Digging Into Autodesk Civil 3D 2007

Level 1 Training

Rick Ellis

PO Box 344
Canby Oregon 97013
www.cadapult-software.com
training@cadapult-software.com
(503) 829-8929

Publisher: Cadapult Software Solutions, Inc.

ISBN 0-9740814-5-0

About the Author

Rick Ellis is the founder and CEO of Cadapult Software Solutions, Inc. At Cadapult Software Solutions, Rick provides training and consulting services to clients around the country helping them to get the most out of their design software investment. He is also an independent consultant and instructor specializing in Autodesk Civil 3D, Autodesk Land Desktop, Autodesk Civil Design, Autodesk Survey, Autodesk Map 3D, Autodesk Raster Design, and AutoCAD. Mr. Ellis is a member of the Autodesk Developer Network, author of "Digging Into Autodesk Land Desktop 2006" and "Digging Into Autodesk Civil 3D 2006", co-author of "Digging Into Autodesk Map 3D 2005" and "Introducing Autodesk Civil 3D", a select author for CAD Digest, and an instructor for the AUGI training program.

Prior to founding Cadapult Software Solutions Rick worked as a CAD Manager and Civil Designer and then as the Technical Services Manager for an Autodesk Reseller. He brings this real world experience and industry knowledge to his training and consulting projects to provide practical examples and solutions to clients.

You can email Rick at: rick@cadapult-software.com

About the Technical Editor

Russell Martin is an independent consultant who has worked with spatial data and cartographic design tools since 1985, on several different platforms and in many different business and academic settings. He served as Staff Geographer, Network Administrator and CAD Manager at a multi-disciplinary engineering firm. More recently, Russell served as the Director of Planning on an innovative urban transportation project, where he coordinated the development of an integrated computerized billing and reservations system, wrote the documentation, supervised installation in satellite offices and trained personnel in its setup, use and maintenance. Russell is a member of the Association of American Geographers, the Autodesk Developer Network, and co-author of "Digging Into Autodesk Map 3D 2005".

You can email Russell at: russell@cadapult-software.com

About Cadapult Software Solutions, Inc.

www.cadapult-software.com

Founded in 2002, Cadapult Software Solutions, Inc. is an independently owned small business located near Portland, Oregon specializing in training, consulting services, and technical support for CAD systems with a focus on the Civil/Survey/GIS industry. Cadapult Software Solutions helps clients maximize the return on their software investment through training classes, consulting services, and support. We offer a wide range of Training options, from standard open enrollment classes to customized on-site training. Our mobile training lab gives us the flexibility to bring classes to our clients regardless of the location. Support options ranging from telephone support to on-site visits help to ensure the continued success of your CAD solution. Although we hold several certifications with Autodesk, Cadapult Software Solutions is an independent company and therefore can provide recommendations and solutions that best fit a clients needs rather than being limited to a specific company's product line. Further affiliations with other consultants and software companies give Cadapult Software Solutions a broad range of experience and industry knowledge to draw from that is not common for a company of its size.

Acknowledgements

Rick Ellis

I would like to thank all the people who made this book possible. There is no way that I can adequately explain the importance of their involvement. But I can simply state that this project could not have been completed without any of them.

Many thanks to Russell Martin for his work as Technical Editor on this project. His input, suggestions, and collaboration on this book were invaluable.

Thank you to Brandt Melick and the City of Springfield, Oregon for providing the project data.

And most of all, thank you to my wife Katie, and our children Courteney, Lucas, and Thomas for their love and support. This project would never have been possible without your encouragement and understanding.

Table of Contents

Chapter 1

Introduction

Welcome to **Digging Into Autodesk Civil 3D 2007**. This book of tutorials is designed to introduce you to many fundamental concepts and procedures commonly used in *Autodesk Civil 3D*. This chapter will help you to become familiar with how the book is laid out and the basics of the Autodesk Civil 3D interface.

- **Overview**
- **Project Overview**
- **Included Data**
- **Style Conventions**
- **Description of Autodesk Civil 3D**

1.1 Overview

Welcome to **Digging Into Autodesk Civil 3D 2007**. This book of tutorials is designed to introduce you to many fundamental concepts and procedures commonly used in *Autodesk Civil 3D*. It is meant to be a resource and a supplement to instructor led training.

It is not meant to, nor is any book able to, replace instructor led training. It is also not a book that will teach you civil engineering, although you may pick up some concepts along the way. This book works through a basic *Civil 3D* project from beginning to end. Showing you many different methods of using *Civil 3D* to accomplish certain tasks and solve problems along the way. Keep in mind that all projects are different so this is not the exact process that you will use to go through every project. It is more important that you understand the individual tools that you are using than focusing on the overall process. This way when you encounter examples that are not covered in this book you can identify and use the proper tool to accomplish your task.

1.2 Project Overview

The exercises in this book lead you through a basic road design project from beginning to end. The project begins by importing data from a variety of sources. You will learn how *Civil 3D* can leverage existing GIS data, aerial photography, and aerial survey data to efficiently add information to your project and manage it with one program.

The project continues with preliminary layout and design by building a surface from the aerial survey data, laying out a preliminary alignment, and exporting points for field verification.

Next you will add survey data to the project and learn to manage points through the use of description keys, point groups, point styles, and point label styles. The survey points will be used to create a more accurate surface and that new surface will be merged with the aerial surface to give you more accurate surface data on the site while using the aerial survey information around the edge of the site where you do not have survey data. Parcels will also be defined based on the survey points.

You will continue on to alignment design where you will lay out and edit the horizontal alignment. You will learn how alignments interact with parcels and you will subdivide and layout new parcels based on several criteria. You will also create and edit existing and finished grade profiles.

The road design will be completed as you create a corridor that combines the alignment, profile, surface, and assembly into a model of the road design. The corridor model is used to create surfaces and sections that you will plot as well as use for quantity takeoffs.

The book concludes with a look at the grading tools. You will create a grading group to grade a pond and automatically generate a surface. The pond surface will be compared with the existing surface to create a volume surface and calculate cut and fill quantities.

1.3 Style Conventions

The specific formatting and use of fonts to denote different things is kept deliberately simple in this book, to keep it easy to follow. No "legend" is required to decipher a page of instructions; we simply use the following conventions.

Pull-down and right-click menus: **Map ⇒ Tools ⇒ Import**
Command Line prompts and entries: **`Command:  Zoom`**
Buttons: **<<OK>>**

Actions to be taken in specific steps appear in **boldface**, to help them stand out when following along in class, and names of dialog boxes and sections within them appear in *italics*, to help distinguish them from instructions.

1.4 Included Data

The data supplied on the CD with this book, as well as the exercises, are designed to work with a standard "out of the box" installation of *Autodesk Civil 3D 2007*. All of the data needed for the exercises in this book will be installed when you run the dataset installer contained on the CD. The dataset installer should automatically run when you insert the CD. However, if it does not you can run the **Setup.exe** file contained on the root of the CD. When extracted to your C drive it will create a folder called *C:\Cadapult Training Data\Civil 3D 2007\Level 1*. There are several folders created below the *Level 1* folder that contain the source data you will use for the exercises in this book. A folder called *Chapter Drawings* is also created that contains a drawing that can be used to begin each chapter as well as each major chapter section. This will allow you to jump in at the beginning of any chapter or major chapter section of the book and do just the specific exercises that you want if you do not have time to work through the book from cover to cover. The drawings in the *Chapter Drawings* folder are not necessary and only need to be used if you want to start in the middle of the book or if you want to overwrite any mistakes that you may have made in previous chapters. A drawing template called *_Cadapult Level 1 Training by Style.dwt* is also installed into your drawing template directory.

If for any reason you have problems using the data installer all of the data needed for the exercises in this book is also contained in a zip file called **"Civil 3D Level 1.zip"** in the support folder on the CD. Just extract this zip file to your C drive and the dataset will be installed. The drawing template is also available in the support folder as well and can be copied manually to your drawing template folder. You can find your current drawing template folder by looking on the Files tab of the AutoCAD Options dialog box.

1.5 Description of Autodesk Civil 3D

Autodesk Civil 3D has all the features of AutoCAD, all of the GIS and drawing management tools in Autodesk Map 3D, and specific civil design tools to work with Points, Surfaces, Parcels, Alignments, Profiles, Corridors, Sections, Grading, and Pipes. Civil 3D creates special objects for each of these features that have the intelligence to interact dynamically with each other.

1.5.1 Program Interface

1. The *Autodesk Civil 3D* interface is essentially the same as AutoCAD 2007, with several new menus and optional toolbars plus two unique features, which you will use frequently throughout this book and in many future projects of your own; the *Toolspace* and the *Task Pane*.

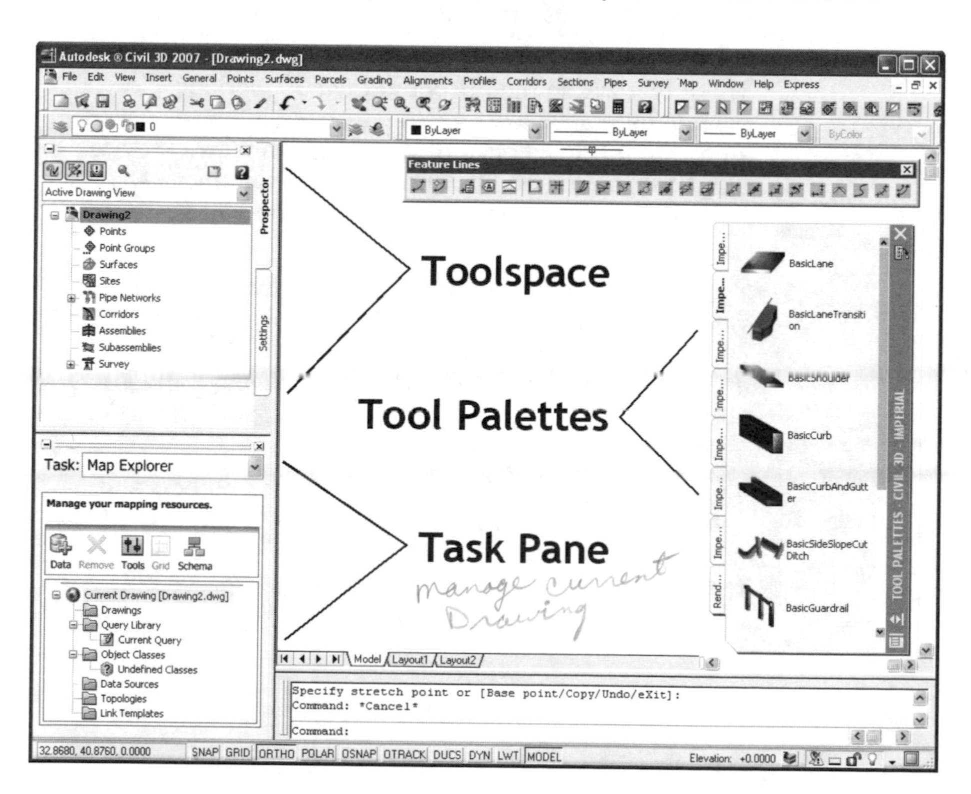

2. The *Toolspace*, where you manage civil objects, has two default tabs, *Prospector* and *Settings*. The *Prospector* tab accesses all of the civil objects in the drawing, while the *Settings* tab manages styles for the display of these civil objects. Tabs called *Toolbox* and *Survey* will also be displayed in the *Toolspace* when you use certain commands. The *Toolbox* tab contains the *Report Manager* as well as third party tools that you may acquire over time. The *Survey* tab is used to manage survey data.

3. The *Task Pane*, where you manage your current drawing, has three options, *Map Explorer*, *Display Manager*, and *Map Book*. The *Map Explorer* shows all attachments to the current drawing; source drawings, queries, feature classes, topologies and link templates. The *Display Manager* is where you manage styles and themes to control how these map objects are displayed. The *Map Book* is where you create map books for hard copy as well as DWF output.

4. These features in the *Autodesk Civil 3D* working environment are shown in the illustration above docked along the left hand edge of the editor window. They can also float, be turned off, and their transparency can be set, to limit their intrusion as you work in the editor.

5. You may be familiar with using *Tool Palettes* in previous version of AutoCAD to manage blocks, hatch patterns, and custom commands. *Civil 3D* adds several *Tool Palettes*, and allows you to create your own, to manage and insert *Subassemblies*.

6. You will become familiar with all of these objects and terms as you work through the tutorials in this book. For now, just familiarize yourself with having these additional features open and part of your working environment.

1.5.2 Workspaces

Workspaces are saved user interface configurations that include the state and positions of menus, toolbars, buttons, and dockable windows. Basically it is what the name implies, a saved snapshot of the user interface, or workspace.

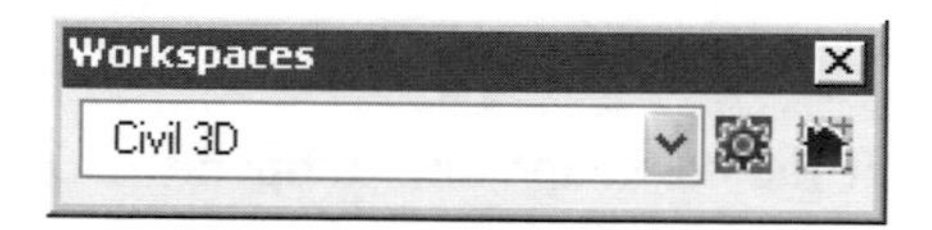

You can switch between workspaces by selecting them from the Workspaces toolbar. Workspaces can be can be created and customized in the CUI command.

1.5.3 Dynamic Input

Dynamic Input was a new feature in *AutoCAD 2006* that allows command input at the cursor rather than the command line. Using *Dynamic Input* in *Civil 3D* can be confusing at times and cause unexpected results. It is recommended by *Autodesk* in the ReadMe file that you turn off *Dynamic Input* when using *Civil 3D*.

1. Confirm that **Dynamic Input** is turned off. If it is on Click the DYN button on the *Status Bar* to disable it.

1.5.4 Selection Preview

Selection Preview was another new feature in AutoCAD 2006. When enabled, it highlights objects in the drawing editor as you move your cursor over them. This can be a nice feature at times, however, in Civil 3D where you are working with many large objects it can get in the way and slow down your computer. This exercise will show you how to turn off Selection Preview.

1. Type **OP** at the command line to open the *Options* dialog box.

2. Select the **Selection** tab.

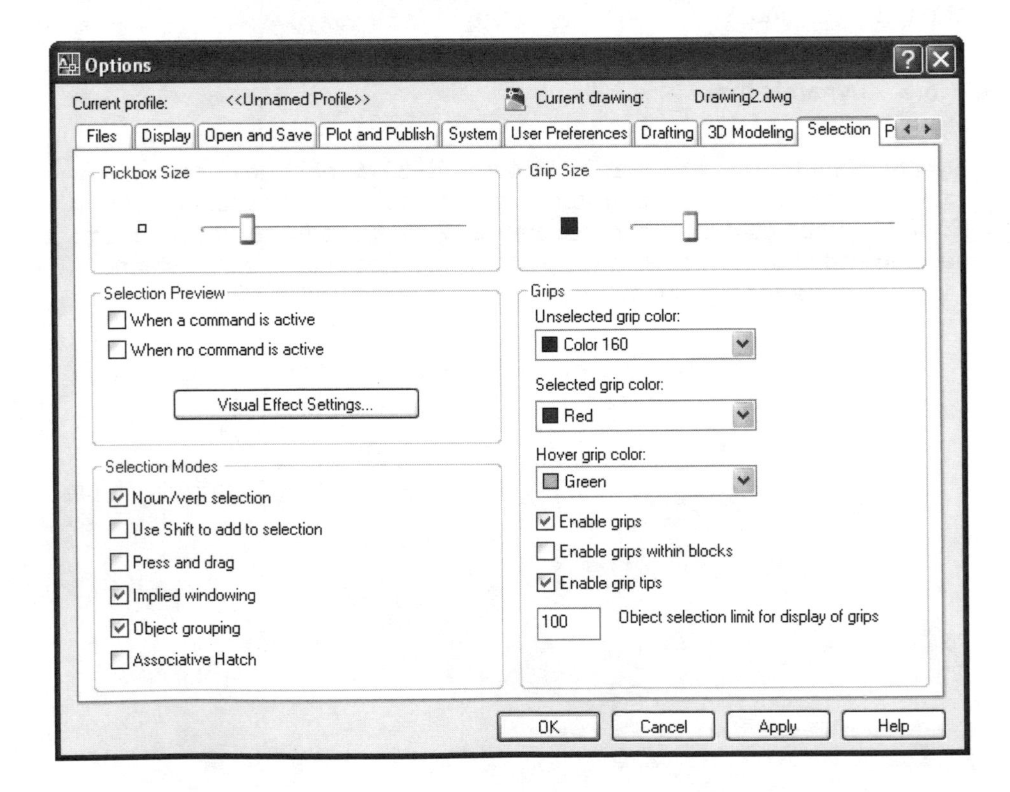

3. **Disable** the **When a command is active** *Selection Preview* option.

4. **Disable** the **When no command is active** *Selection Preview* option.

5. Click **<<OK>>** to save your changes and exit the dialog box.

1.5.5 Data Management

Civil 3D 2007 is a drawing based program and this book will cover working with *Civil 3D* in a standalone manner. However, *Civil 3D 2007* does include an option to install *Autodesk Vault*. *Vault* is a powerful document management tool and you should give your implementation of *Vault* careful consideration. This is a big change if you are a Land Desktop user who learned to leverage its project to share data and reduce drawing size. Since the data is stored in the drawing it is important to Save often. It also means that all of your settings and styles are saved in the drawing and should also be saved in a drawing template so they are available when you create new drawings.

1.5.6 Styles

One of the most important features in *Civil 3D* is *Styles*. *Styles* control the display, labeling, and in some cases the geometric properties of *Civil 3D* objects. This book will expose you to many different object and label styles as you work through the project. You will find that in many cases the interface to create and edit these styles is similar from object to object. This consistency will help as you learn to work with styles because, for example, creating an alignment style is not that different than creating a profile style.

Styles control the display of the components that make up an object. For example a *Surface Style* has components for triangles, contours, points, elevation bands, slope arrows, and much more. The style controls the visibility, color, layer, linetype, linetype scale, lineweight, and plot style of these components. One of the choices that you will need to make as you set up your own styles is if the styles will be configured to place each component on a different layer so that you can manage the display of the components with standard layer commands; or, if the styles will be configured to manage the display of the components themselves, independent of layers. This book will use styles to manage the display of the components, independent of layers, to expose you to these new options and limit the number of layers in the drawing.

Chapter 2

Data Collection and Base Map Preparation

This chapter shows how to build base maps with existing map information. It covers importing GIS data, inserting registered ortho photography, as well as attaching and querying specific layers from planimetric map sheets and Digital Terrain Model (DTM) source sheets.

- **Importing GIS Data**
- **Using Queries to Manage and Share Data**
- **Inserting Images And The Project Boundary**
- **Importing Point And Breakline Information From Aerial Mapping**

Dataset:

To start this chapter you need to install the dataset from the CD that came with the book. The dataset installer should automatically run when you insert the CD. However, if it does not you can run the **Setup.exe** file contained on the root of the CD. The dataset will create a folder named **C:\Cadapult Training Data\Civil 3D 2007\Level 1** which contains the base data that you will use for the exercises in this book as well as drawings to begin each chapter if you choose to start in the middle of the book. A drawing template called *_Cadapult Level 1 Training by Style.dwt* is also installed into your drawing template directory.

2.1 Overview

One of the strengths of *Civil 3D* is its ability to bring together data from a variety of sources for use in your project. This chapter shows several ways to bring data from different sources together to build base maps with existing map information. It covers importing GIS data, inserting registered ortho photography, as well as attaching and querying specific layers from planimetric map sheets and Digital Terrain Model (DTM) source sheets. Several of the tools covered in this chapter are part of *Autodesk Map 3D*, which is the foundation for *Civil 3D*.

2.1.1 Concepts

In this chapter you will import GIS data in the form of ArcView Shapefiles. This type of information is commonly available from local city and county government agencies. You will also learn to use Queries to manage and share large data sets. You will learn how to manage registered aerial photos. Finally, you will use Queries to extract point and breakline data from several drawings for use in Chapter 3 to create a surface.

2.1.2 Terms

Attribute - Any data attached to an object. In AutoCAD you may be familiar with block attributes, which is a way to attach data to an AutoCAD block. Autodesk Map 3D expands your ability to attach data to any object, not just a block, with object data and external database links. This means that in Autodesk Map 3D the term Attribute Data applies not only to blocks but also to object data and external database links.

Coverage - is a term used in GIS applications like ArcInfo to define a layer of information such as parcels, soils or street centerlines.

Shapefile - An ArcView or ESRI shapefile consists of a main file, an index file, and a dBASE table. The main file describes a shape with a list of its vertices. The index file contains the offset of the corresponding main file record from the beginning of the main file. The dBASE table contains feature attributes. At least these three files with extensions of shp, shx, and dbf must be present for you to have a complete and usable shapefile.

Coordinate system - in Autodesk Map 3D and Civil 3D, a "Coordinate System" includes the projection method, the datum AND the base coordinates (Cartesian, Lat-Long, UTM, etc.), - whereas in cartography it only refers to the X,Y coordinates themselves. There are many systems used throughout the world.

Object Data - attributes attached to drawing objects and stored in object data tables.

Object Data Table - table saved within a drawing that stores text and data that is attached to the drawing objects.

2.2 Importing GIS Data

The use of the import command is very similar for all the different types of GIS data file formats that it supports.

In this chapter you will add parcel and street data from the GIS department as well as an aerial photo of the project area. You will also add point data and breakline data that you will use later to build a surface.

2.2.1 Importing ESRI Shapefiles

ArcView Shapefiles are one of the most common types of GIS data files that you will encounter. In this exercise you will add parcel and street information from the city, which is in a Shapefile format, to your drawing for use as background information. This is an example of how you can use Civil 3D to bring data together from a variety of sources to meet the needs of your project.

1. Select **File ⇒ New.**

2. Select the drawing template **acad.dwt**.

3. Click <<**Open**>>.

4. From the main menu select **Map ⇒ Tools ⇒ Import.**

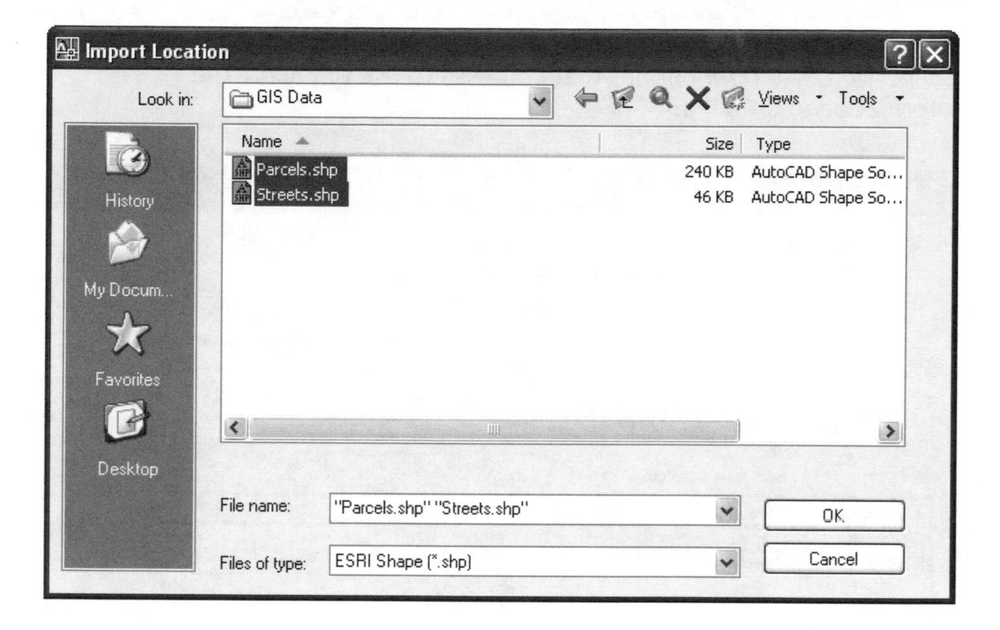

5. Browse to:

   C:\Cadapult Training Data\Civil 3D 2007\Level 1\GIS Data

6. Set the Files of type to **ESRI Shape**.

7. Select the files **Parcels.shp** and **Streets.shp**.

8. Click **<<OK>>** to continue.

In the *Import* dialog box you can configure the *Layer, Coordinate Conversion*, and *Data* options that you wish to use to Import the information into *Civil 3D*.

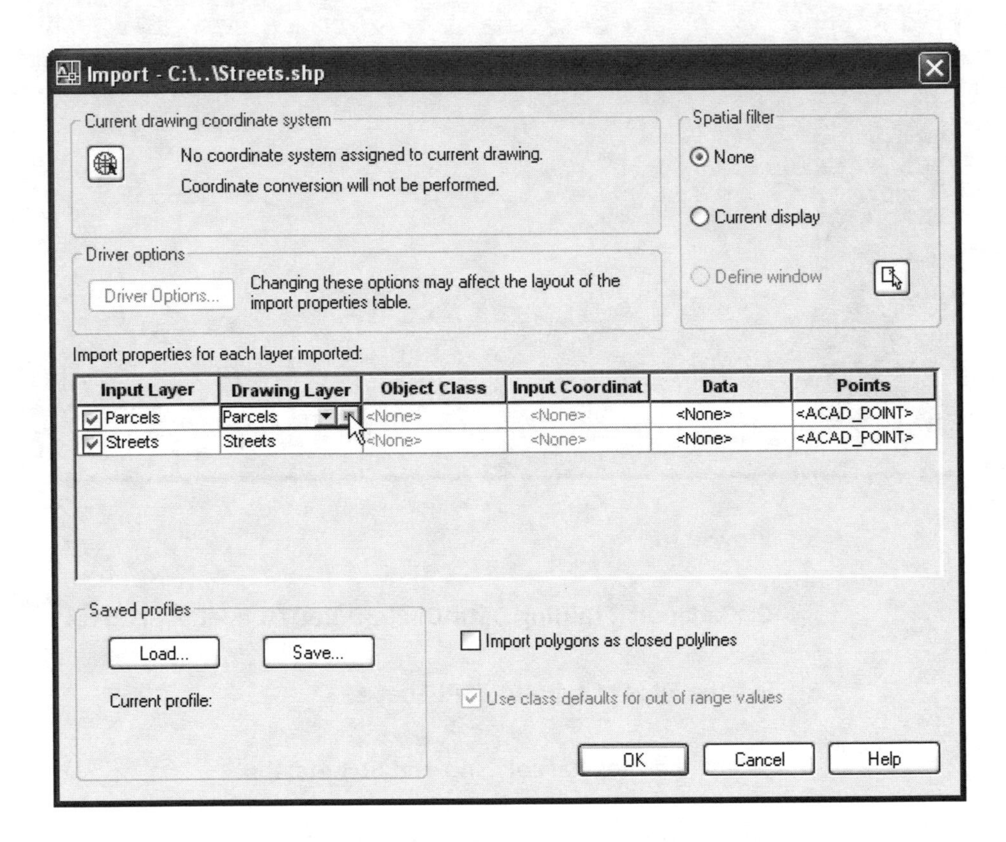

9. Click on the **Drawing Layer** field in the *Parcels* row to activate the **ellipses << ... >>** button. Click this button to bring up the *Layer Mapping* dialog box.

Here you can choose to import the drawing objects onto an existing layer, create a new layer, or select a column of data from the file that you are importing to determine the layer names. This last option will allow you to do some basic thematic mapping during the import of the objects. For example, if you were importing parcel data and that data set had a column for zoning. You could have the import command create a new layer for each zoning type and place each parcel on the appropriate layer for its zoning designation.

10. Choose **Create on new layer** and name it "**EX-Parcels**".

11. Click **<<OK>>**.

12. Repeat the process for the *Streets ShapeFile*, creating it on a new layer named "**EX-Streets**".

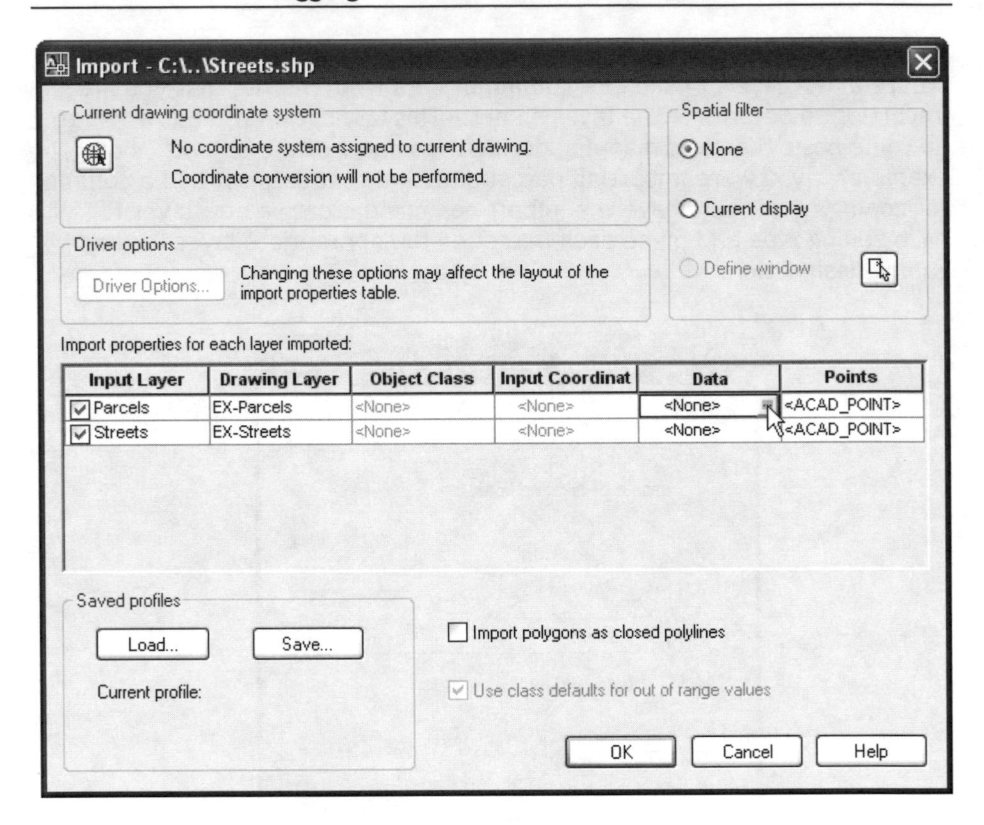

13. Back in the *Import* dialog box, click on the **Data** field in the *Parcels* row to activate the **ellipses << ... >>** button. Click this button to bring up the *Attribute Data* dialog box.

Here you will create *Object Data* from the shapefile's attribute data. You can enter the desired name for the *Object Data Table* and select the desired fields to import. This is the step that allows you to bring the intelligence of the GIS file along with the geometry into *AutoCAD*. By creating the object data table and populating it with the information provided in the shape file you will be able to click on a parcel and find the owner name, address, zoning, and any other information that was added by the GIS department. This will also allow you to edit the geometry and data from the GIS file in *AutoCAD* and then export it back to any of the supported GIS formats without losing any of the attached data. If you leave the *Data* option set to *None* or *Do not import attribute data*, then you will only import the geometry of the file and you will lose all of the attached information.

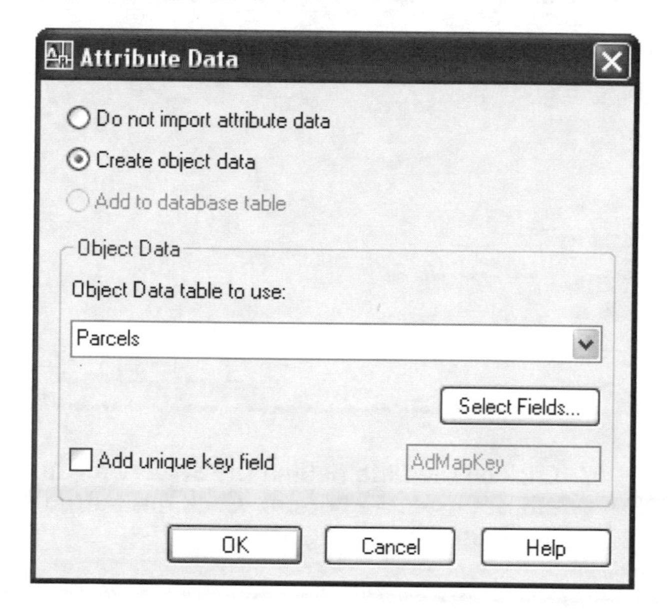

14. Choose **Create object data.**

15. Use the default name **Parcels** for the *Object Data table*.

16. Click **<<OK>>**.

This returns you to the *Import* dialog box.

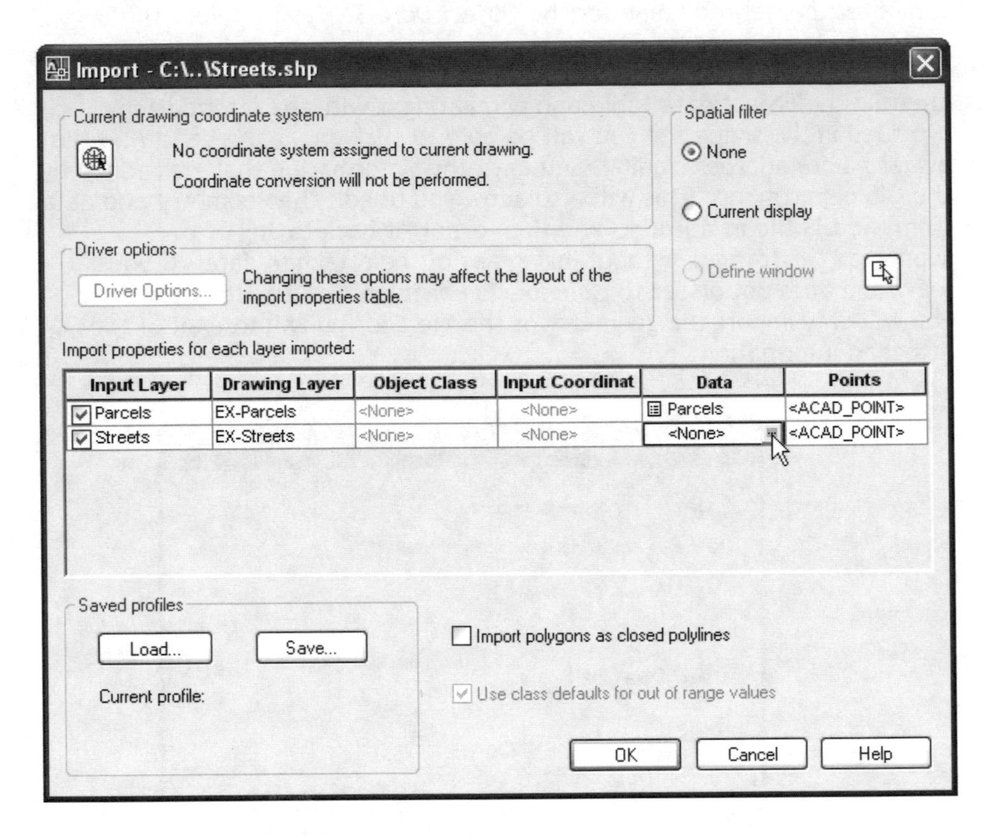

17. Click on the **Data** field in the *Streets* row to activate the ellipses << ... >> button. Click this button to bring up the *Attribute Data* dialog box.

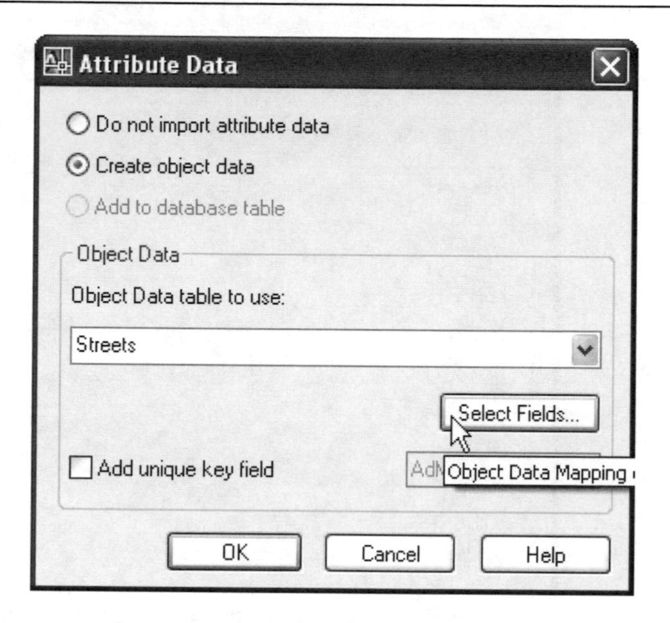

18. Choose **Create object data**. Use the default name **Streets** for the *Object Data table*.

19. This time, click the **<<Select Fields>>** button. This will allow you to select only the attribute data you wish to import rather than the entire database.

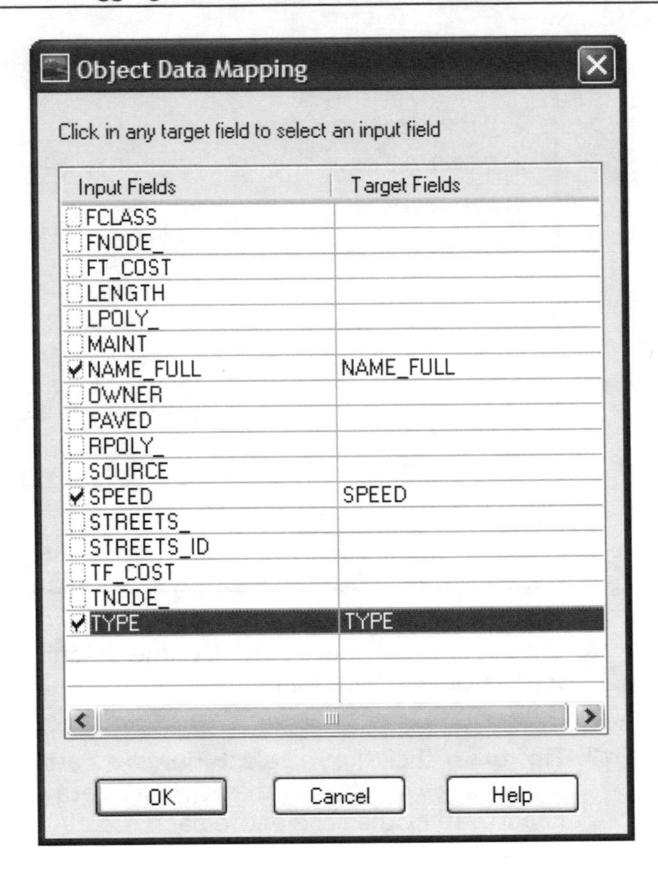

20. Deselect all fields except **"NAME_FULL"**, **"SPEED"**, and **"TYPE"**.

This option allows you to be selective about what information you import into *AutoCAD*. If you don't want to import all of the attached data, you have control over the columns of information that you import from the shape file.

21. Click <<**OK**>> to return to the *Import* dialog box. The completed dialog box should look like the one below.

If you have assigned a coordinate system to the drawing it will display at the top of the Import dialog box. You can also assign a coordinate system to the drawing at this time by selecting the *Assign Global Coordinate System* button. If a coordinate system has been assigned to the drawing the *Input Coordinate System* column will be activated. This allows you to assign a coordinate system to the files that you are importing. If the input coordinate system is different than the current drawing coordinate system the geometry will be converted to the current drawing coordinate system as it is imported.

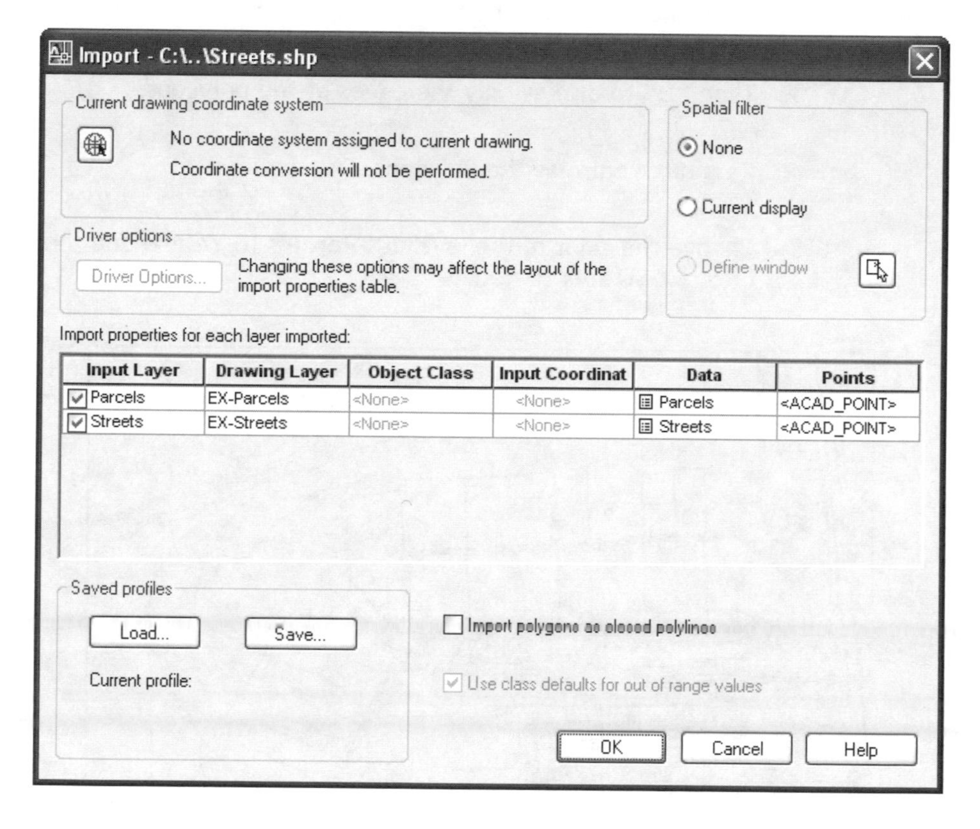

22. Click **<<OK>>** to begin importing the data.

23. **Zoom to Extents** to view the imported data.

2.2.2 Controlling the Display of Polygons

Notice that the polygons are all displayed with a solid hatch fill. This display option is a feature of the *MPOLYGON* object. To display just the edges of the polygons you need to set the polygon display mode.

1. At the command line enter:
2. `Command: POLYDISPLAY`
3. Type "`E`" to display only the edges of the polygons.
4. **Regen** to redisplay the Polygons.
5. Change the color of the layer **EX-Parcels** to **Yellow** and the later **EX-Streets** to **Red**.

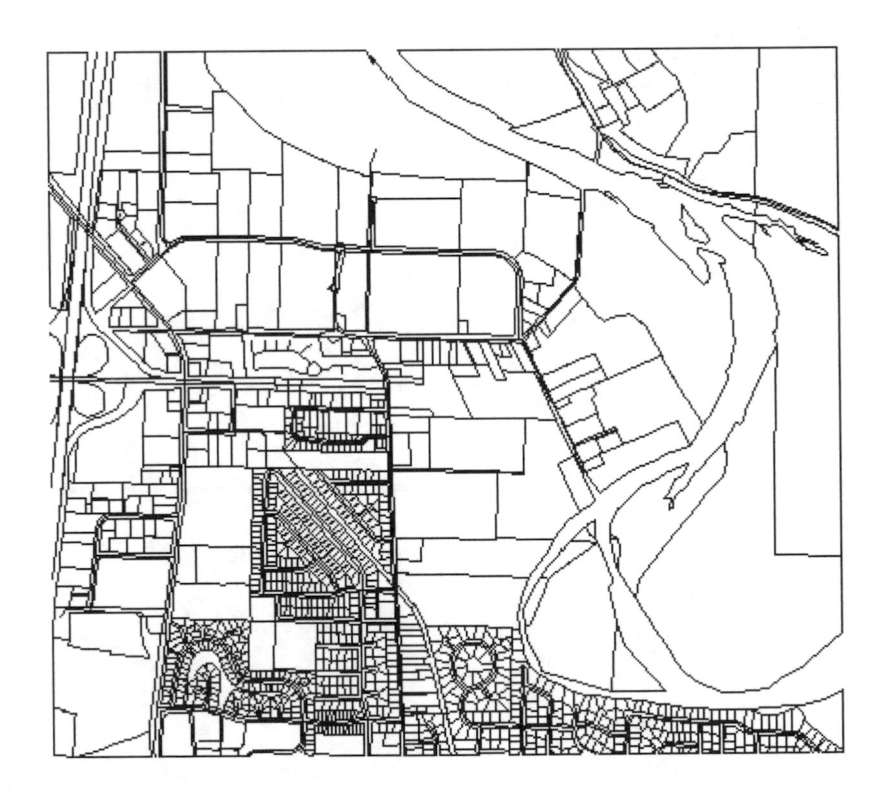

2.2.3 Viewing GIS Attributes in *AutoCAD*

When talking about GIS data the term *attributes* refers to any type of attached data, not just block attributes as in *AutoCAD* terminology. In *Autodesk Map 3D*, attribute data can be object data, external database data, or block attributes. Within *AutoCAD*, the *Map* functions enable you to view and edit this attribute data, including attributes imported from GIS files, as well as attributes created as object data in Map. *Civil 3D* is built on and includes *Autodesk Map 3D* so all of that functionality is available to you in *Civil 3D* as well. To view the attributes imported above follow these steps.

1. Select **Map ⇒ Object Data ⇒ Edit Object Data.**

2. Then select the desired parcel or street to display its object data.

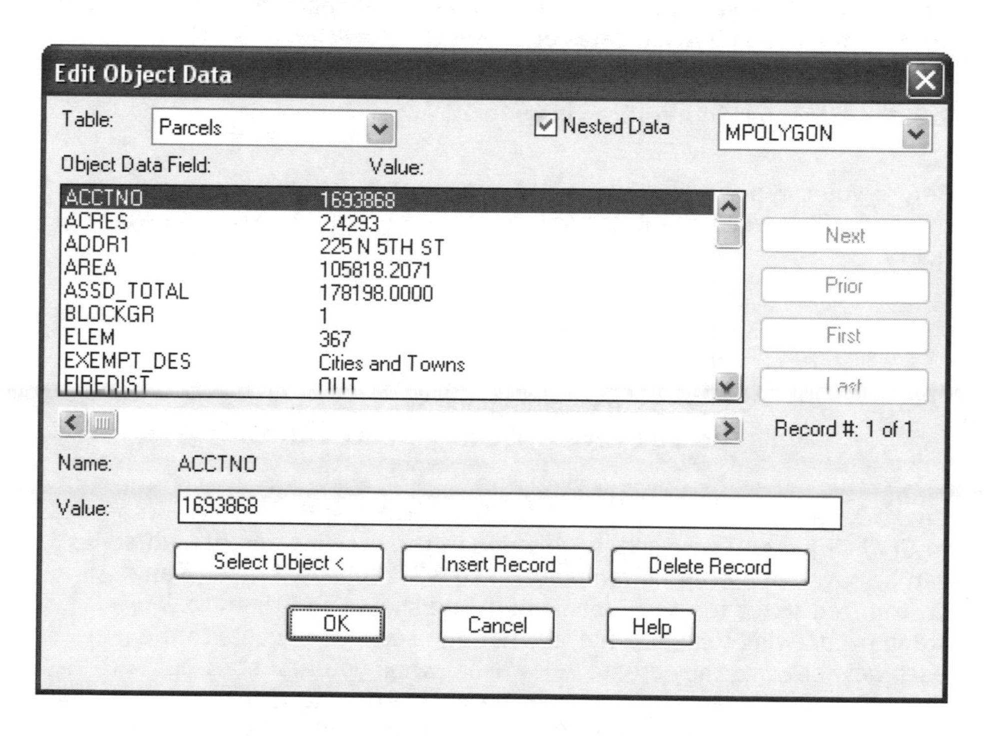

3. Here you can view and edit any object data as needed. Click **<<Cancel>>** to leave the *Edit Object Data* dialog box without making any changes.

You can also display and edit object data with the *AutoCAD Properties* command.

4. Save the drawing as **"Base Map.dwg"**, in the folder **C:\Cadapult Training Data\Civil 3D 2007\Level 1\.**

5. Close the drawing.

2.3 Using Queries to Manage and Share Data

You will use *Autodesk Map's* ability to query information from other drawings to import, edit, and save back specific parcel and street information. This gives you the ability to work with small, specific, pieces of larger data sets as well as allowing you to share the data with others that may need to be using it at the same time. You will also use queries in a later exercise to extract spot elevation and breakline data for a specific area from a large citywide data set.

2.3.1 Attaching Source Drawings

Before you can perform a query, you must attach at least one source drawing. This will become the source of the information you find with the query.

1. Select **File ⇒ New.**

2. Select the drawing template: **_Cadapult Level 1 Training by Style.dwt.**

Civil 3D makes extensive use of styles to control the display and properties of *Civil Objects*. These styles are saved in the drawing and default styles can be saved in a drawing template. You will probably create many customized styles for your organization and should add them to your standard drawing template file, so that all users have access to them in each new drawing they create. However, when you first start out with *Civil 3D* you will want to start with an existing template like the one provided with this book, *_Autodesk Civil 3D Imperial By Style.dwt*, so that you can see some example styles and modify them to fit your needs. If you start with a blank template like "acad.dwt" you will only have the *standard* style available for all objects.

3. Click **<<Open>>**.

4. If the *Map Explorer* is not visible on your screen, At the command line enter:

5. `Command: MAPWSPACE`

6. Type "`ON`" to display the Task Pane which includes the Map Explorer.

7. Set the **Task** at the top of the *Task Pane* to **Map Explorer**.

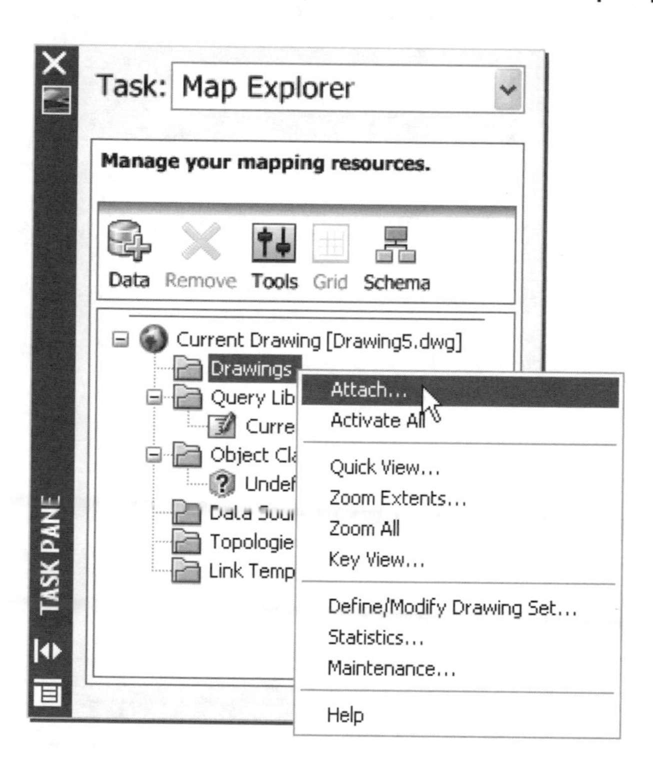

8. In the **Map Explorer** right-click on the **Drawings** folder and select ⇒ **Attach** from the fly-out menu.

To attach *Source Drawings*, you must have a *Drive Alias* defined. There is a default alias for the C drive. A drive alias is a named shortcut to a drive or folder on your local machine or network. Its purpose is to aid in sharing data with others whose directory structure may be different from yours. You can share a common named drive alias and each have your own specific paths to the data.

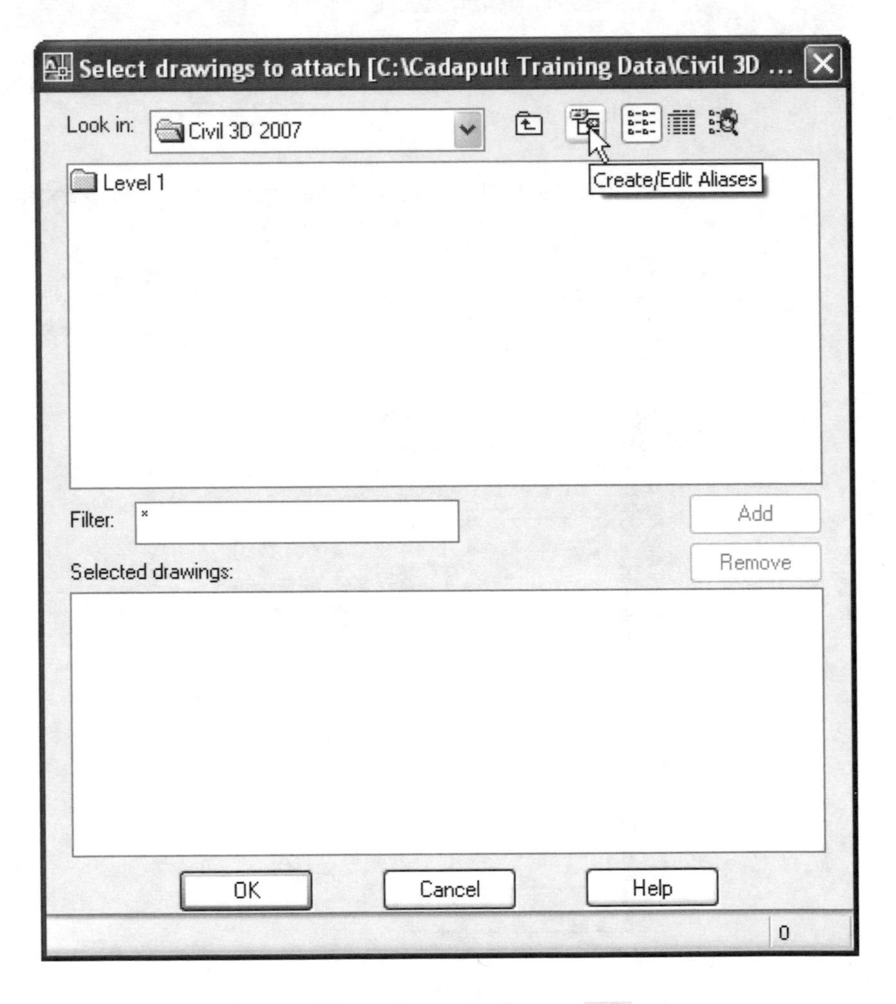

9. Click the **Create/Edit Aliases** button at the top of the dialog box to create a drive alias.

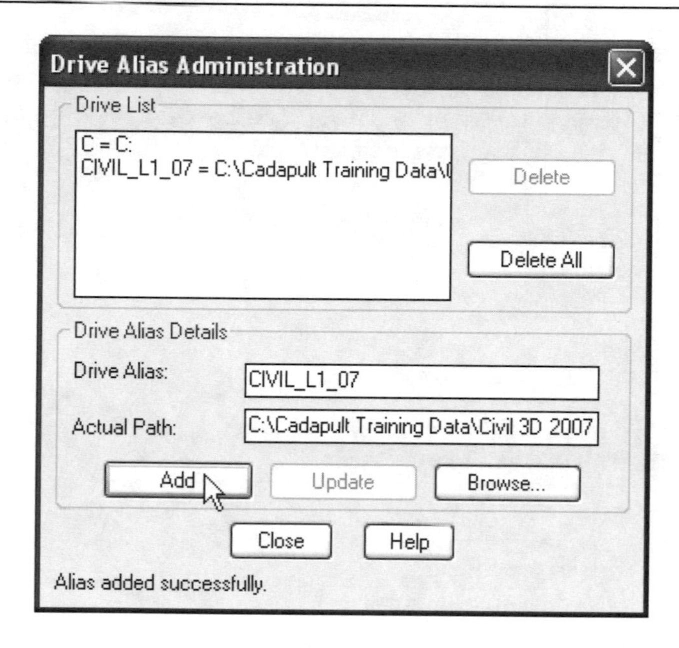

10. In the *Drive Alias Administration* dialog box enter **"CIVIL_L1_07"** for the alias name.

Aliases cannot contain any spaces and must be in upper-case.

11. Click <<**Browse**>> to add the *Actual Path*. For all the exercises in this book the path will be: **C:\Cadapult Training Data\Civil 3D 2007\Level 1\.**

12. Click <<**Add**>> and this alias will appear in the *Drive List*.

If you click <<**Close**>> before you click <<**Add**>>, the alias will not be saved.

13. After you have added the alias, click <<**Close**>>.

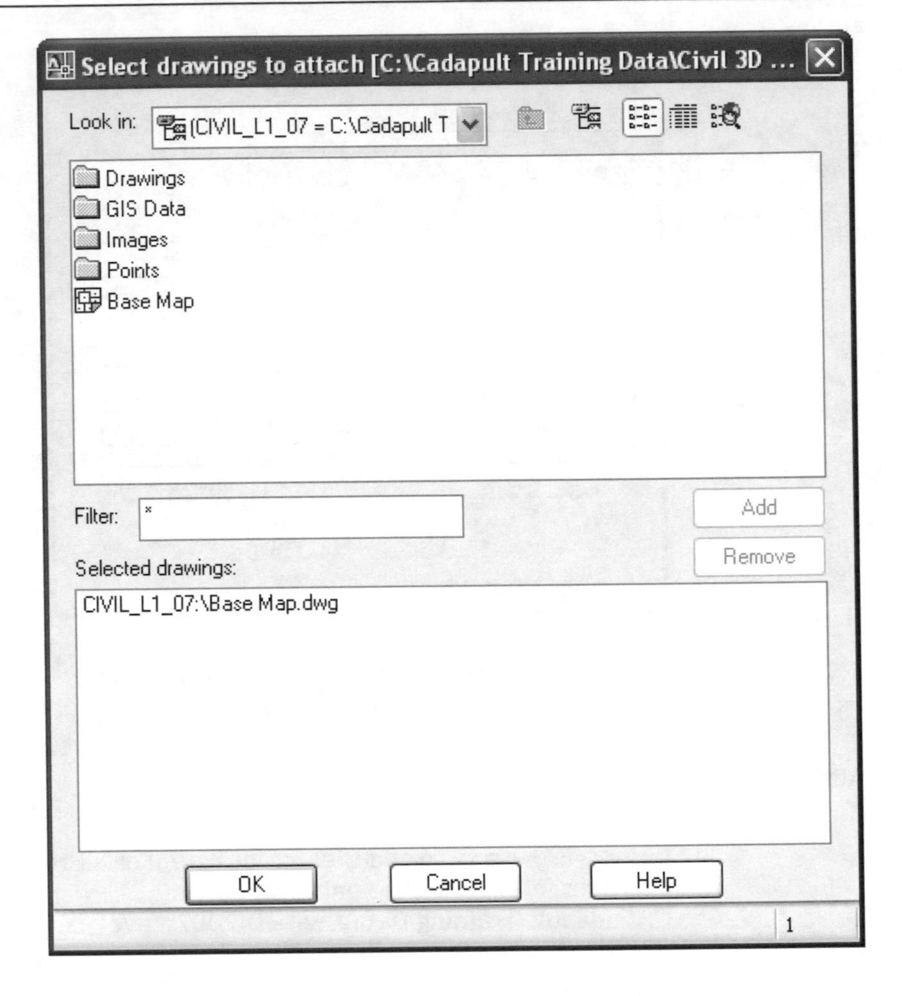

14. Back in the *Select Drawing* dialog box select the **CIVIL_L1_07 alias** from the drop-down list.

15. Double-click on the **Base Map** drawing file, to add it to the *Selected drawings* list.

16. Click **<<OK>>**.

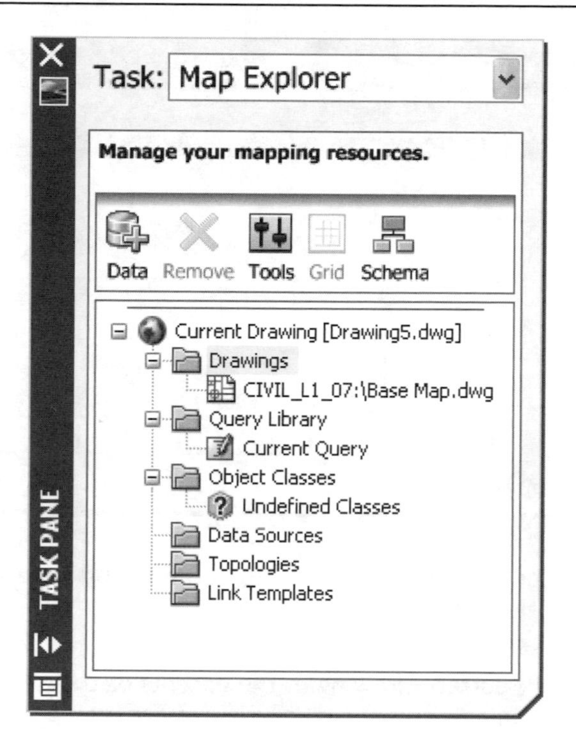

You will now see the drawing listed in the *Map Explorer*. *Active drawings* will be used for all *Query* and *Quick View* commands. Right-clicking on attached drawings in the *Map Explorer* and choosing deactivate can deactivate them. *Deactivated drawings* are still attached to the drawing but will not be used in Queries or Quick Views.

17. Right-click on **Drawings** and select ⇒ **Quick View.**

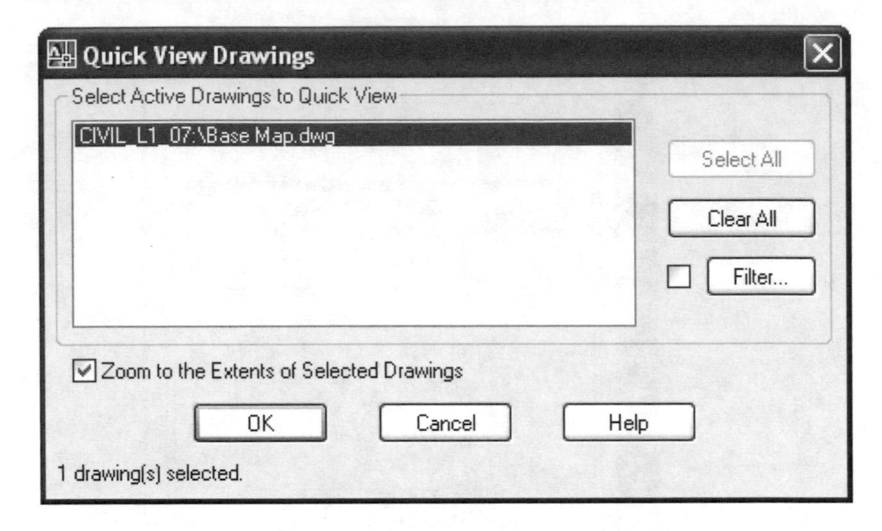

18. Confirm that the option to **Zoom to the Extents of Selected Drawings** is Enabled.

This will zoom your current drawing to the extents of the attached drawings. If you don't do this the quick view image may appear in an area that is not shown on your screen.

A Quick View will show you a preview image of what is contained in any attached and active drawings. The quick view image will go away when you use a Redraw or during an AutoSave. If you lose the quick view just run the quick view command again to redisplay it.

19. Save the drawing as **Design,** in the folder **C:\Cadapult Training Data\Civil 3D 2007\Level 1**

2.3.2 Defining a Query

You can use any combination of four different Query Types to create your query and filter through the attached, active, drawings to select the exact objects that you want to bring into your drawing. The four types are Location, Property, Data, and SQL.

- **Location** allows you to limit your Query to a specific area which you define within the source drawings.

- **Property** allows you to select the object to be queried by any AutoCAD property such as Layer, Linetype, Color, etc.

- The **Data** and **SQL** options allow you to select options for your Query according to their attached data; both object data and external database data.

By using the above Query types together you can limit the objects that are Queried into your drawing giving you a very specific selection set and keeping your drawing sizes smaller.

1. Right-click on **Current Query** under the *Query Library* in the *Map Explorer* and select ⇒ **Define** to open the *Define Query* dialog box.

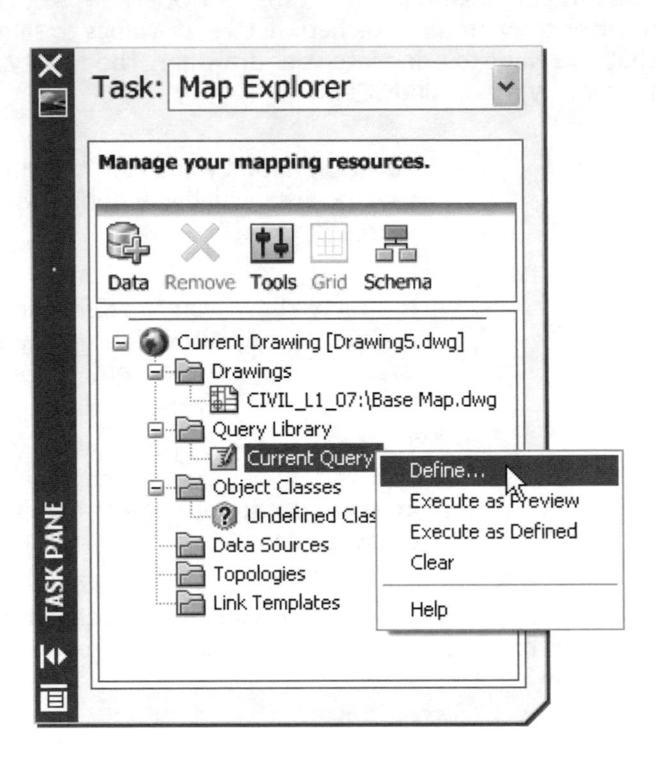

Here, you set up the parameters of your query.

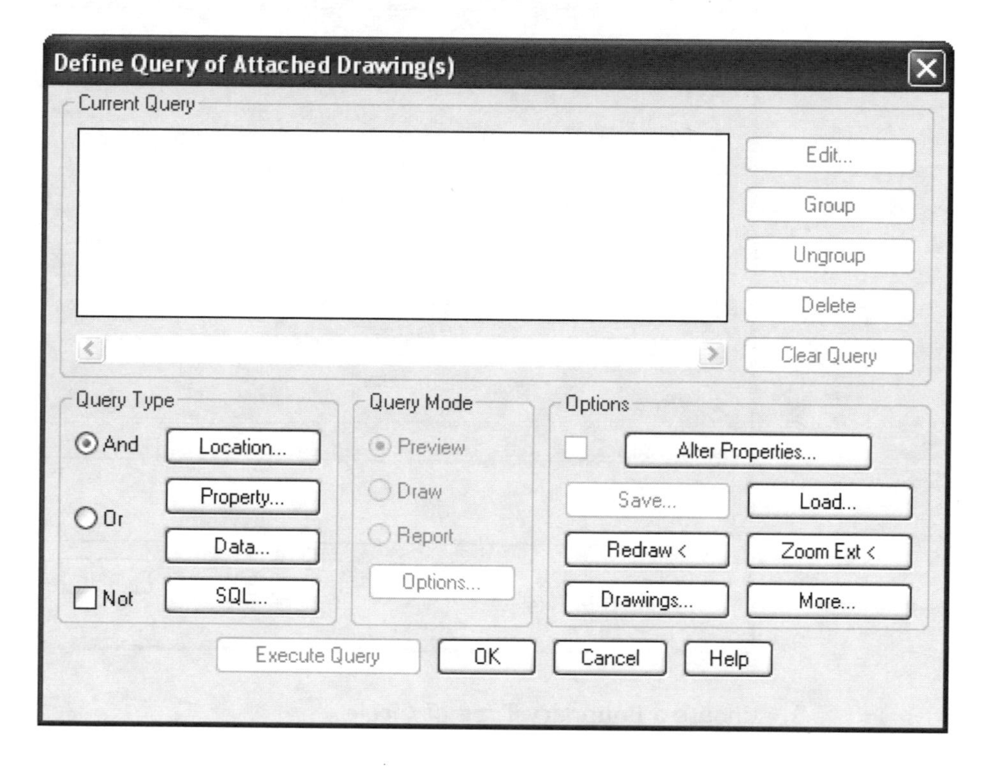

2. Click **<<Location>>** in the **Query Type** section.

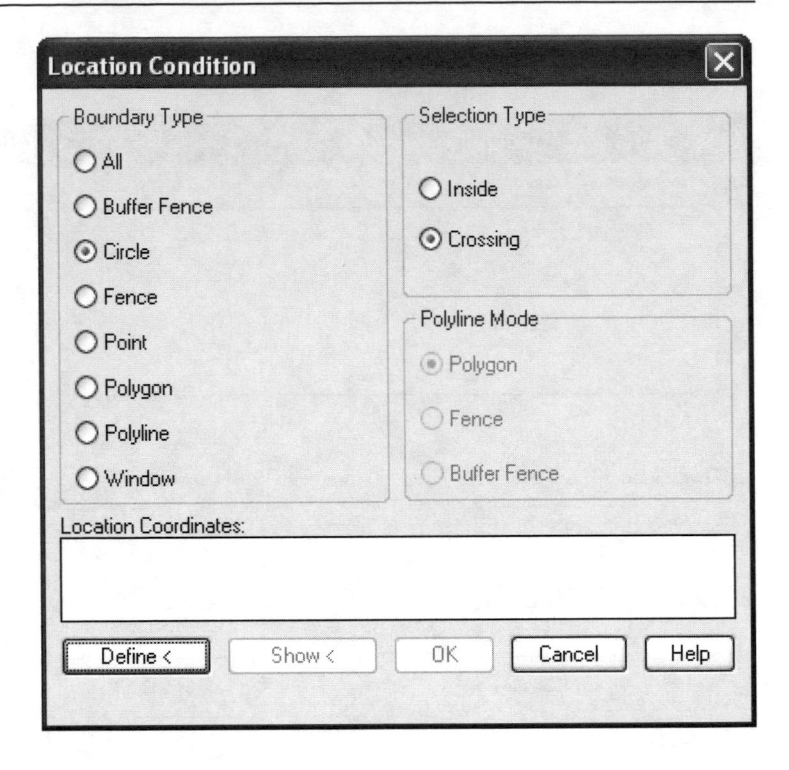

3. Choose a **Boundary Type** of **Circle**.

4. Click **<<Define>>**.

5. At the command line enter `1336700,892000` for the center of the circle.

6. Enter `3500` for the radius.

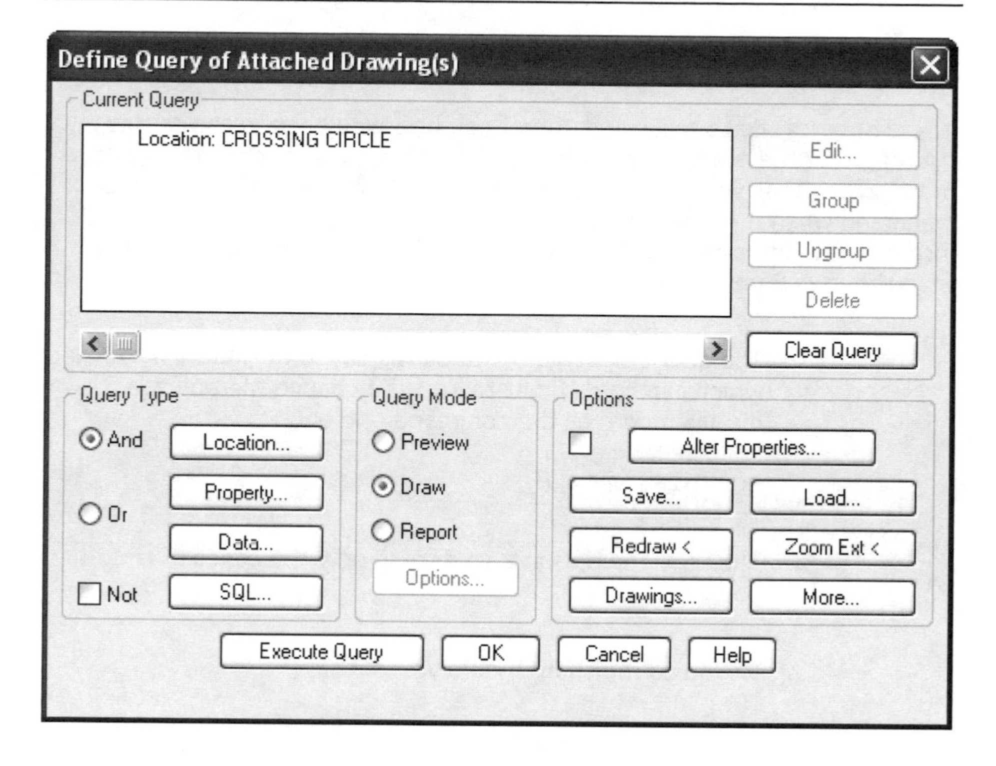

7. Back in the *Define Query* dialog box; change the **Query Mode** to **Draw** and click **<<Execute Query>>**.

This will bring all of the objects in the attached drawing or drawings that meet the query criteria into the current Drawing. They are brought in as regular *AutoCAD* objects so you can edit them using any of your *AutoCAD* commands. Any changes can be saved back to the source drawings.

2.3.3 Saving Changes Back To the Source Drawings

Any changes made to a queried object can be saved back to the source drawing that the object was queried from. This is a very powerful feature but also one that you need to be very careful with. Using the Save Back command when it is not appropriate can edit base data and cause you to lose valuable information. Remember, the Undo command does not work with the Save Back command.

1. The queried polygons will be displayed as filled polygons, which can be difficult to edit. To change the polygon display mode, at the command line enter:

   `POLYDISPLAY`

2. Then use the option **`E`** to display only the edges of the polygons.

3. **`Regen`** to redisplay the Polygons.

4. Edit any of the Queried objects using standard *AutoCAD* commands, like stretch or move. Make sure it is a large enough change that it is easy to see.

5. Once you finish the edit you will be asked if you want to add object(s) to the save Set.

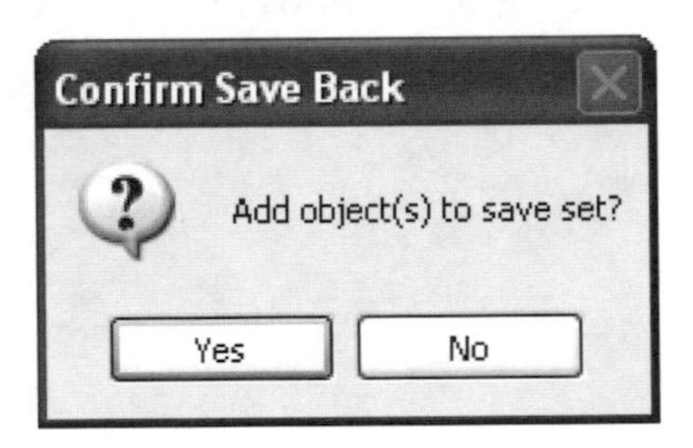

6. Click **<<Yes>>** to add the object(s) that you just edited to the save set.

This has not saved the edited objects back to the source drawing yet. It has only added them to a group of objects that can be saved back when you are ready.

7. Select **Map ⇒ Save Back ⇒ Save to Source Drawings.**

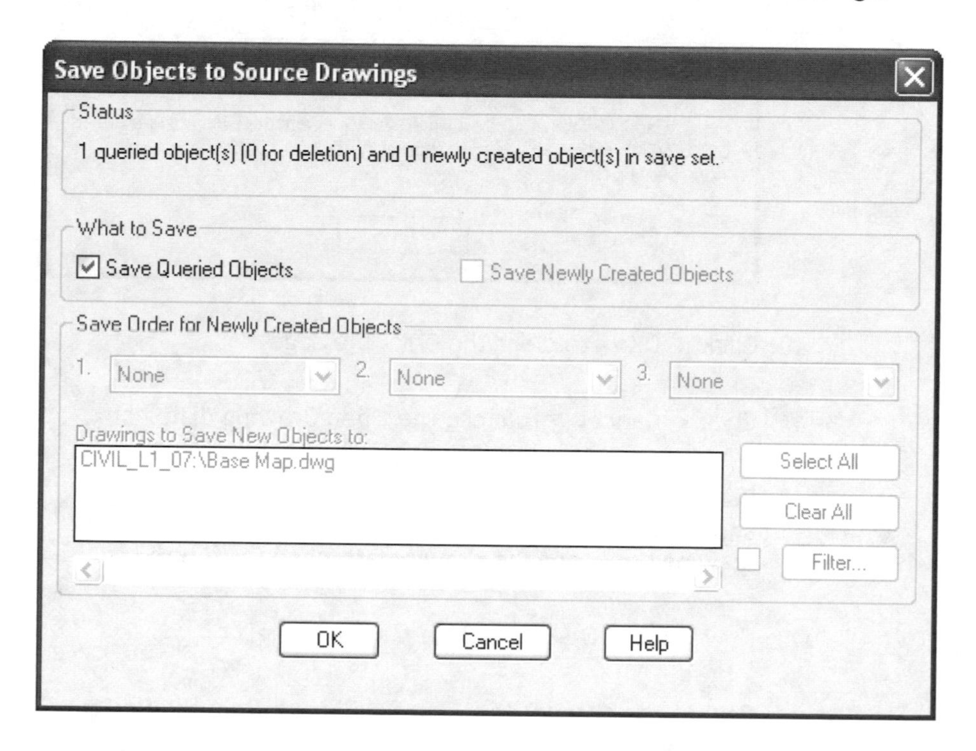

8. Click **<<OK>>** to save the object in the save set back to the source drawings.

Notice that the object(s) that you saved back have disappeared from your drawing. To bring them back you can just run the query again.

9. Now open the drawing **Base Map.dwg** to see that your changes were saved back to their source.

When you try to open a drawing that is attached and active in your current drawing you will receive the following error message.

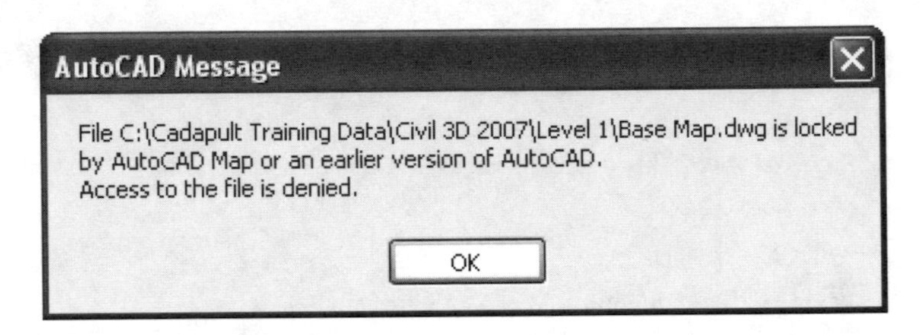

10. Click <<**OK**>> to clear the error message.

11. Click <<**Cancel**>> to close the Open Drawing dialog box.

12. Right-click on the **Base Map** drawing in the *Map Explorer* and select ⇒ **Deactivate**. This removes the lock on the drawing but keeps the drawing attached for future use.

13. Now you can **Open** the **Base Map** drawing.

14. Review the changes that you saved back from the Design drawing.

Don't worry that the Base Map drawing is now inaccurate because of your edits. You made these changes to illustrate the usage of the Save Back command. You will not be using this drawing in any future exercises.

15. Select **File** ⇒ **Close** to close the drawing *Base Map.dwg* and return to the drawing *Design.dwg*.

You do not need to save Base Map.dwg if you are prompted when closing the drawing.

16. Right-click on the **Base Map** drawing in the Map Explorer and select ⇒ **Detach**.

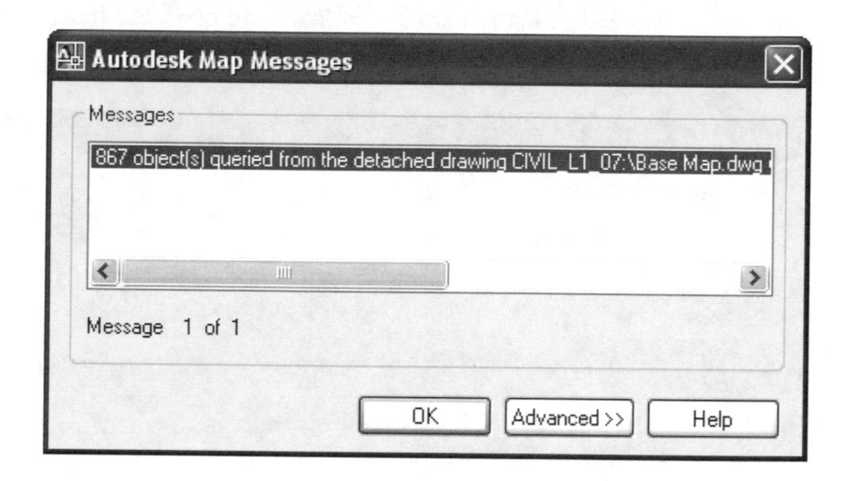

Detaching a source drawing after you have queried objects from it will trigger the above message from Map. This is not an error it is just an informational message explaining that if you detach the source drawing the objects that you have queried will no longer know that they were queried from a source drawing. If you make any edits to these objects the program will not prompt you to add them to the save set. You can always erase these objects, attach the source drawing again, and rerun the query to get a current copy of the object in the drawing so you will be able to save edits to the source drawing.

2.4 Inserting Images and the Project Boundary

These exercises explore ways to add raster images and a block containing the project boundary to your drawing.

2.4.1 Inserting a Registered Image (Rectified Aerial Photography)

It is important to use the **Map** ⇒ **Image** ⇒ **Insert** command when inserting a registered image, rather than the commands in the *AutoCAD Insert* pull-down. This is because the Map command utilizes correlation information, in our example World Files (.tfw), to automatically place your image correctly in your coordinate system. Otherwise you would need to register each image manually.

1. Create a layer called **"Image"** and set it **Current**.

2. Select **Map** ⇒ **Image** ⇒ **Insert** to open the Insert image dialog box.

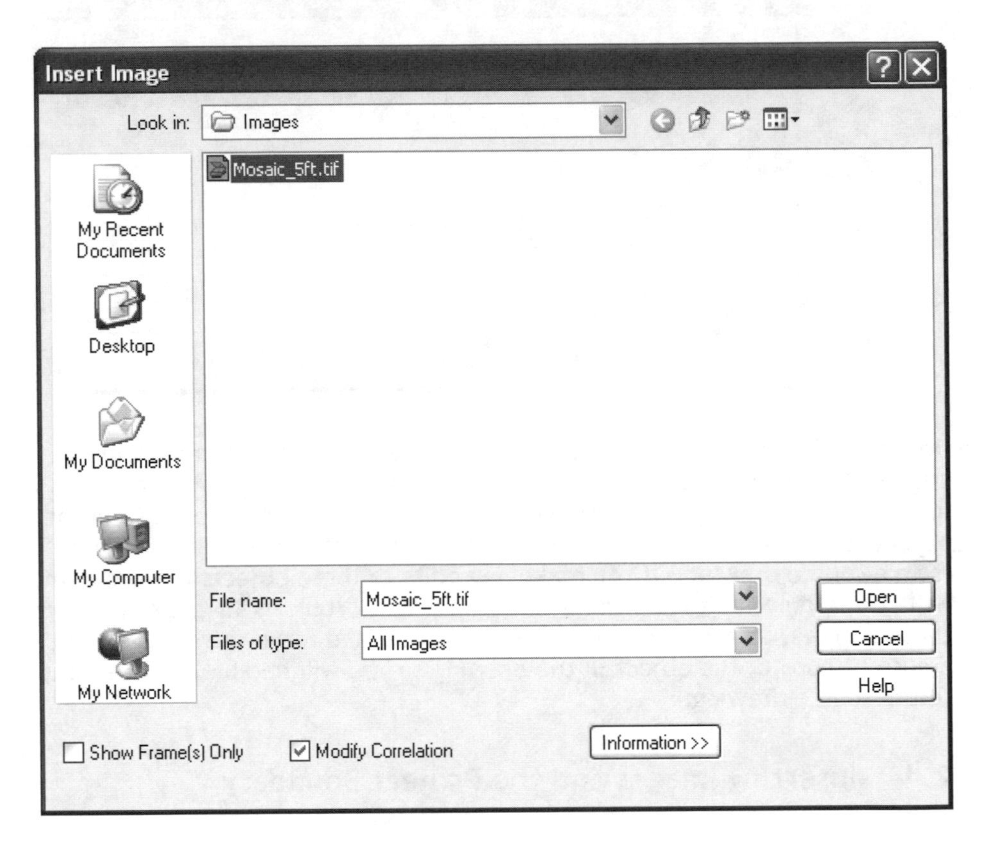

3. Browse to:
   C:\Cadapult Training Data\Civil 3D 2007\Level 1\Images.

4. Confirm that the *Files of type:* is set to **All Images**.

5. Select the file **Mosaic_5ft.tif**.

6. Click **<<Open>>**.

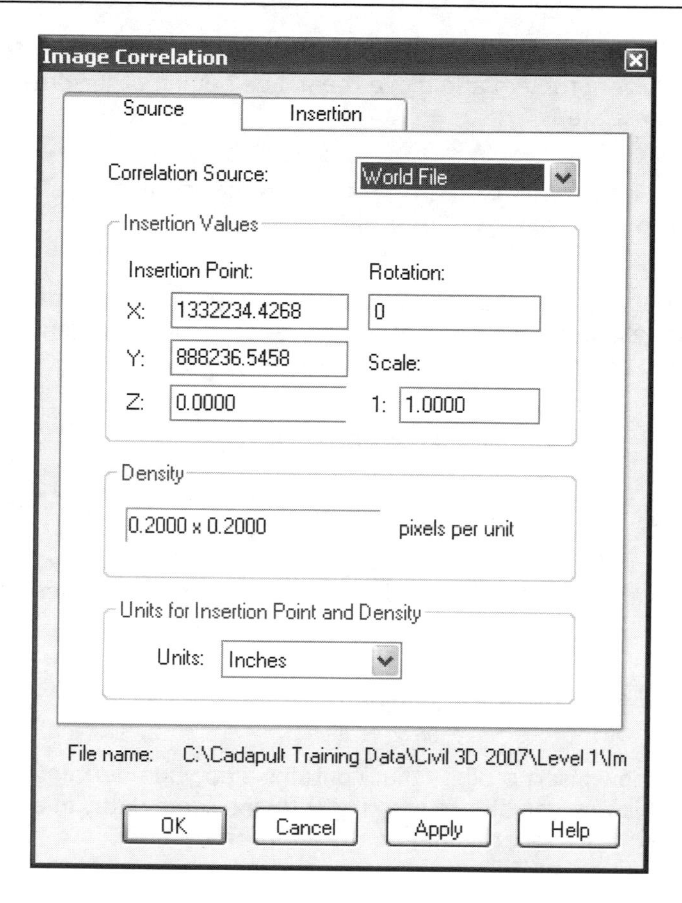

You will see that *Civil 3D* has automatically found the *World File* and set the *Correlation Source* and *Insertion Values*. You will also see a frame in your *AutoCAD* drawing editor showing the proposed location of the image.

7. Click **<<OK>>** to insert the image.

By default, the image will cover everything in your drawing. To correct this *display order* problem and move the image behind other drawing entities, follow these steps.

8. Select the Image by its frame, right-click and select **Display Order ⇒ Send to Back.**

The aerial photographs provide useful background information, however they do take a lot of system recourses to process and regenerate. So, when you are not using them it is best to turn them off.

9. Set layer **0** current.

10. Freeze the layer **Image**.

The image is still attached and will be displayed again when the layer is thawed.

2.4.2 Adding the Project Area

You will now insert a block that contains a polyline defining our project area. This way you all are working with the same data, in a normal project you would just draw this line in the appropriate area.

1. Select **Insert ⇒ Block.**

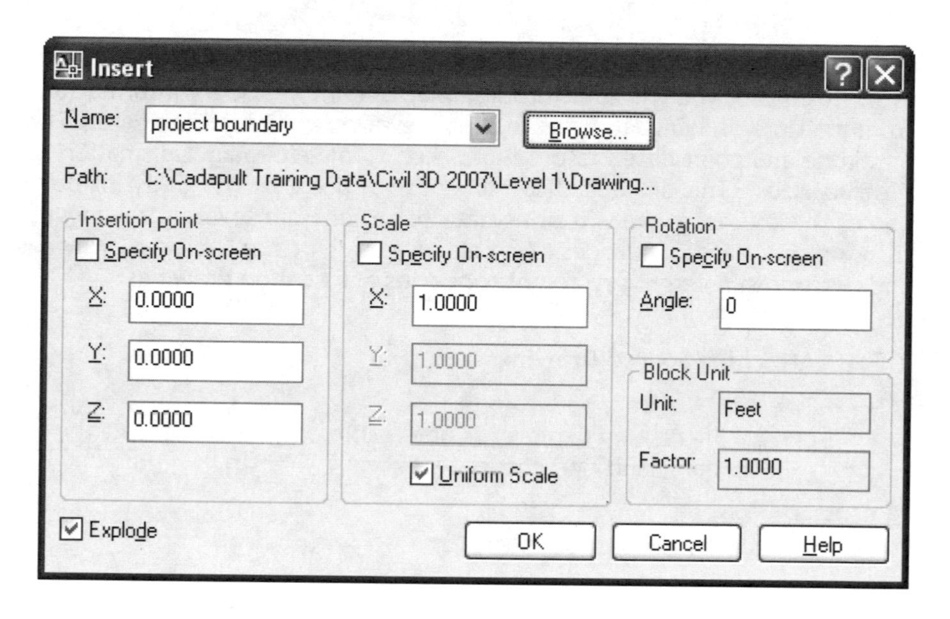

2. Browse to the file:

C:\Cadapult Training Data\Civil 3D 2007\Level 1\Drawings\project boundary.dwg

3. Disable the option to **Specify the Insertion Point** On-screen.

4. Enable the option to **Explode** the block.

5. Click **<<OK>>** to insert the project boundary.

6. Freeze the **Ex-Parcels** and **Ex-Streets** layers.

7. Save the drawing.

2.5 Importing Point and Breakline Information from Aerial Mapping

In this exercise, you will use *Autodesk Map's* ability to query information from multiple drawings to bring in spot elevation and breakline data. The breakline and point data is stored in a tiled format to maintain smaller drawing sizes. This project, like many others, does not fit neatly inside of one of these tiles, so we will bring data together from several drawings to fit our project area. *Autodesk Map* allows you to attach multiple drawings and perform simple or very complex queries across tiled drawings.

2.5.1 Attaching Source Drawings

1. If the *Map Explorer* is not visible on your screen, At the command line enter:

2. `Command:` **`MAPWSPACE`**

3. Type "**`ON`**" to display the Task Pane which includes the Map Explorer.

4. Set the **Task** at the top of the *Task Pane* to **Map Explorer**.

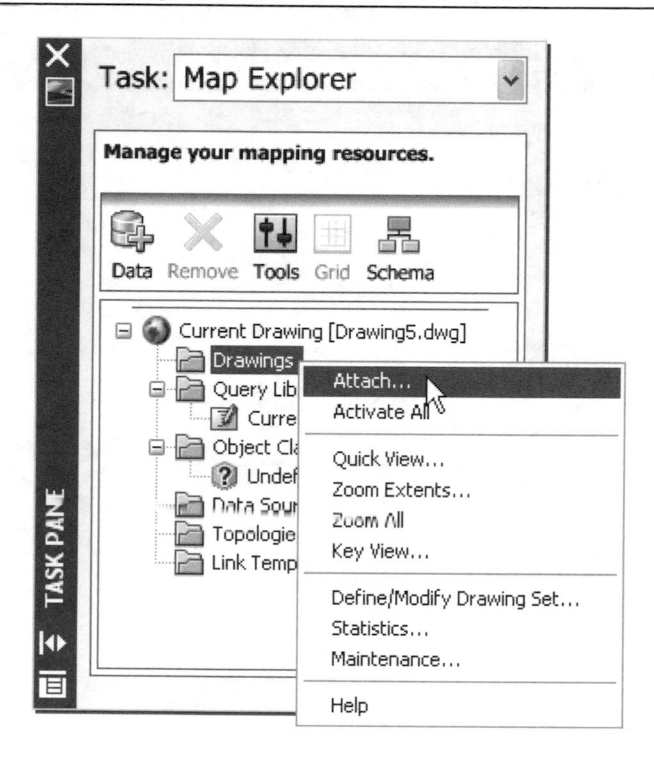

5. Right-click on **Drawings** and select ⇒ **Attach**.

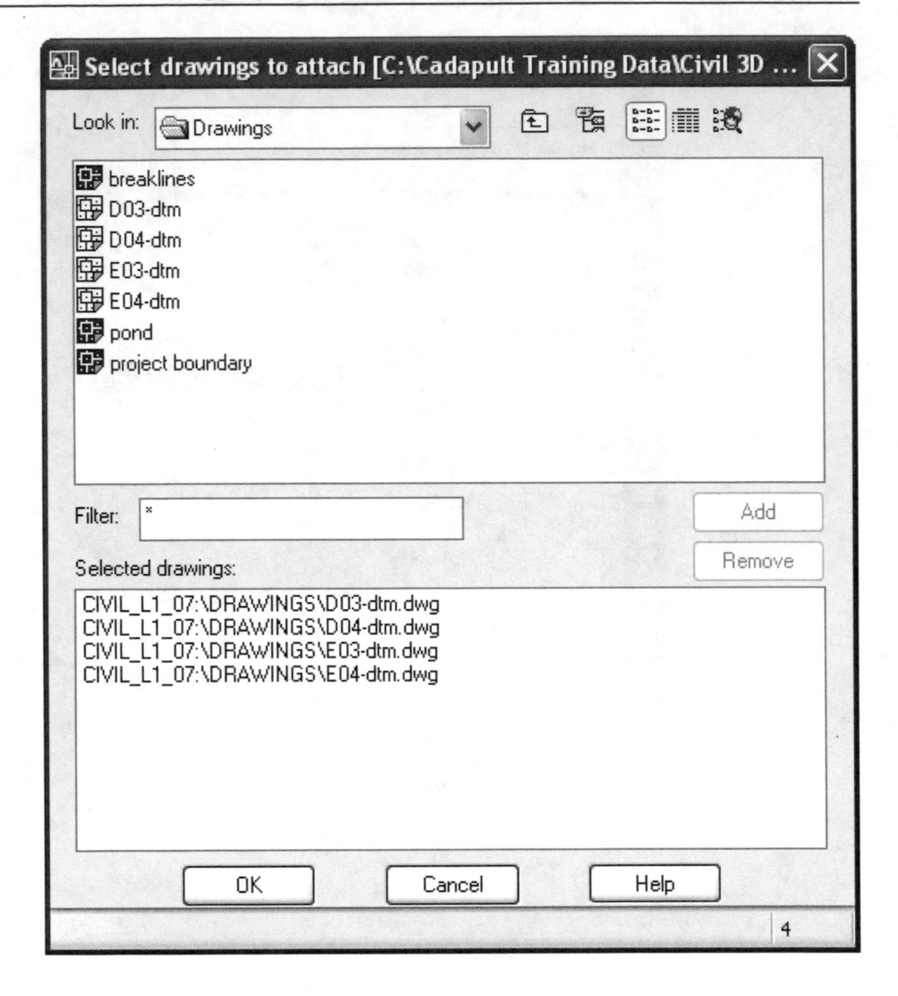

6. In the *Select Drawing* dialog box, select the **CIVIL_L1_07** alias at the top of the dialog box.

7. Open the **Drawings** folder.

8. Hold down the **Ctrl key**, select **D03-dtm, D04-dtm, E03-dtm** and **E04-dtm,** and then click **<<Add>>**.

9. Click **<<OK>>**.

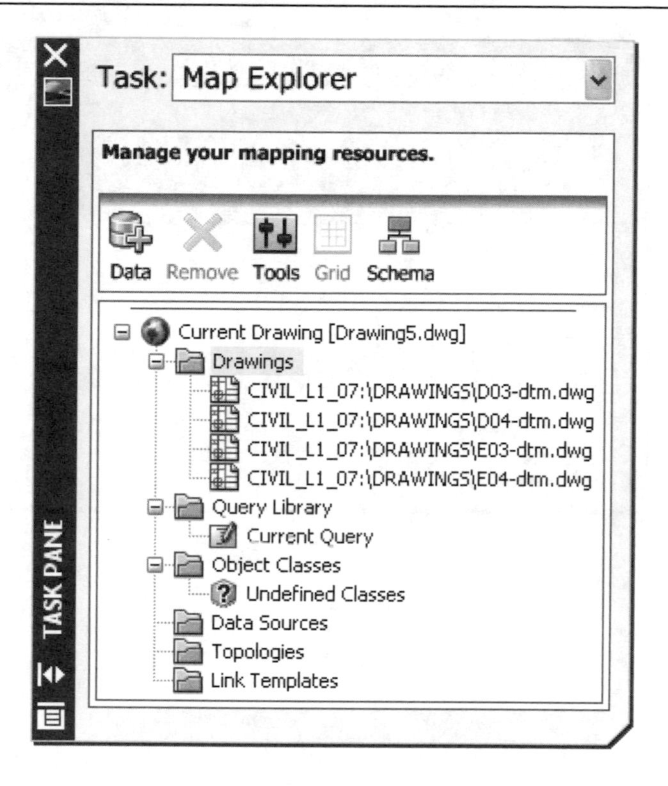

You will now see the four drawings listed in the *Map Explorer*.

10. Right-click on the drawings folder and select ⇒ **Quick View**.

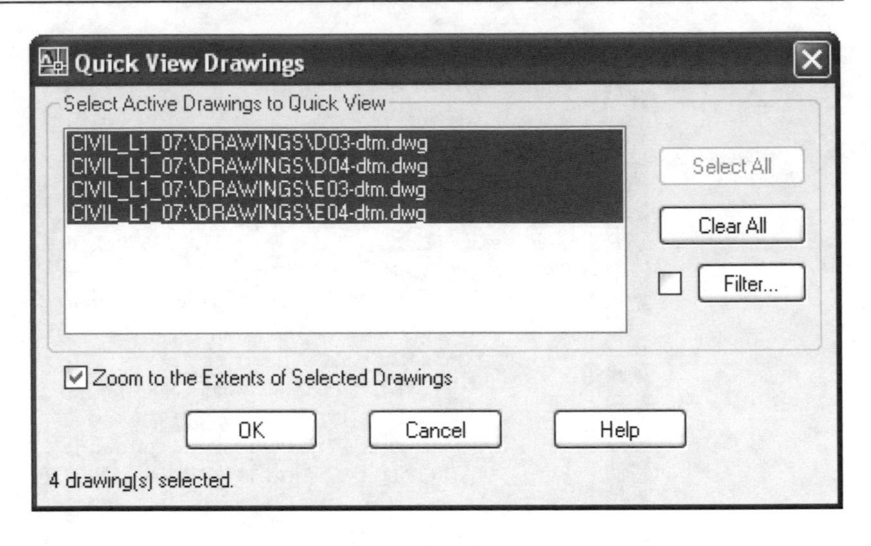

11. Click <<**Select All**>>.

12. Click <<**OK**>> to display the *Quick View* of all the attached drawings.

This will show you a preview image of what is contained in the attached drawings. The Quick View image will go away when you use a Redraw.

13. After you finish reviewing the contents of the attached drawings enter **R** at the command line to Redraw the display and clear the *Quick View*.

2.5.2 Defining a Compound Query

You can use any combination of four different query types to create your query. The query types are *Location, Property, Data*, and *SQL*, which are detailed in section 2.3.2.

By using several query types together you can limit the objects that are queried into your drawing giving you a very specific selection set and keeping your drawing sizes smaller. This is typically referred to as a *Compound Query*.

In this exercise you will use the query to select just the points and breaklines from all four attached drawings containing aerial survey data.

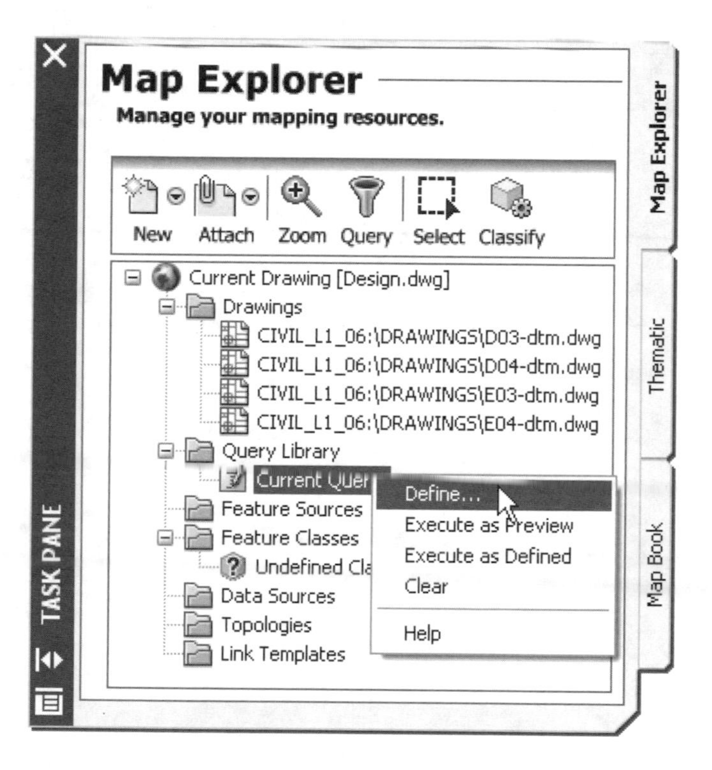

1. Expand the **Query Library** node in, then right-click on **Current Query** and select ⇒ **Define.**

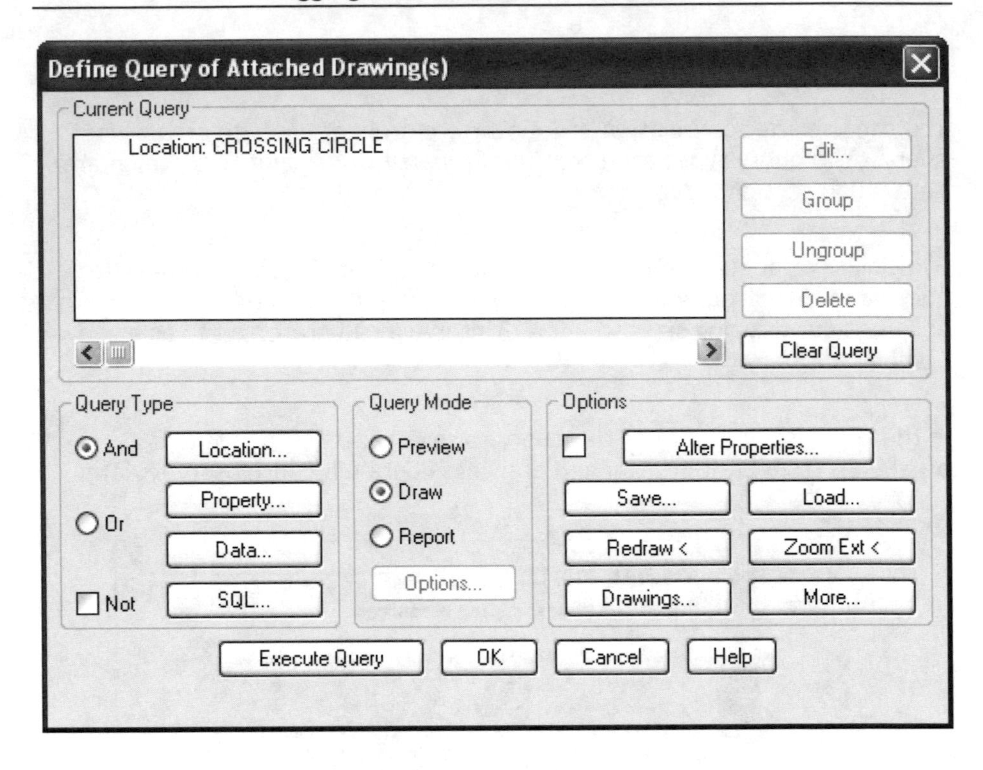

2. The previous query is still saved in the query dialog box. Click **<<Clear Query>>** to clear the previous query criteria and start a new one.

3. Click **<<Location>>** in the "Query Type" section.

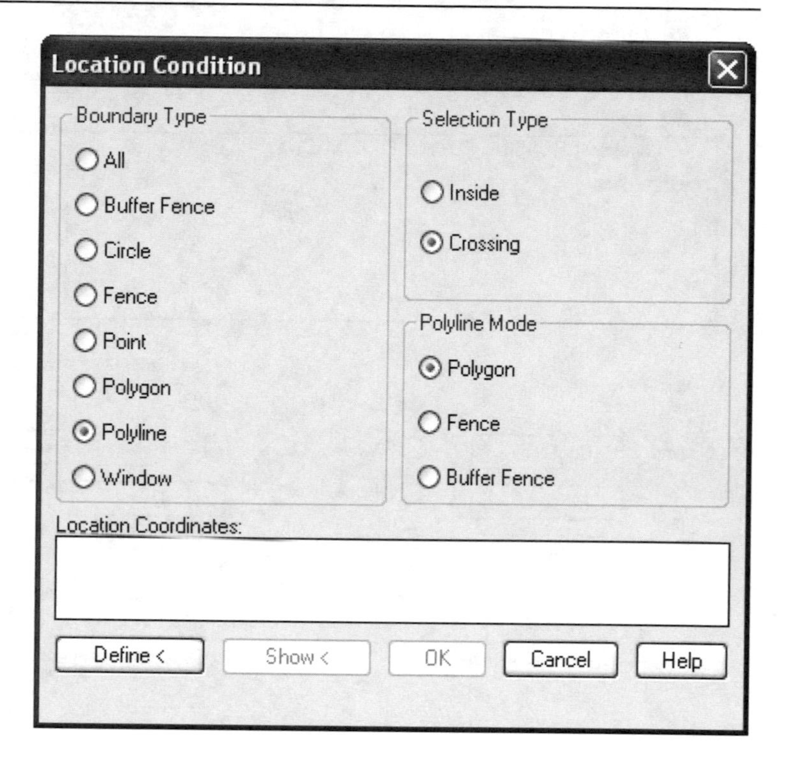

4. Choose a **Boundary Type** of **Polyline**.

5. Verify that the **Selection Type** of **Crossing** is chosen.

6. Verify that the **Polyline Mode** of **Polygon** is chosen.

7. Click **<<Define>>**.

8. When asked to select the polyline, pick the magenta site boundary that you just inserted on the layer *Project Boundary*. You will see a preview of the boundary you selected on the screen.

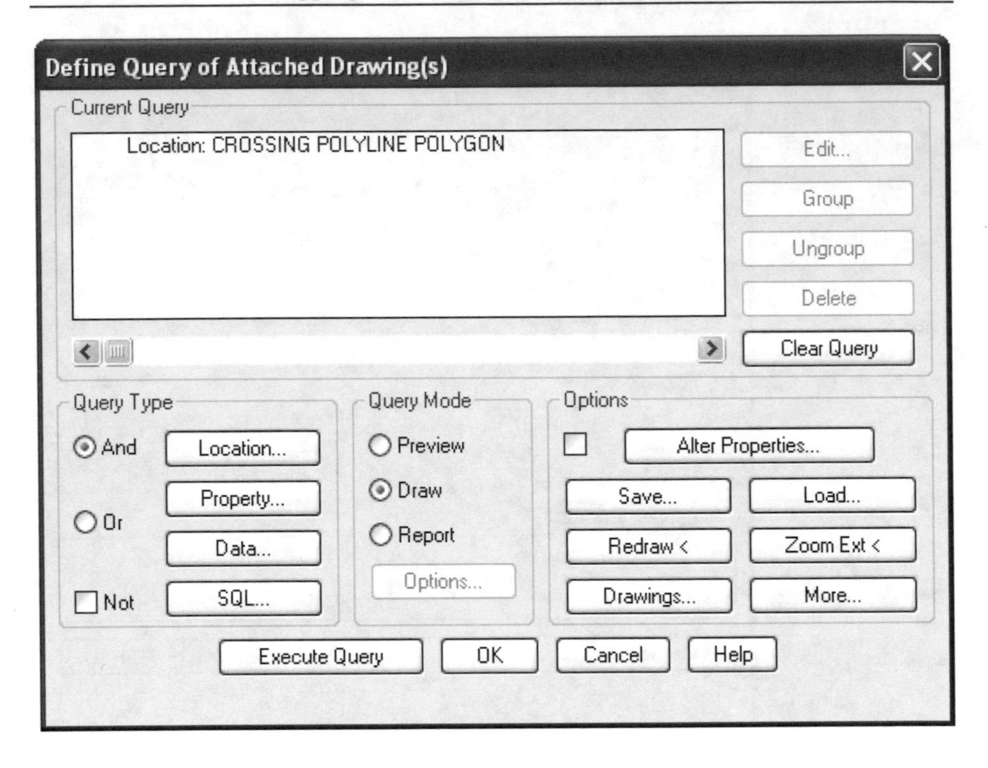

9. Back in the *Define Query* dialog box, in the **Query Type** section click **<<Property>>**.

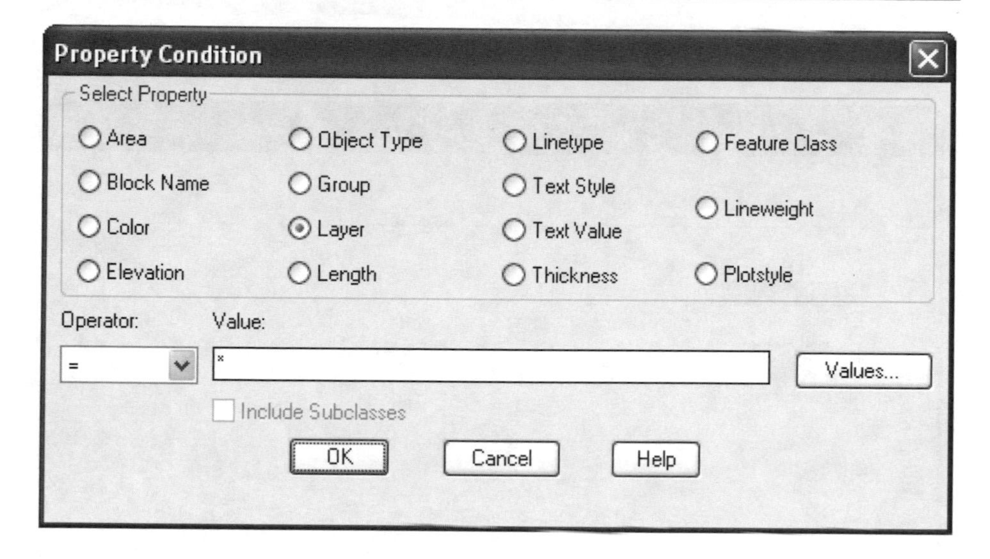

10. Select the **Property** of **Layer**.

11. Click **<<Values>>**.

You will see a list of all of the layers in all of the active attached drawings.

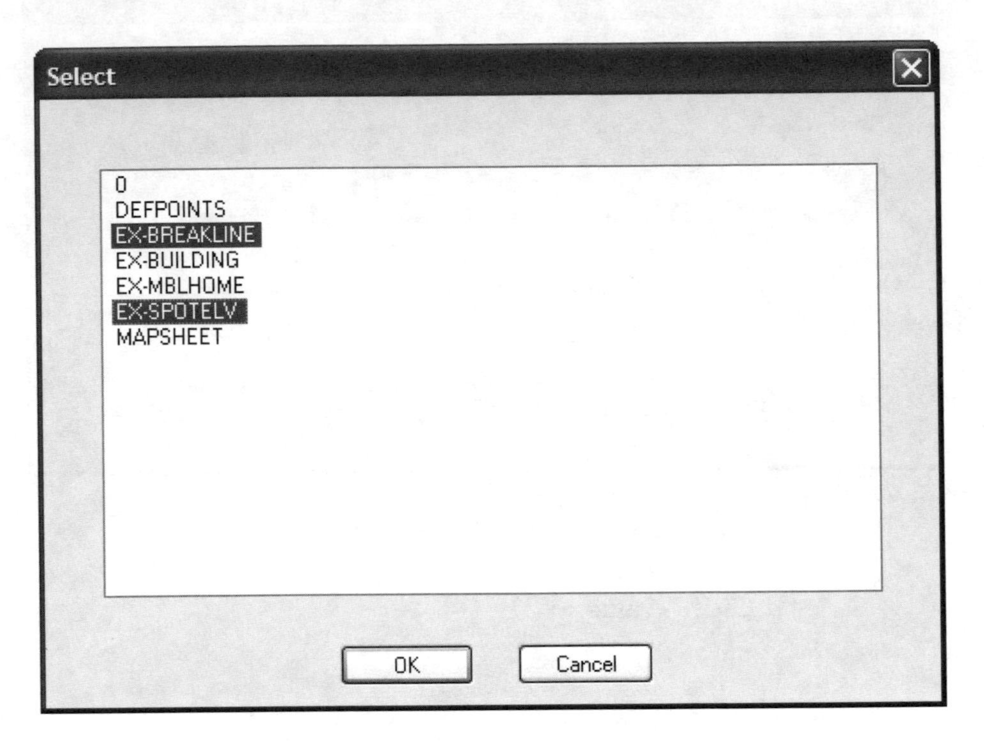

12. Hold down the control key, and select the layers: **EX-BRKLINE, EX-SPOTELV**.

13. Click **<<OK>>**.

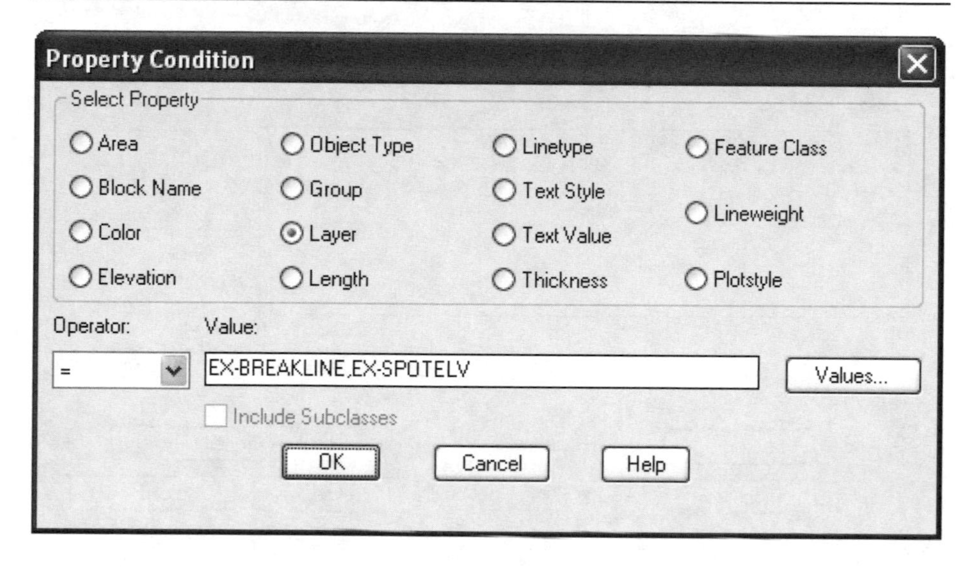

14. Click **<<OK>>** to save your changes to the **Property Condition** dialog box.

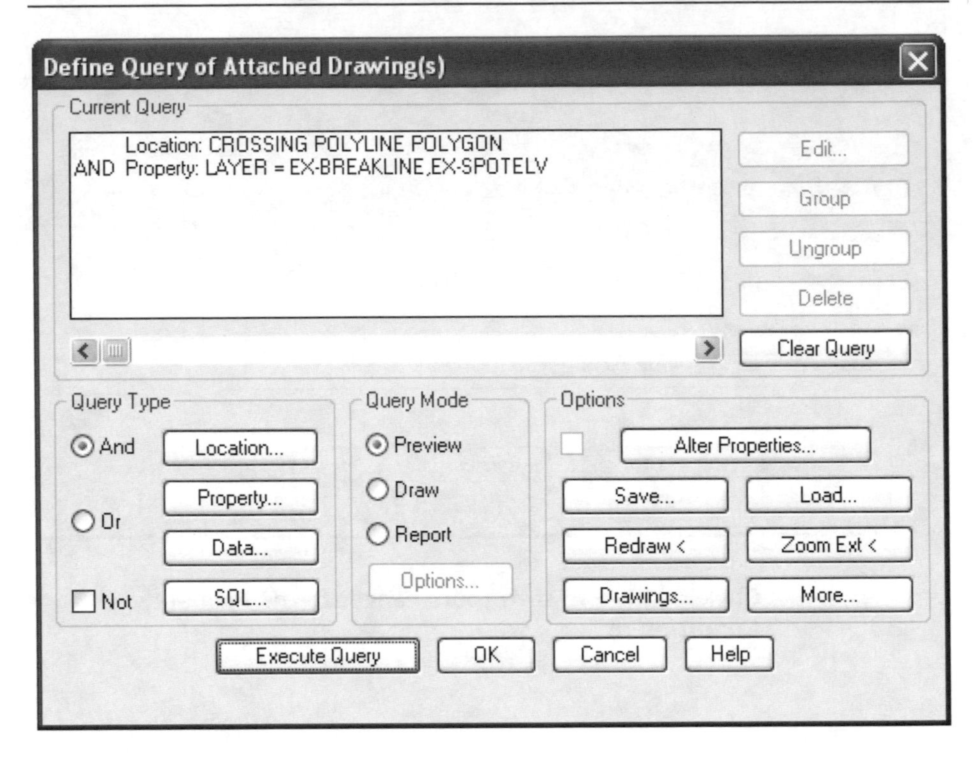

15. Back in the *Define Query* dialog box, Set the **Query Mode** to **Preview** and click **<<Execute Query>>**.

Preview mode will display a *Quick View* of the objects that meet the criteria you set in the *Query*. This gives you a chance to confirm that you have defined the query correctly and it is selecting the objects that you expected it to. You can clear that display with a redraw. If you do not see the points and breaklines displayed, or if you see too much information, you will need to check the definition of the query before you proceed.

16. Right-click on the **Current Query** in *Map Explorer* and select ⇒ **Define**.

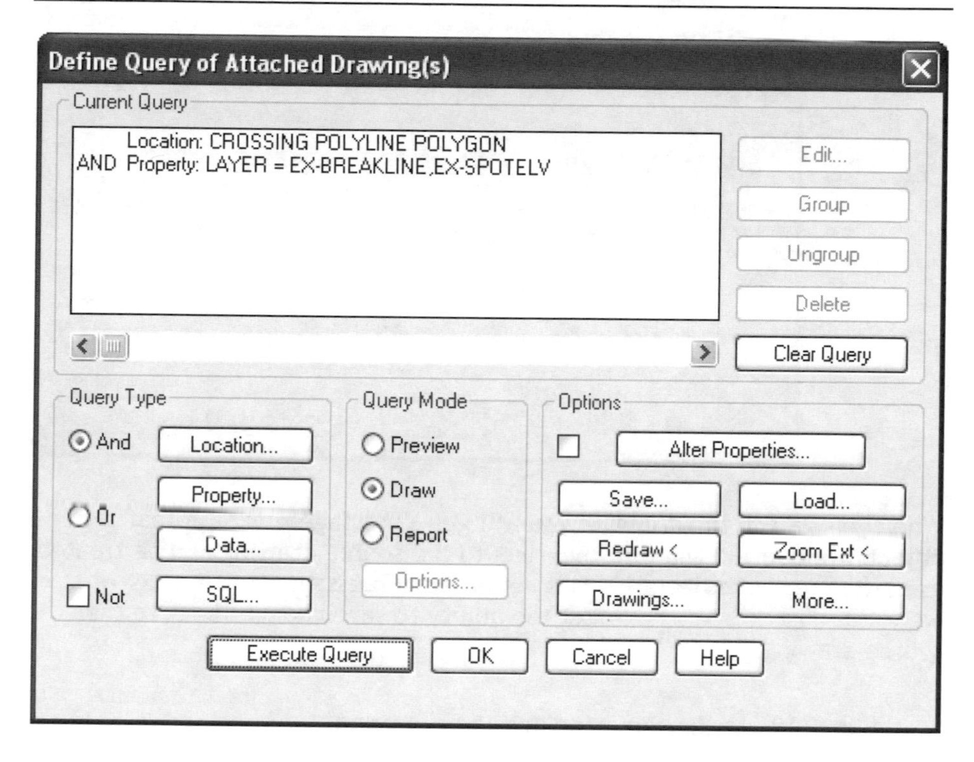

All of the information from the previous query is retained.

17. Change the **Query Mode** to **Draw** and click **<<Execute Query>>**.

This will bring all of the objects that you saw in the preview into the drawing. They are brought in as regular *AutoCAD* objects so you can edit them using any of your AutoCAD commands. Any changes can be saved back to the source drawings. Be careful with *Savebacks*. In many situations, like yours here, you would never want to save changes back to the source data. That would change and corrupt the information from your aerial survey.

18. To prevent any changes from being saved to the source drawing select all four of the source drawings in the *Map Explorer* using the shift key. Then right-click on the selected source drawings and select ⇒ **Detach**.

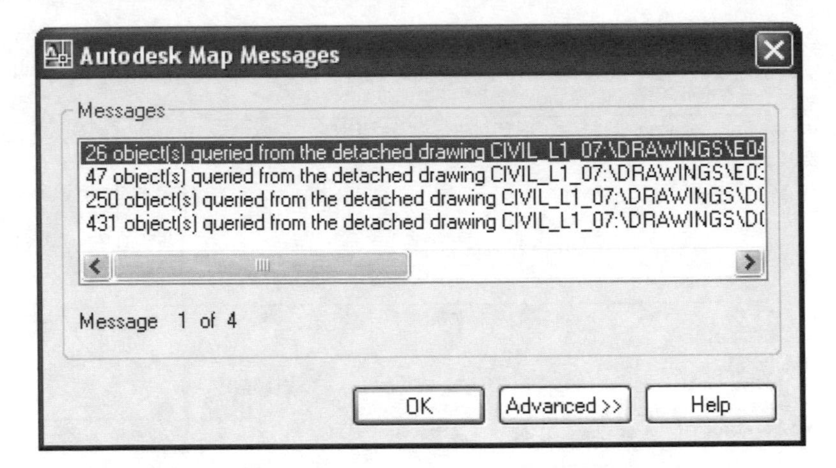

When you detach the drawings you will get a message that says that the objects we queried will lose their link to the source drawing and be treated as newly created objects. This is not an error message; it is exactly what we wanted to do. You have lost the ability to save changes back to the source drawing.

19. At the command line, type "**PDMODE**".

20. Set the point mode to "**3**", then **<<Enter>>** to end the command.

The spot elevations will now be displayed with an X rather than just dots. The size is relative to the screen so as you zoom in and out they will change size after a Regen.

21. Save the drawing.

2.6 Chapter Summary

In this chapter you have used *Civil 3D*, and the *Map 3D* components of *Civil 3D* in particular, to collect base data for your project from a variety of sources. It is important to keep in mind that every project is different and an important first step is to think about what type of data that you have available and what you want to collect. You can then use a variety of different commands, as they are appropriate; to compile that data similar to the way we have in this chapter.

Chapter 3

Preliminary Layout

This chapter will look at creating a preliminary surface of the existing ground from the *AutoCAD* objects that you queried into the drawing in the previous chapter. In later chapters you will build a surface from survey data and merge it with the preliminary surface using the preliminary surface to add buffer data around your survey.

- **Creating a Preliminary Existing Ground Surface**
- **Creating a Preliminary Alignment**
- **Creating Points from an Alignment**
- **Creating a Point Group**
- **Exporting the Points for Field Verification**

Dataset:

To start this chapter you will continue working in the drawing named **Design.dwg**. You can continue with the drawing that you currently have from the end of the previous chapter or, if you are starting in the middle of the book, you can open the drawing **CH-03.dwg** located in the folder **C:\Cadapult Training Data\Civil 3D 2007\Level 1\Chapter Drawings.** Opening the drawing from the dataset provided will ensure that you have the drawing set up correctly for the exercises in the following chapter and overwrite any mistakes that you may have made in previous exercises.

3.1 Overview

This chapter will look at creating a preliminary surface of the existing ground from the *AutoCAD* objects that you queried into the drawing in the previous chapter. In later chapters you will build a surface from survey data and merge it with the preliminary surface using the preliminary surface to add buffer data around your survey. You will also create a preliminary alignment and export points based on that alignment for field verification.

3.1.1 Concepts

In this chapter you will be introduced to several concepts that will be covered in more detail in later chapters of this book. You will learn how to create surfaces from *AutoCAD* objects and control surface display. You will use transparent commands to layout a preliminary alignment and you will create points based on that alignment. In later chapters *Points*, *Surfaces*, and *Alignments*; and the *Styles* that control them, will be covered in much greater detail with specific examples. In this chapter just try to get comfortable working with the objects.

3.1.2 Styles

Styles are saved in the drawing and can be created and edited on the Settings tab of the *Toolspace* or on the fly during many object creation and editing commands.

Surface Styles control the display of surfaces. You have individual control over many components of the surface including contours, triangles, points, borders, slope analysis, elevation bands, watersheds, and flow arrows.

Alignment Styles control the display of the alignment geometry. You have individual control over many components of the alignment including lines, curves, spirals, direction arrows, and PI points.

Alignment Label Sets control the group of styles used to label the alignment. This includes displaying stationing, ticks, geometry information, and much more.

Point Styles control the display of the point marker. They can use an *AutoCAD Point*, a custom marker similar to *Land Desktop*, or an *AutoCAD Block* for the marker. *Point Styles* can be applied to individual points, point groups, or point group overrides.

Point Label Styles control the display of the point label text and can be customized to show any type of point related information. *Point Label Styles* can be applied to individual points, point groups, or point group overrides.

3.2 Creating a Preliminary Existing Ground Surface

This set of exercises will look at creating a preliminary surface of the existing ground from the *AutoCAD* objects that you queried into the drawing in the previous chapter. In later chapters you will build a surface from survey data and merge it with the preliminary surface using the preliminary surface to add buffer data around your survey.

3.2.1 Creating a Surface

A *Surface* in *Civil 3D* is an *Object*. This surface object contains the surface definition and is saved in the drawing on a layer just like any other *AutoCAD* object. The surface object can be displayed many different ways by changing the surface style. As an example the surface can be displayed as triangles, contours, or elevation bands.

1. Continue working in the drawing **Design.dwg**.

2. If it is open, **Close** the **Map Task Pane**.

3. If the *Toolspace* is not visible, turn it on by selecting **General ⇒ Show Toolspace**.

4. On the *Prospector* tab of the *Toolspace*, right-click on **Surfaces** and select ⇒ **New**.

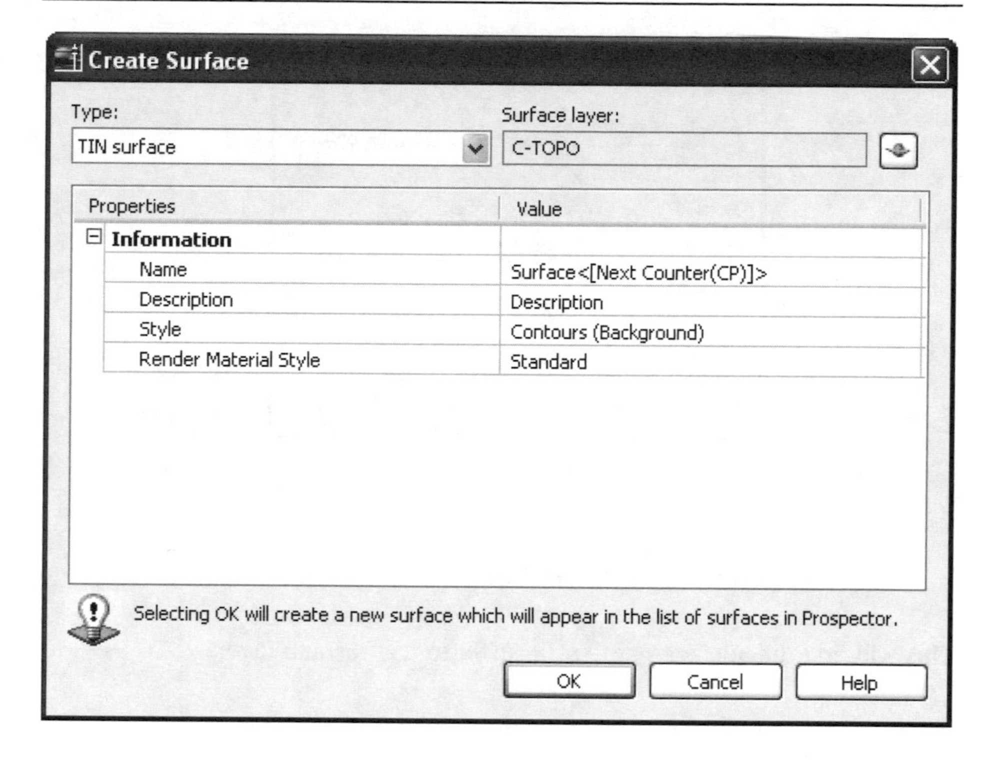

5. Confirm that **TIN surface** is selected as **Type**.

The *TIN Surface* type is the most common type of surface. It can be created from many different types of data that form a triangulated irregular network.

You can also create *Grid Surfaces* that are optimized for data on a regularly spaced grid. A common use of the Grid Surface is for building a surface from DEM data.

TIN Volume and *Grid Volume* surfaces can also be created for volume calculations between two surfaces. These are similar to the Composite Volume and Grid Volume calculations in *Land Desktop*.

6. Click the **Surface Layer** button to open the *Object Layer* dialog box.

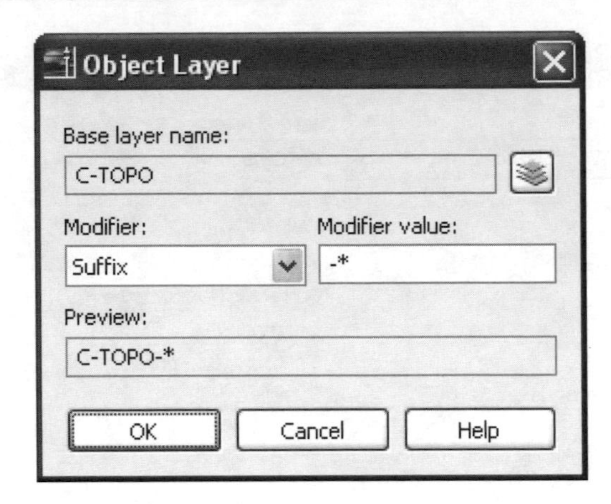

7. Set the **Modifier** to **Suffix**.

8. Enter "**-***" as the **Modifier value.**

This will add the surface name as a suffix to the surface layer.

9. Click **<<OK>>** to close the **Object Layer** dialog box.

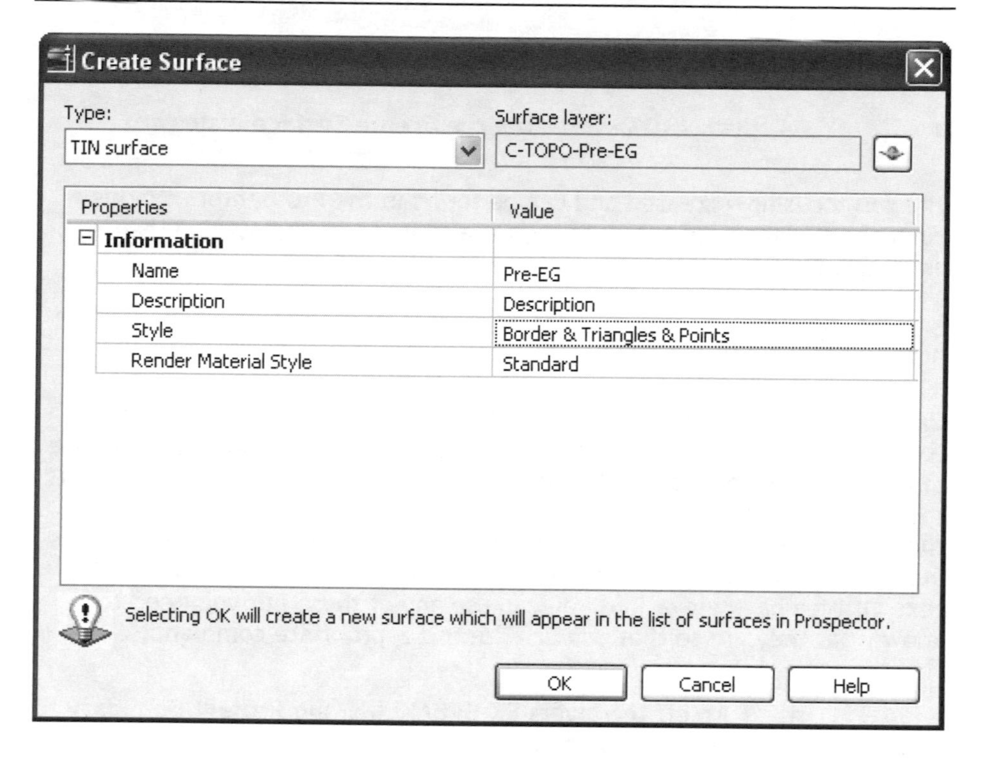

10. Enter **"Pre-EG"** for the Surface **Name**.

11. Click in the *Style Value* field to activate the **Ellipses** <<...>> **button**, and then click it to select a *Style*.

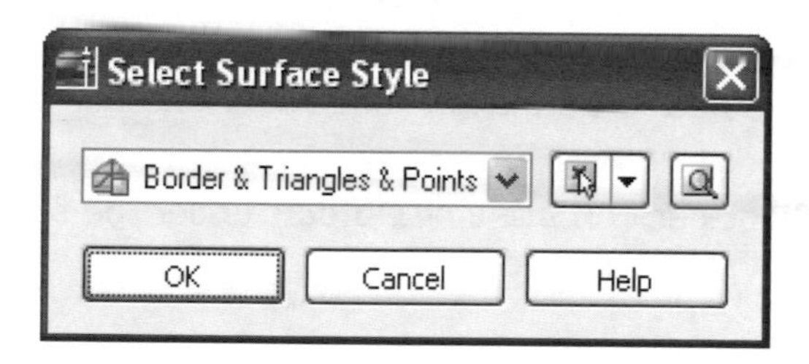

12. Select the *Style* **Border & Triangles & Points** from the drop-down list.

This will set your surface to display a standard looking TIN as soon as surface data is added.

13. Click <<**OK**>> to close the *Select Surface Style* dialog box.

14. Click <<**OK**>> to close the **Create Surface** dialog box.

The surface is now created and can be found in the Prospector. At this point the surface does not have any data so it is not displayed in the drawing editor.

3.2.2 Adding Surface Data

Once a surface is created you can add many different types of data to construct the model. The first step in this process is to find out what type of data you have available. You may want to List the properties of objects that you intend to use in the surface to find out what type of an object they are and if they have an elevation assigned to them. For example it is important to know if your spot elevations are *AutoCAD* points, blocks, or text. If they have elevations you can use any of these but you need to know what they are so that you can use the appropriate command.

1. Turn off the layers **EX-BREAKLINE** and **Project Boundary** to isolate the spot elevations layer.

2. On the *Prospector* tab of the *Toolspace*, expand the **Surfaces** node.

3. Expand the *Surface* **Pre-EG**.

4. Expand the **Definition** node under *Pre-EG*.

5. Right-click on **Drawing Objects** under the *Definition* node and select ⇒ **Add**.

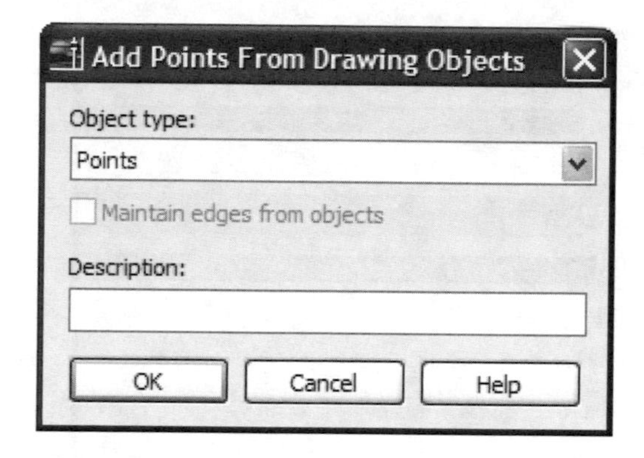

6. Confirm the **Object** type is set to **Points.**

7. Click **<<OK>>.**

8. Select all the points with a crossing window, then **<<Enter>>** to end the command.

The spot elevations are added to the surface and the surface is updated. The surface will display on your screen as triangles according to the surface style. If the surface is not visible on your screen, then check to see if the layer *C-TOPO-Pre-EG* is off or frozen. If it is, turn it on to display the surface. You may need to **Regen** the drawing after changing the layer state to see the surface.

9. Turn off the layer **EX-SPOTELEV.**

10. Turn on the layer **EX-BREAKLINE.**

11. Confirm that the **Definition** node under the Surface **Pre-EG** is expanded on the *Prospector* tab of the *Toolspace*.

12. Right-click on **Breaklines** under the **Definition node** and select ⇒ **Add.**

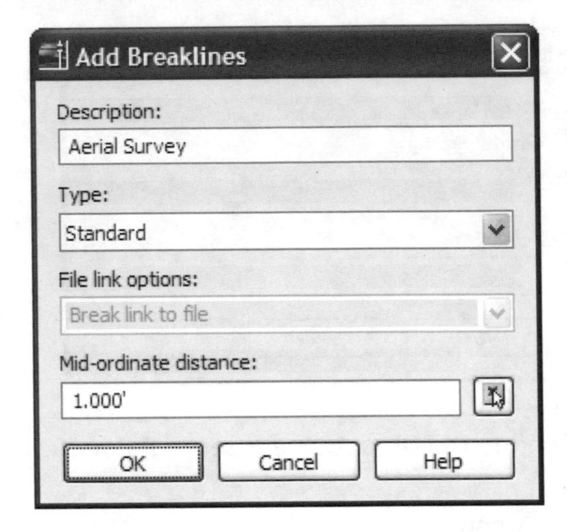

13. Enter **"Aerial Survey"** for the **Description**.

14. Confirm that the **Type** is set to **Standard.**

15. Click **<<OK>>**.

16. Pick the **breaklines** with a crossing window.

The breaklines are added to the surface and the surface is updated.

17. Turn **on** the layer **Project Boundary**.

18. Confirm that the **Definition** node under the Surface **Pre-EG** is expanded on the *Prospector* tab of the *Toolspace*.

19. Right-click on **Boundaries** under the **Definition node** and select ⇒ **Add.**

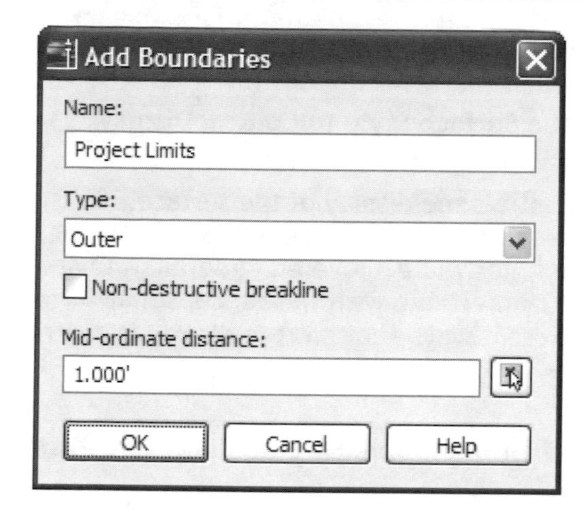

20. Enter **"Project Limits"** for the **Name.**

21. Confirm that the **Type** is set to **Outer.**

22. Confirm that the option for **Non-destructive breaklines** is **Disabled.**

Enabling the *Non-destructive breaklines* option trims the surface at the boundary while disabling it erases any surface lines that cross the boundary. This option is useful when you have good surface data on both sides of the boundary and are using the boundary to limit the extents of the surface. It is typically not used for outer boundaries.

23. Click **<<OK>>.**

24. Pick the **Boundary.**

The boundary is added to the surface and the surface is updated. The surface is dynamically linked to all of its surface data. So if you edit a point, breakline, or boundary the surface and the corresponding node under the surface definition will display as "Out of date" in the *Prospector*. If you right-click on the surface name in the *Prospector* and select *Rebuild* the surface will be updated to reflect the changes in the surface data. The undo command can be used to undo any changes to the surface or the surface data.

25. Save the drawing.

3.2.3 Changing the Surface Style to Control Display

The surface style controls the display of the surface. By changing the surface style you can display the surface in many different ways. Creating and editing surface styles will be covered in detail in *Chapter 5*. For now you should just get comfortable with the concept that surfaces can be displayed many different ways and that the display is controlled by the surface style.

1. Turn off the layers **EX-BREAKLINE** and **Project Boundary.**

2. On the *Prospector* tab of the *Toolspace*, right-click on Surface **Pre-EG** and select ⇒ **Properties.**

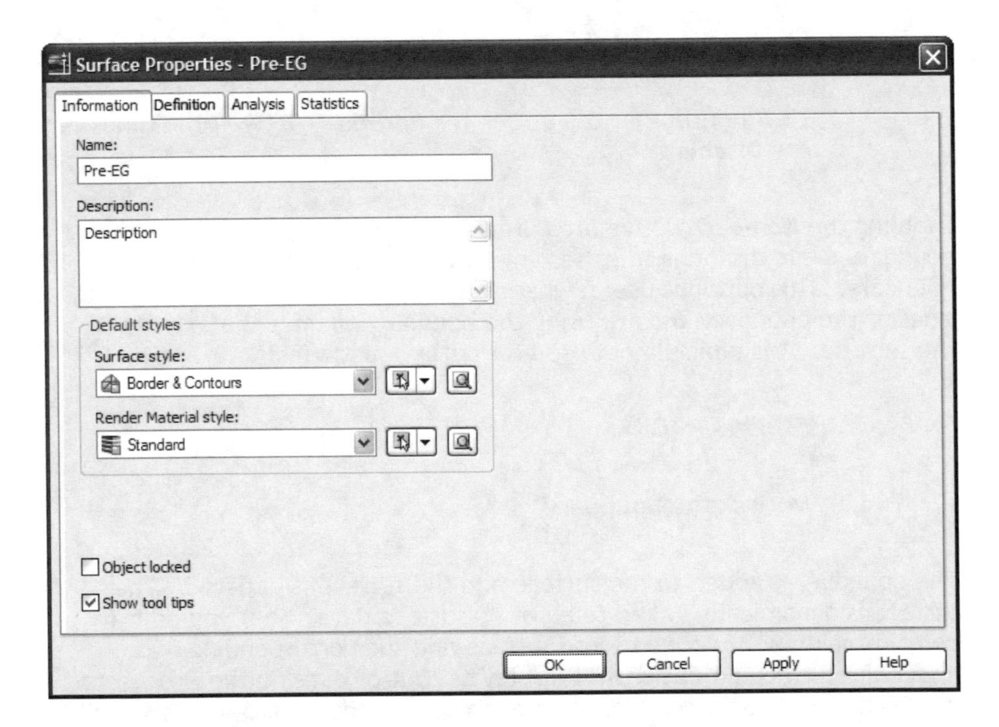

3. Select **Border & Contours** from the **Object style** list.

4. Click <<**OK**>>.

The surface is now displayed showing contours. The contour interval, color, smoothing, and much more are all part of the surface style.

3.2.4 Using the Object Viewer

The *Object Viewer* is a separate window that will allow you to view a selected object or objects in 3D and rotate them in real-time.

1. Pick one of the **contours** to highlight the entire surface.

2. **Right-click** and select ⇒ **Object Viewer**.

3. In the **Object Viewer**, click and drag while holding down the left mouse button to rotate the surface in 3D.

Once you rotate to a 3D view the contours will change to 3D faces. This is controlled by the surface object style.

4. In the *Object Viewer*, right-click and select ⇒ **Visual Styles** ⇒ **3D Wireframe.**

You will now see the wireframe triangles representing the surface.

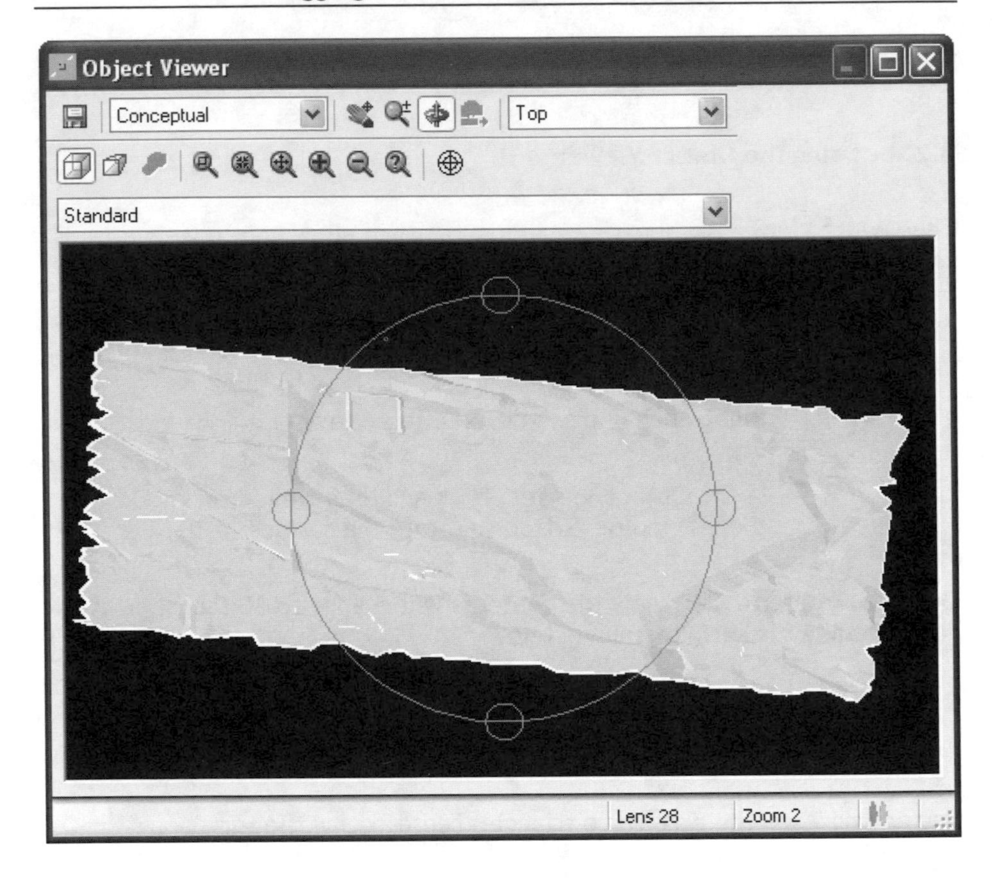

5. In the *Object Viewer*, right-click and select ⇒ **Visual Styles** ⇒ **Conceptual.**

You will now see shaded triangles representing the surface.

6. Continue to rotate the surface to examine it from different angles.

7. When you are finished viewing the surface, close the object viewer window to return to the drawing editor.

8. Save the drawing.

3.3 Creating a Preliminary Alignment

In this section you will layout a preliminary version of the horizontal alignment for the road in our project. This will be based on our aerial survey data and will need to be checked, and most likely modified, by the survey crew when they go on site and conduct their survey.

3.3.1 Drafting the Preliminary Alignment Using Transparent Commands

An *Alignment* in *Civil 3D* is an *Object*. This alignment object contains the alignment information and parameters like stationing, design speed, and superelevation information. The alignment is saved in the drawing on a layer just like any other *AutoCAD* object. The display of the alignment object is controlled by the alignment style and the labeling is controlled by the alignment label set.

1. Confirm that **Dynamic Input** is turned off. If it is on Click the DYN button on the *Status Bar* to disable it.

Using *Dynamic Input* with the *Transparent Commands* can be confusing and cause unexpected results.

2. Select **Alignments ⇒ Create By Layout.**

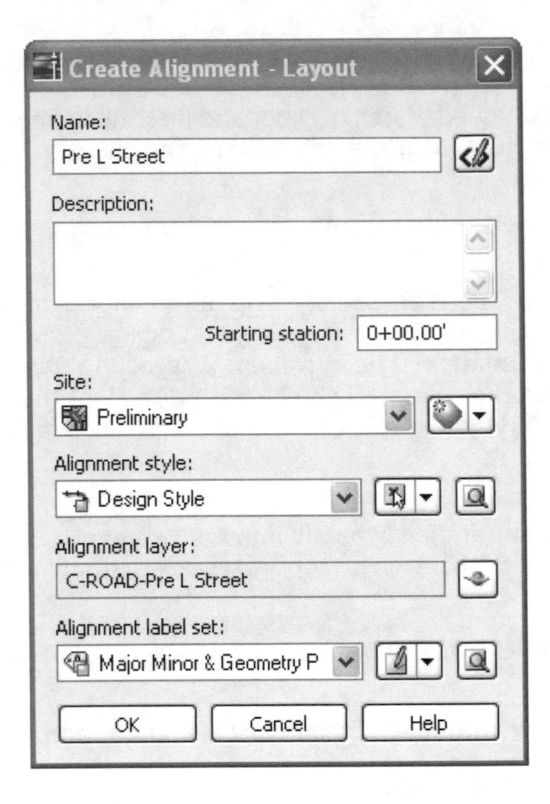

3. Change the default **Name** to "**Pre L Street**".

In *Civil 3D* a *Site* is a collection of objects that relate to and interact with each other. *Alignments*, *Parcels*, and *Grading Groups* can all be contained in a *Site*. By using multiple *Sites* in your drawing you can control which of these objects interact with each other. You may want to use different sites for different phases of a project or for different design alternatives.

4. Click the **Create New Site** button to open the *Site Properties* dialog box.

5. **Name** the new Site **"Preliminary"**.

6. Click **<<OK>>** to create the new site and return to the *Create Alignment - Layout* dialog box.

7. Confirm that the **Alignment Style** is set to **Design Style**.

The alignment style controls the display of the alignment geometry. This particular style will use continuous lines with the tangents colored red and the curves colored blue. This may be helpful during the design process because you can easily identify where the tangents and curves begin and end. However, it may not work well for plotting. In that case you can change the alignment style when you are ready to plot to display the colors and linetypes that you prefer to see on your plots.

8. Click the **Alignment Layer** button to open the *Object Layer* dialog box.

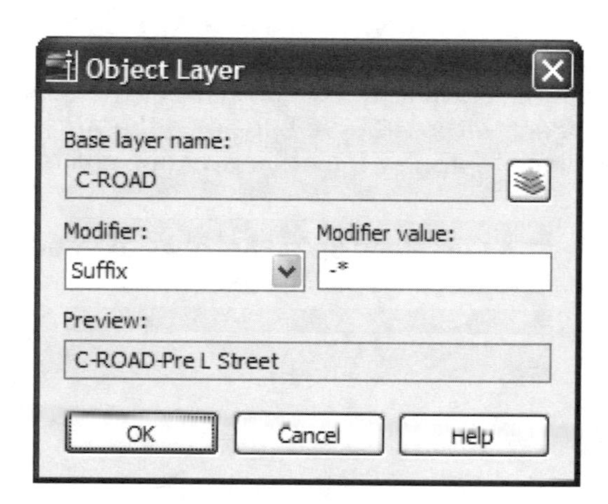

9. Set the **Modifier** to **Suffix**.

10. Enter **-*** as the *Modifier value.*

This will add the alignment name as a suffix to the alignment layer.

11. Click **<<OK>>** to close the *Object Layer* dialog box.

12. Confirm that the Alignment label set is set to **Major Minor & Geometry Point**.

The alignment label set controls the group of styles used to label the alignment. This includes displaying stationing, ticks, geometry information, and much more. Alignment Labels will be discussed in more detail in *Chapter 6*.

13. Click **<<OK>>** to create the alignment and open the *Alignment Layout* toolbar.

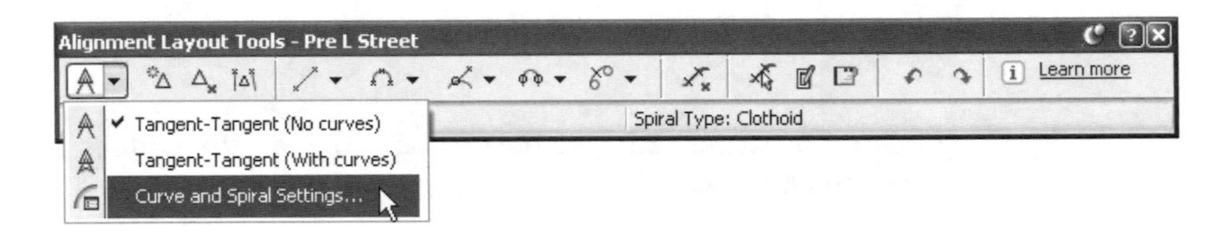

14. Click the down arrow next to the **Draw Tangent-Tangent without Curve** button on the *Alignment Layout* toolbar, to display the other creation commands.

15. Select ⇒ **Curve and Spiral Settings** to display the *Curve and Spiral Settings* dialog box.

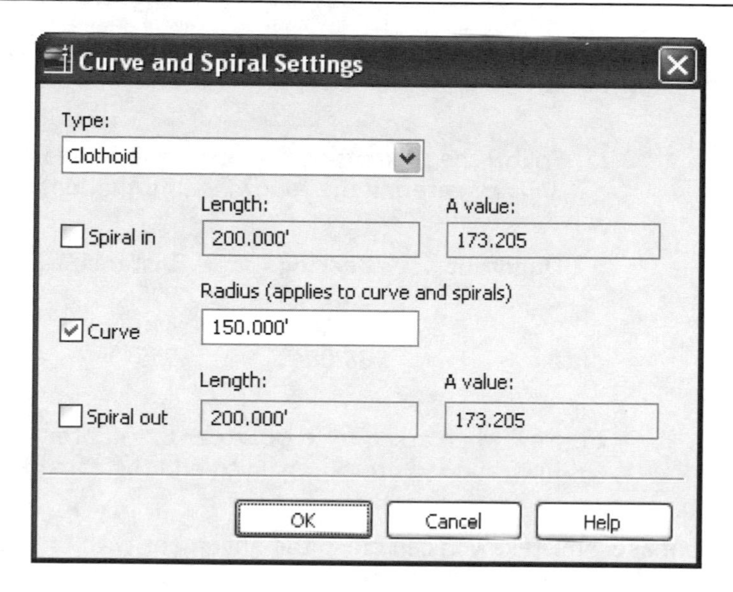

16. Set the default Curve **Radius** to **150.**

17. Click **<<OK>>.**

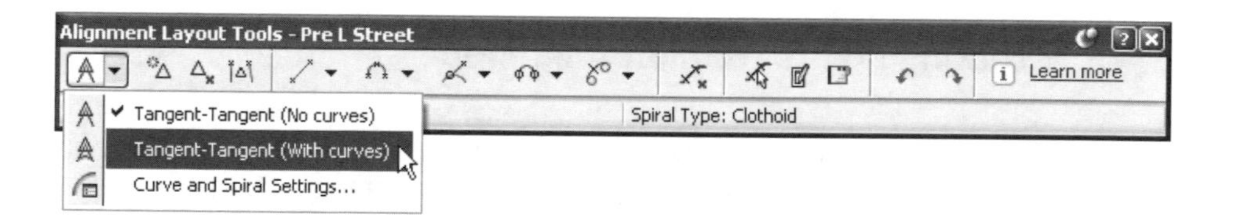

18. Click the down arrow next to the Draw **Tangent-Tangent (No curves)** button on the *Alignment Layout* toolbar to display the other creation commands.

19. Select ⇒ **Draw Tangent-Tangent (With curves).**

20. Enter "**1336096.7, 891291.3**" for the start point.

21. Enter `'bd` to start the bearing and distance transparent command.

You can also click the Bearing and Distance button from the *Transparent Commands* toolbar.

22. Follow the prompts at the command line to create the next 3 PIs by entering the following information.

Quadrant	Bearing	Distance
2	88.2435	995
3	03.5411	410
4	88.0843	955

23. Now when asked for a `Quadrant`, **<<Enter>>** to end the line, and **<<Enter>>** again to end the command.

If you make a mistake you can erase the alignment with the *AutoCAD* erase command or by right-clicking on it in the *Prospector*. You can also grip edit the alignment or edit it with the edit alignment command. All of these options will be covered in detail in Chapter 6.

24. Save the drawing.

3.4 Creating Points from an Alignment

Points can be created based on the alignment geometry. These points can be used to field check the preliminary layout of the alignment. This section will explore working with default settings for creating points and creating points based on an alignment.

3.4.1 Establishing the Point Settings

The *Create Points* toolbar contains all of the commands for creating points as well as an area to control all of the point settings. You may need to change the point settings often depending on the command that you are using and the type of points that you are creating.

1. Select **Points** ⇒ **Create Points.**

2. Select the **large down arrows** button on the right of the *Create Points* toolbar to expand it and show the point settings.

3. Expand the **Points Creation** parameters.

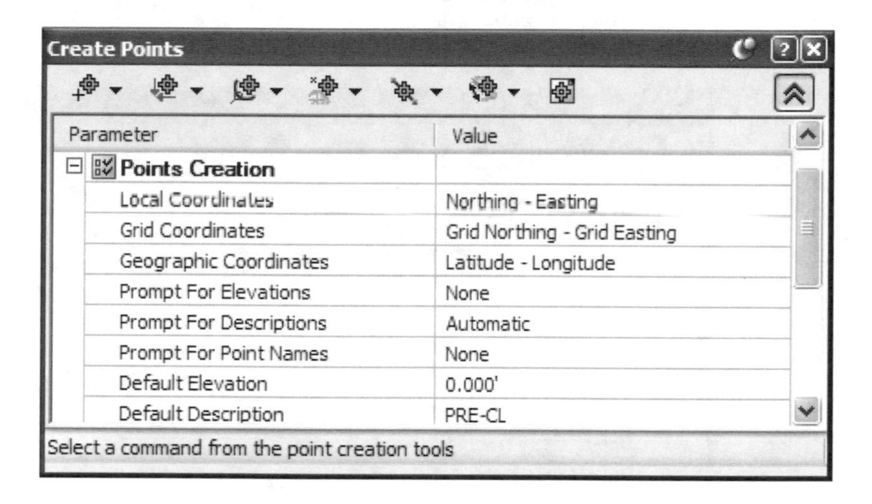

4. Set **Prompt For Elevations** to **None**.

This will automatically give the points a NULL elevation as they are created. This is important because if points with a NULL elevation are included in a surface they are ignored, while points with a zero elevation create a hole in the surface going down to elevation zero.

5. Set **Prompt For Point Names** to **None.**

6. Set **Prompt For Descriptions** to **Automatic.**

7. Enter a **Default Description** of **PRE-CL.**

This will automatically give the points a description of *PRE-CL* as they are created, rather than prompting you to enter the description of each point as it is created.

3.4.2 Setting Points on the Alignment

In this exercise you will create points based on the alignment geometry. These points are not linked to the alignment. So if the alignment is changed you will need to delete and recreate the points.

1. Click the **down arrow** next to the third button from the left on the *Create Points* toolbar, to display the other alignment options.

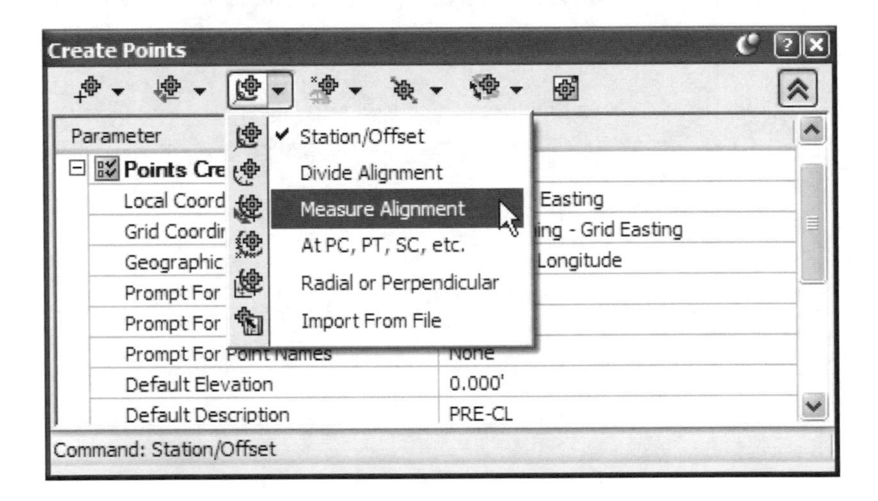

2. From the drop-down list, select the **Measure Alignment** button.

3. Select the **Pre L Street** alignment from the screen.

4. Accept the **default** beginning and ending stations.

5. Accept the **default** offset of **"0"**.

This will place the points on the centerline.

6. Use a station interval of **"50"**.

7. **<<Enter>>** when asked to select another Alignment to end the command.

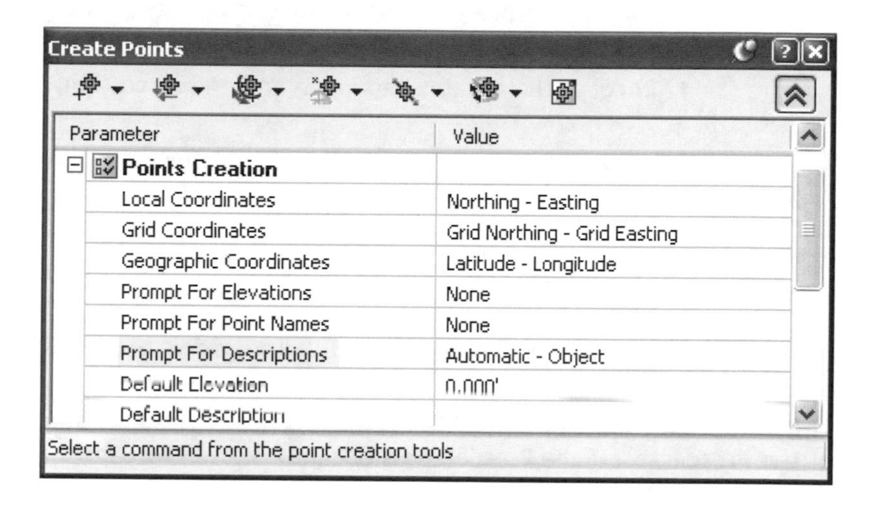

8. In the Points Creation Parameters, Set **Prompt For Descriptions** to **Automatic - Object.**

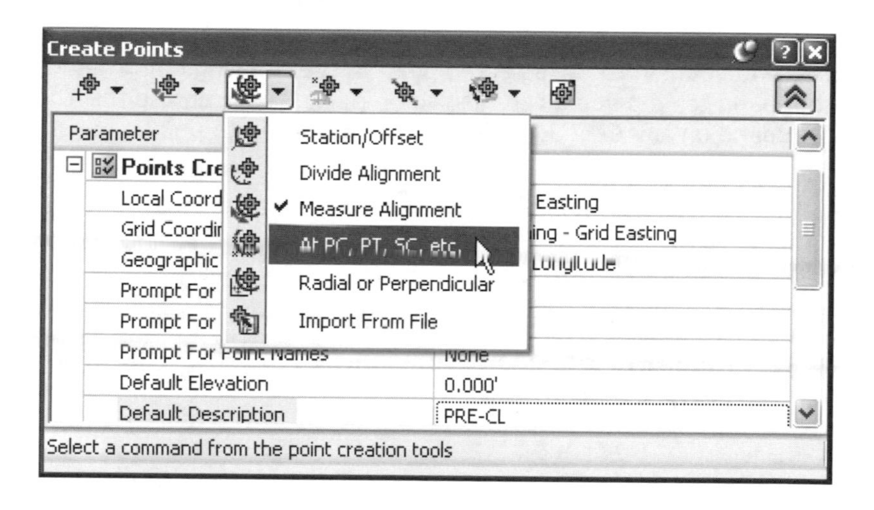

9. Click the **down arrow** next to the third button from the left on the *Create Points* toolbar, to display the other alignment options.

10. From the drop-down list, select **At PC, PT, SC, etc...**

11. Select the **Pre L Street** alignment from the screen.

12. Accept the **default** beginning and ending stations.

13. **<<Enter>>** when asked to select another Alignment to end the command.

14. Close the **Create Points** toolbar.

15. On the *Prospector* tab of the *Toolspace*, pick **Points**.

This will display a list of all the points in the preview window at the bottom of the Prospector, if the Prospector is docked. If the Prospector is not docked it will display on the side.

16. Right-click on any point in the preview window of the *Prospector* and select ⇒ **Zoom to.**

This will zoom in to the selected point in the drawing editor and is a great command to help you locate a specific point. You can use the *Zoom to* command on any *Civil 3D* object, it is not limited to points.

17. Zoom Extents to view the entire drawing.

18. Save the drawing.

3.5 Creating a Point Group and Exporting the Points for Field Verification

Civil 3D creates a default point group called *_All Points*. As you might expect by the name this point group contains every point in the drawing. You will want to create other, more specific, point groups so that you can easily select the desired points for different commands. Point groups can also control the display of the points by assigning a *Point Style* and a *Point Label Style* to the group. *Point Styles* and a *Point Label Styles* will be covered in detail in *Chapter 4*.

3.5.1 Creating a Point Group

Create a *Point Group* of the *Preliminary Centerline Points*. This point group will be used to export these specific points to the surveyors.

1. On the *Prospector* tab of the *Toolspace*, right-click on **Point Groups** and select ⇒ **New.**

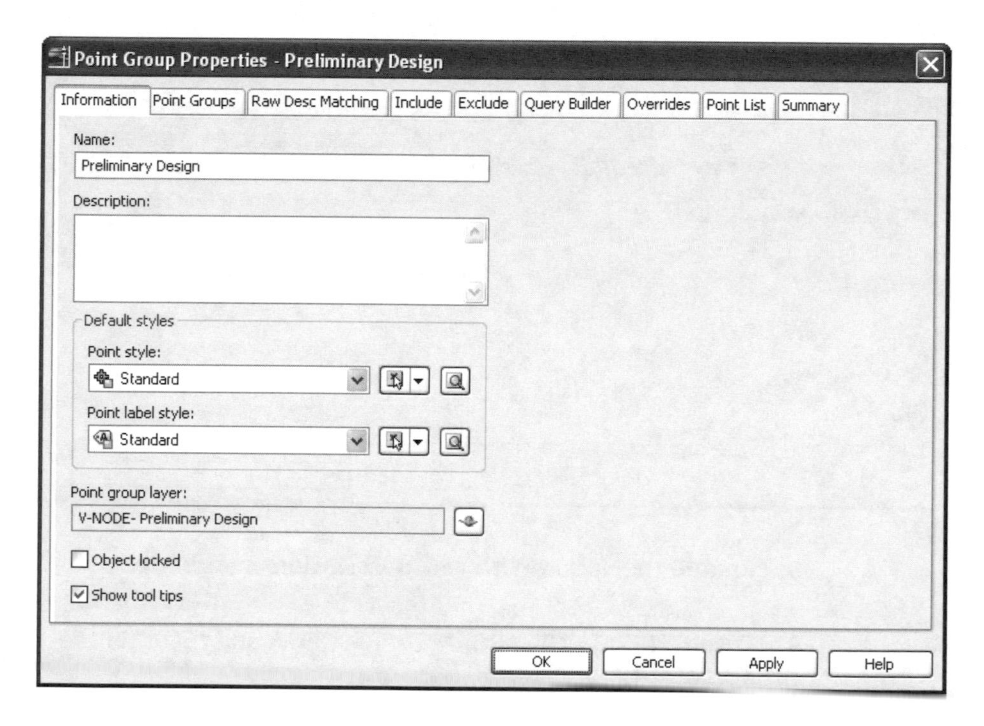

2. Enter **"Preliminary Design"** for the **Name.**

3. Click the **Point group Layer** button to open the *Object Layer* dialog box.

4. Set the **Modifier** to **Suffix.**

5. Enter **-*** as the **Modifier** value.

6. Click **<<OK>>** to close the *Object Layer* dialog box.

7. Select the **Include** tab in the *Point Group Properties* dialog box.

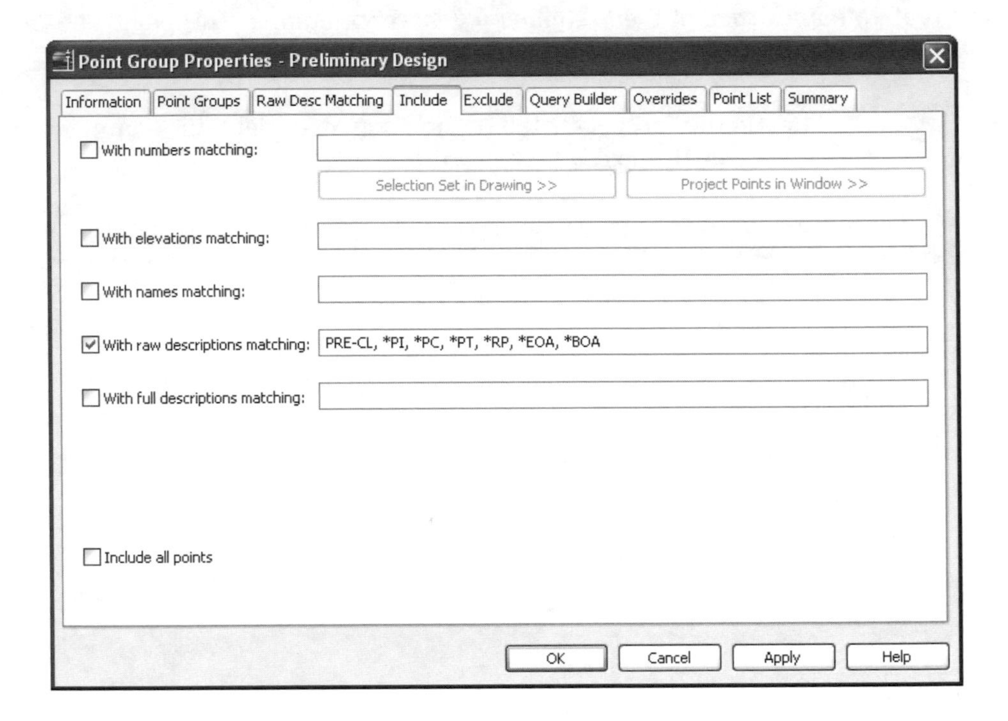

8. **Enable** the option **With raw descriptions matching.**

9. Enter "**PRE-CL, *PI, *PC, *PT, *RP, *EOA** and ***BOA**" in the adjacent field.

The asterisks (*) works as a wild card and in this example will select any points that have a description the ends with PI or any of the other descriptions using the wildcard.

10. Click **<<OK>>**.

3.5.2 Creating a Point Import/Export Format

In this exercise you will create an *Import/Export* format. *Civil 3D* comes with many standard import/export formats predefined. However, at some point you may need to define a new format for a specific project need. Following is the process of creating your own custom import/export format.

1. On the *Prospector* tab of the *Toolspace*, expand the **Point Groups** node.

2. Right-click on **Preliminary Design** and select ⇒ **Export Points.**

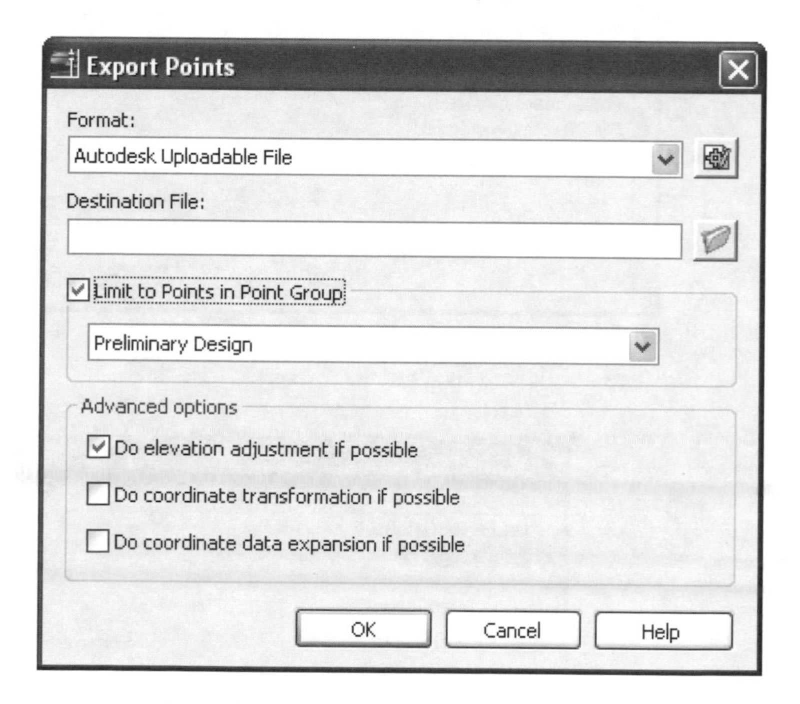

3. Click the **Format** button to open the *Point File Formats* dialog box.

Here you can see a list of all the available formats in your drawing. Your new format can be saved to a drawing template so that it is added to all new drawings.

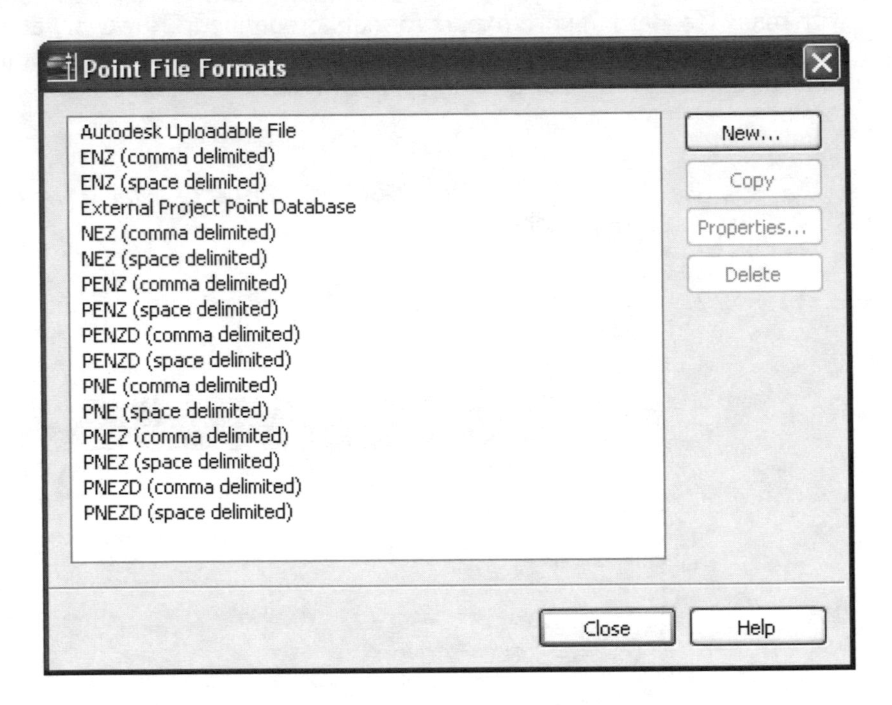

4. Click <<**New**>>.

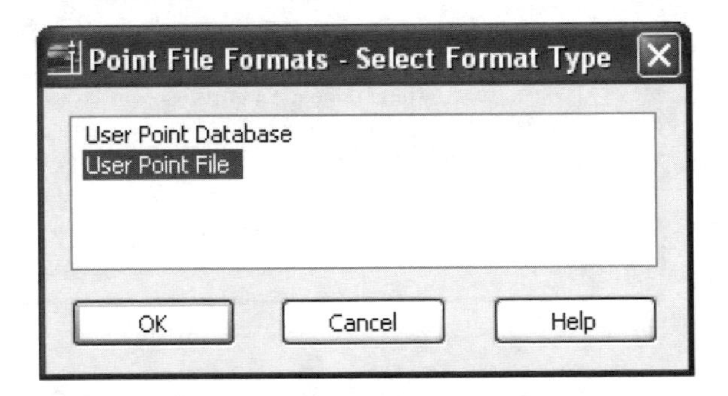

5. Select **User Point File** and click <<**OK**>>.

6. Name the format **"Training"**.

7. Set the format to be **delimited by a comma.**

8. Enter the "#" symbol as a **Comment Tag.**

Civil 3D will ignore any line that begins with the *Comment Tag* symbol during the import of the file. So if you want to have information in the header of the point file, as in this example, you must use a *Comment Tag*.

When picking any of the column headers at the bottom of the dialog box you can select the type of data in that column, by default all display as unused.

9. Create a format that contains columns for **Point Number, Northing, Easting, Point Elevation**, and **Raw Description,** in that order.

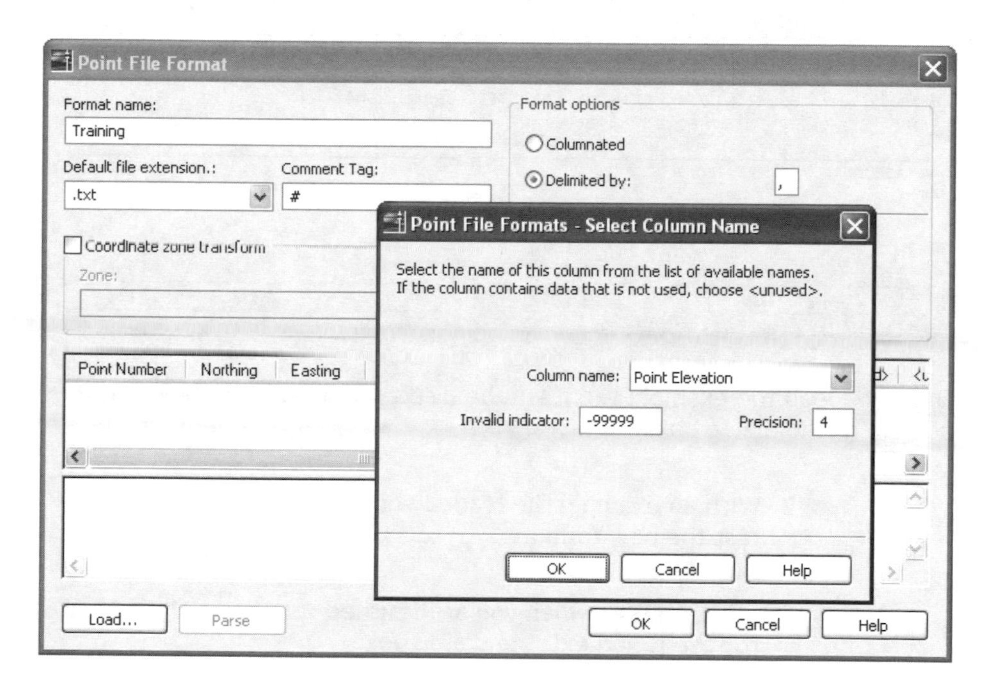

10. When setting up the *Point Elevation* field, set the **Invalid Indicator** to **-99999**.

You can use the Invalid Indicator option to set the value that your data collector uses for null entries.

11. To test your new format you can <<**Load**>> an example file that would use the new format. This example has loaded the file:

C:\Cadapult Training Data\Civil 3D 2007\Level 1\Points\Site.txt

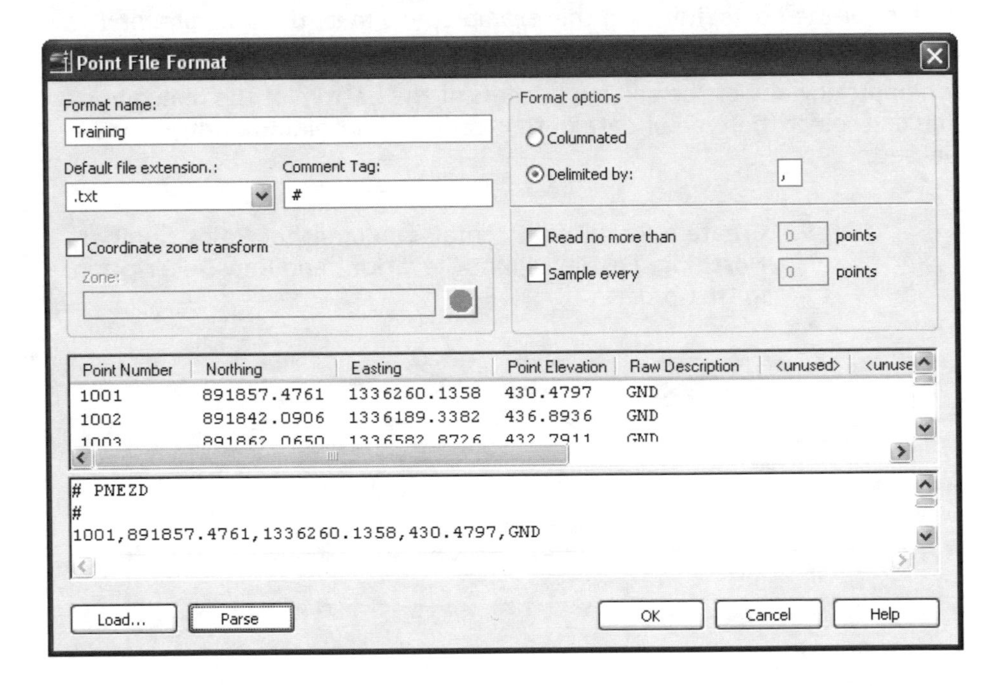

Once you load the example file it will be displayed in the text box at the bottom of the dialog box.

12. With an example file loaded you can click <<**Parse**>> to test the new format.

13. Click <<**OK**>> when you are finished setting up the format to save it and exit the dialog box.

14. <<**Close**>> the main *Point File Formats* dialog box.

3.5.3 Exporting Points to an ASCII File

1. Back in the *Export Points* dialog box, confirm you have set the **Format** to **Training.**

If you are not currently in the *Export Points* dialog box just right-click on the point group that you want to export and select *Export Points*.

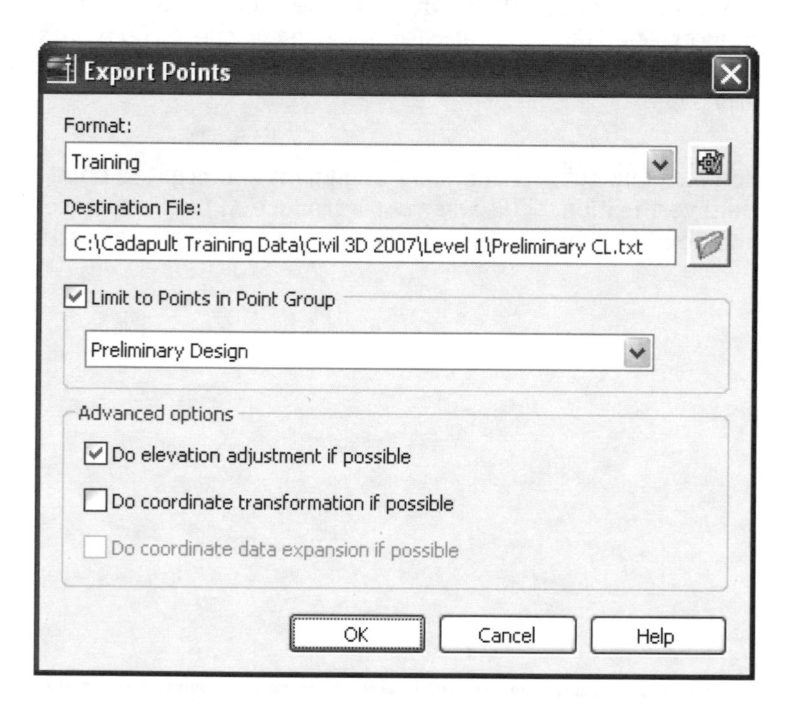

2. Click **the Destination File browse** button to select the location you would like to save the exported point file. Browse to the path: **C:\Cadapult Training Data\Civil 3D 2007\Level 1**

3. Enter a file name of **"Preliminary CL.txt"**

4. Confirm that the option is enabled to limit to points in the *Point Group* **"Preliminary Design"**.

5. Click **<<OK>>** to export the points.

3.6 Chapter Summary

In this chapter you used *Civil 3D* to create a surface from *AutoCAD* objects. This is a process that you will often use if you are receiving surface data from a photogrammetrist, surveyor, or anyone else that doesn't happen to be using *Civil 3D.* Many programs now have the option to import and export surface and other project data through the *LandXML* format. However, there are still a lot of people using software that does not support *LandXML.* In that case, exchanging the surface data as drawing objects is one of your only options, which is why this is an important process.

You also laid out a preliminary alignment and created points along it for field verification. This was your introduction to the surface, alignment, and point commands, which you will use again in detail later.

Chapter 4

Creating a Survey Plan

In this chapter you will work with description keys and point groups to organize the drawing as you import survey data. You will also work with styles to control the display and labeling of points and parcels. The chapter finishes by working with parcels and parcel segment labeling.

- **Importing Survey Points**
- **Working with Point Groups**
- **Controlling Point Display**
- **Drawing Linework Using Transparent Commands**
- **Working with Parcels**
- **Labeling Linework**

Dataset:

To start this chapter you will continue working in the drawing named **Design.dwg**. You can continue with the drawing that you currently have from the end of the previous chapter or, if you are starting in the middle of the book, you can open the drawing **CH-04.dwg** located in the folder **C:\Cadapult Training Data\Civil 3D 2007\Level 1\Chapter Drawings.** Opening the drawing from the dataset provided will ensure that you have the drawing set up correctly for the exercises in the following chapter and overwrite any mistakes that you may have made in previous exercises.

4.1 Overview

In this chapter you will work with description keys and point groups to organize the drawing as you import survey data. You will also work styles to control the display and labeling of points and parcels. There are several exercises devoted to controlling point display through the use of *Description Keys, Point Groups,* and *Layers.* Take your time and don't be afraid to experiment with them on your own. Managing points can be a very powerful, and complex, process and there are several ways to approach it. After you become familiar with how the commands work you can make decisions about the method that makes the most sense for your projects. The chapter finishes by working with parcels and parcel segment labeling.

4.1.1 Concepts

This chapter focuses on importing and managing survey data. If you are a Land Desktop user you will see some concepts that look similar to Land Desktop and some that look very different. As you work through the exercises take your time and keep an open mind. This will change the workflow that you may be used to. However, it will also allow you to do things that you could not do before.

4.1.2 Styles

Styles are saved in the drawing and can be created and edited on the Settings tab of the *Toolspace* or on the fly during many object creation and editing commands.

Point Styles control the display of the point marker. They can use an *AutoCAD Point*, a custom marker similar to *Land Desktop*, or an *AutoCAD Block* for the marker. *Point Styles* can be applied to individual points, point groups, or point group overrides.

Point Label Styles control the display of the point label text and can be customized to show any type of point related information. *Point Label Styles* can be applied to individual points, point groups, or point group overrides.

Parcel Styles control the display of the parcel segments and the parcel area fill. *Parcel Styles* also control the *Parcel Name Template* which can be used to automatically name the parcels as they are created or renumbered.

Parcel Label Styles control the display of the parcel label text. This typically includes things like the parcel name and/or number, area in square feet and/or acres, and perimeter. *Parcel Labels* can be configured to display many other types of information as well, such as tax lot number, address, and user defined fields.

Parcel Segment Label Styles control the display of basic line and curve labels. To label lines and curves they must first be converted to *Parcel Segments*.

Parcel Table Styles control the display of line, curve, segment, and area tables, including the columns of data displayed.

4.2 Importing Survey Points

In this section you will import survey points from an ASCII file. You will also learn ways to manage and display these point using *Point Groups* and *Description Keys*.

4.2.1 Creating a Description Key File

Description Keys control points as they are created or inserted into the drawing. They filter points based on their *Raw Description* and can control the layer points are inserted on, the full description displayed on the point, the point style, and the point label style as well as scale and rotation.

You are allowed to have more than one description key file in a drawing and the *Description Key Sets Search Order* controls the priority that different description key files are given if the same code is contained in more than one description key set. The *Description Key Sets Search Order* is found by right-clicking on the Description Key Sets node on the *Settings* tab of the *Toolspace*.

If you are familiar with using description keys in *Autodesk Land Desktop* you will notice a difference in *Civil 3D*. There is not an option for inserting a block or symbol with the point. This is now controlled by the *Point Style*, which means that you no longer have a separate *Point* and *Symbol*. Instead, the symbol is part of the point object. You do have new options in the description key to control the *Point Style* and *Point Label Style*. However, you also have the ability to control the *Point Style* and *Point Label Style* with *Point Groups*. Since Point Groups give you the ability to change the *Point Style* and *Point Label Style* of large sets of points quickly and easily this book will use *Point Groups* rather than *Description Keys* to control these styles.

Many of the functions that could only be done with *Description Keys* in *Land Desktop* are now also accomplished by using point groups. The features that are unique to description keys are converting the *Raw Description*, or *Code*, to the *Full Description* as well as scaling and rotating point symbols based on description parameters.

Definitions of the terminology used to create *Description Keys*:

Code: *Raw Description* or the description entered in the field by the surveyors.

Format: *Full Description* or the Description that will be shown on the *Point Object* in the drawing.

Layer: *Layer* in the drawing that the *Point Object* will be inserted on.

1. Continue working in the drawing **Design.dwg**.

2. On the *Settings* tab of the *Toolspace*, expand the **Point** node.

3. Under the *Point* node, right-click on **Description Key Sets** and select ⇒ **New**.

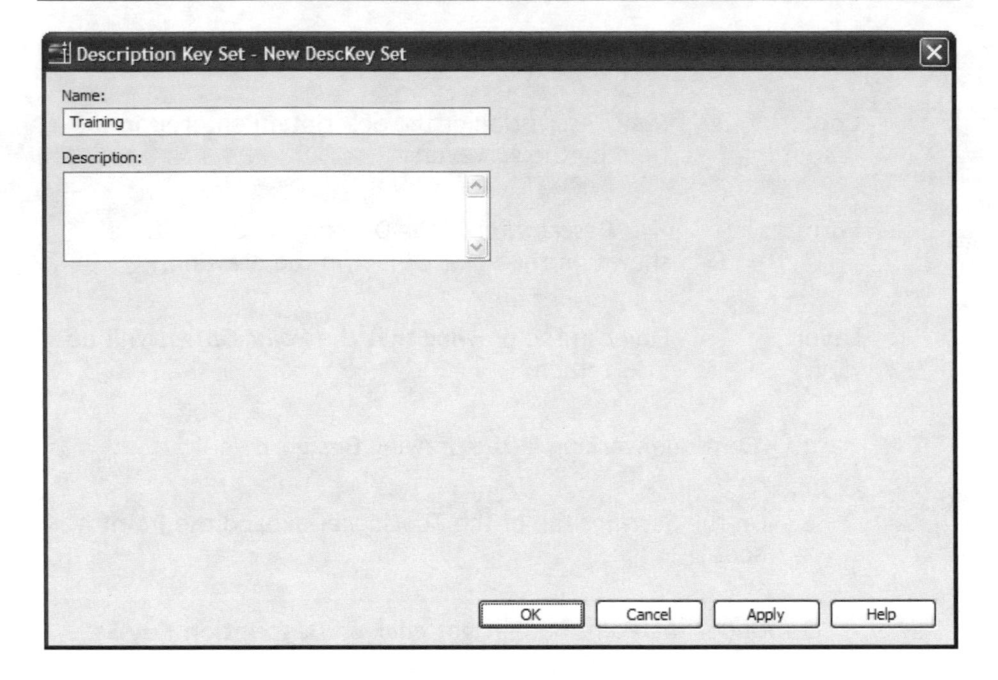

4. Enter the **Name** to **"Training"** and click **<<OK>>**.

5. On the *Settings* tab of the *Toolspace*, expand the **Description Key Sets** node under the *Point* node to display the new set.

6. Right-click on the **Training** *Description Key Set* and select ⇒ **Edit Keys** to open the DescKey Editor.

The first line contains the default *New DescKey* with default values which you will edit. You will need to expand some of the columns to read the complete headers.

In this exercise you are only using *Description Keys* to convert the Raw Description to a Full Description. The Layers and the Styles of the Points will be controlled by *Point Groups*.

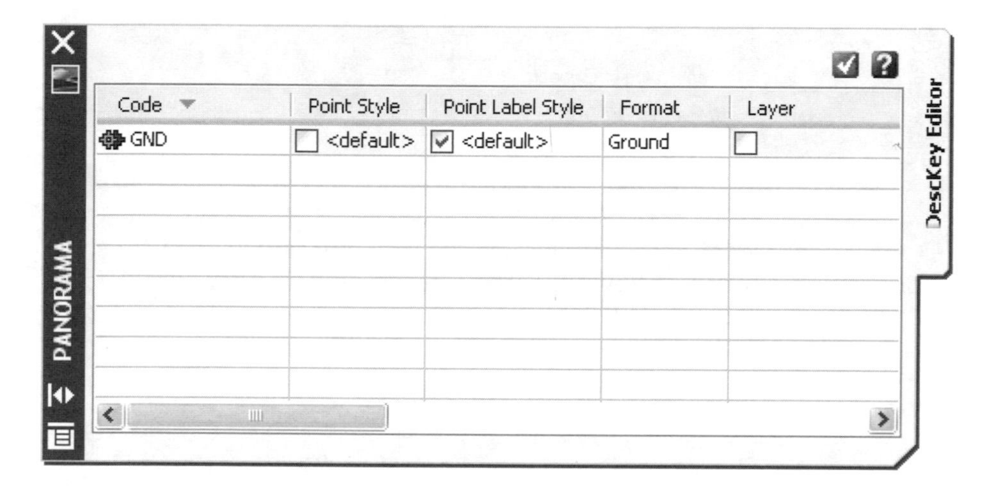

7. Change the **Code** to “**GND**”.

The *Code* is case sensitive.

8. Change the **Format** to “**Ground**”.

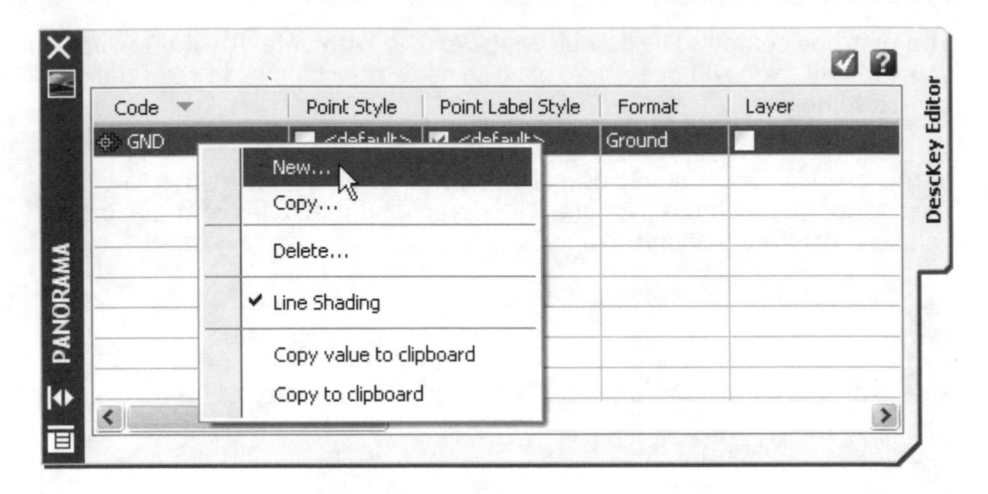

9. Right-click anywhere on the new description key *GND* and select ⇒ **New** to create a new description key with default values below it.

10. Change the **Code** to "**BC**".

11. Change the **Description Format** to "**Building Corner**".

12. Repeat this process for the rest of the keys until the *DescKey* editor table looks like the table below.

Code	Point Style	Point Label Style	Format	Layer
GND	<default>	<default>	Ground	
BC	<default>	<default>	Building Corner	
CRNR	<default>	<default>	Prop Corner	
DT	<default>	<default>	Tree	
AEC	<default>	<default>	Edge of Asphalt	
DWYRK	<default>	<default>	Rock Driveway	
DWYAC	<default>	<default>	Asphalt Driveway	
LP	<default>	<default>	Light Pole	
TOP	<default>	<default>	$*	
TOE	<default>	<default>	$*	
CL	<default>	<default>	Center Line	

The format $* will create a Full Description that is exactly the same as the Raw Description.

13. Close the *Description Key Editor* when finished.

14. Save the drawing.

4.2.2 Importing Points from an ASCII File

In this exercise you will import survey points from an ASCII text file using the format that you created in *Chapter 3*. As these points are created in the drawing the description keys will sort them to specific layers and convert the *Raw Descriptions* provided by the surveyors in the ASCII file to a more explanatory *Full Description*.

1. Freeze the layers **C-ROAD-Pre L Street** and **V-NODE-Preliminary Design**.

You should only see the surface displayed as contours on your screen.

2. Select **Points ⇒ Create Points.**

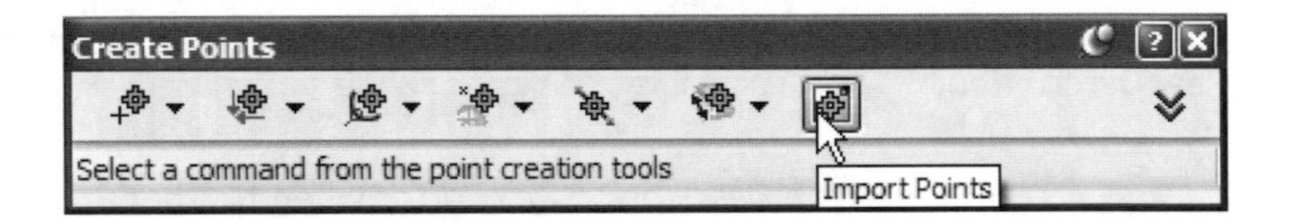

3. Click the **Import Points** button on the *Create Points* toolbar.

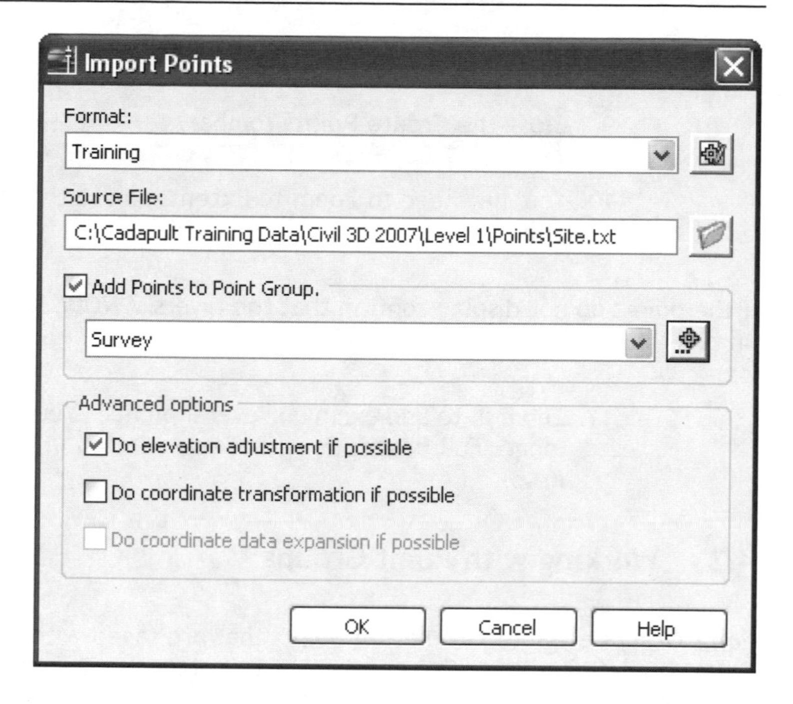

4. In the *Import Points* dialog box, select "**Training**" as the **Format.**

5. Click the **Browse** button to select *the Source File* you want to import:

C:\Cadapult Training Data\Civil 3D 2007\Level 1\Points\Site.txt

6. **Enable** the **Add Points to Point Group** option.

7. Click the Add Points button to the right of the drop down list, to enter a new point group called **"Survey"**.

This will create a *Point Group* for all of the survey points and allow you to easily track which points are survey points collected in the field and which points are design points or other points generated in the office. You might also consider adding a date to the name of the Point Group if you have multiple sets of survey data coming in as the project progresses.

8. Click <<**OK**>> to import the points.

9. Close the **Create Points** toolbar.

10. You may need to **Zoom** to **Extents** to see the imported points.

If the points do not display confirm that the layers *V-NODE* layer is turned on and thawed.

11. Zoom in to and examine several points. You should see the longer Full Descriptions that you entered in the *Description Keys*.

4.3 Working with Point Groups

Point Groups are selection sets of points that are saved with the drawing. Once you define a *Point Group*, whether it is simple or complex, you can then select points by that group while using any point related command. *Point Groups* can also control the display of the points by assigning a *Point Style* and a *Point Label Style* to the group.

4.3.1 Managing Point Group Layers

Point Groups are objects and they each reside on a layer in your drawing. By placing each *Point Group* on its own individual layer you will be able to turn the *Point Group* on and off with standard layer commands. In this exercise you will modify the point group *Survey* so that it is placed on a unique layer.

1. On the *Prospector* tab of the *Toolspace*, expand the **Point Groups** node.

2. Right-click on the *Point Group* **Survey** and select ⇒ **Properties.**

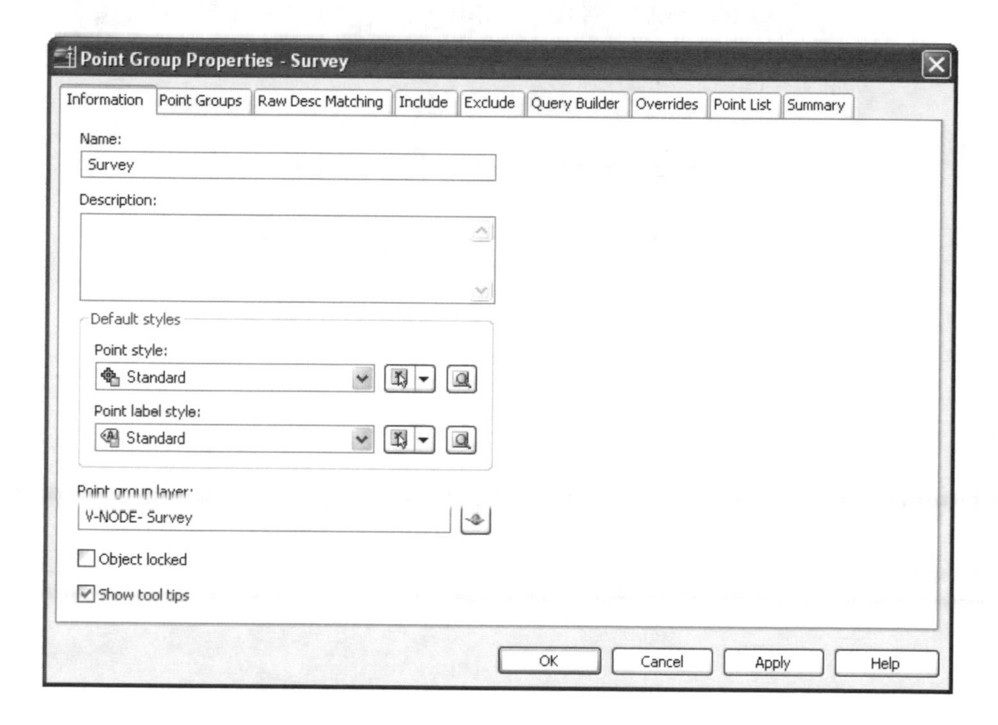

3. Click the **Point group Layer** button to open the *Object Layer* dialog box.

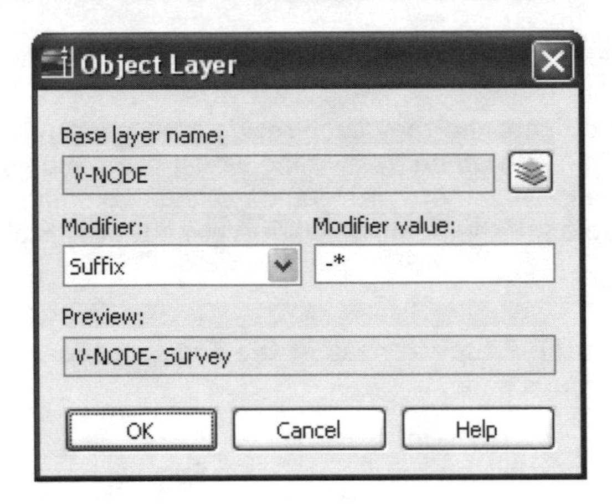

4. Select **Suffix** as the **Modifier**.

5. Enter "**-***" as the **Modifier value.**

6. Click <<**OK**>> to close the **Object Layer** dialog box.

7. Click <<**OK**>> to save the changes to the group.

This will create a new layer called V-NODE-Survey and move the *Point Group* to that layer.

4.3.2 Locking Points and Group Properties

1. On the *Prospector* tab of the *Toolspace*, right-click on the **Survey** point group and select ⇒ **Lock Points.**

This locks all the points that are in the selected group so that they cannot be edited.

2. On the *Prospector* tab of the *Toolspace*, right-click on the **Survey** point group and select ⇒ **Lock**.

This locks the *Properties* of the point group so that points cannot be added or removed from the group. This is a good precaution to take with the survey group to ensure that it contains all the points that came from the surveyor and no more. You are changing this from a dynamic to a static group. Once a point group's *Properties* are locked the icon changes to include a locked symbol.

4.3.3 Creating a Point Group for Property Corners

In this exercise you will create a *Point Group* for the property corner points. This group will search for all the points in our drawing that use a certain description key. This is an example of a *Point Group* where you could save the group properties to a drawing template. This way when you start a new drawing the point group is automatically created. Then when points are imported to the drawing that fit into the properties of the point group they are automatically added to the group.

1. On the *Prospector* tab of the *Toolspace*, right-click on **Point Groups** and select ⇒ **New.**

2. Enter **"Property Corners"** for the **Name**.

3. Click the **Point group Layer** button to open the *Object Layer* dialog box.

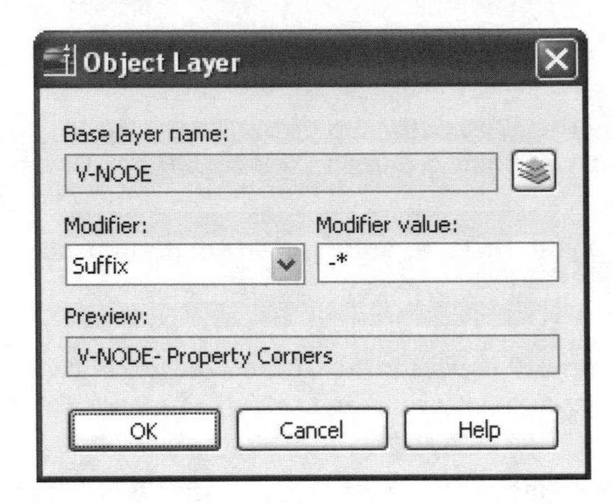

4. Select **Suffix** as the **Modifier**.

5. Enter "**-***" as the **Modifier value.**

6. Click **<<OK>>** to close the **Object Layer** dialog box.

7. Select the **Raw Desc Matching** tab in the *Point Group Properties* dialog box.

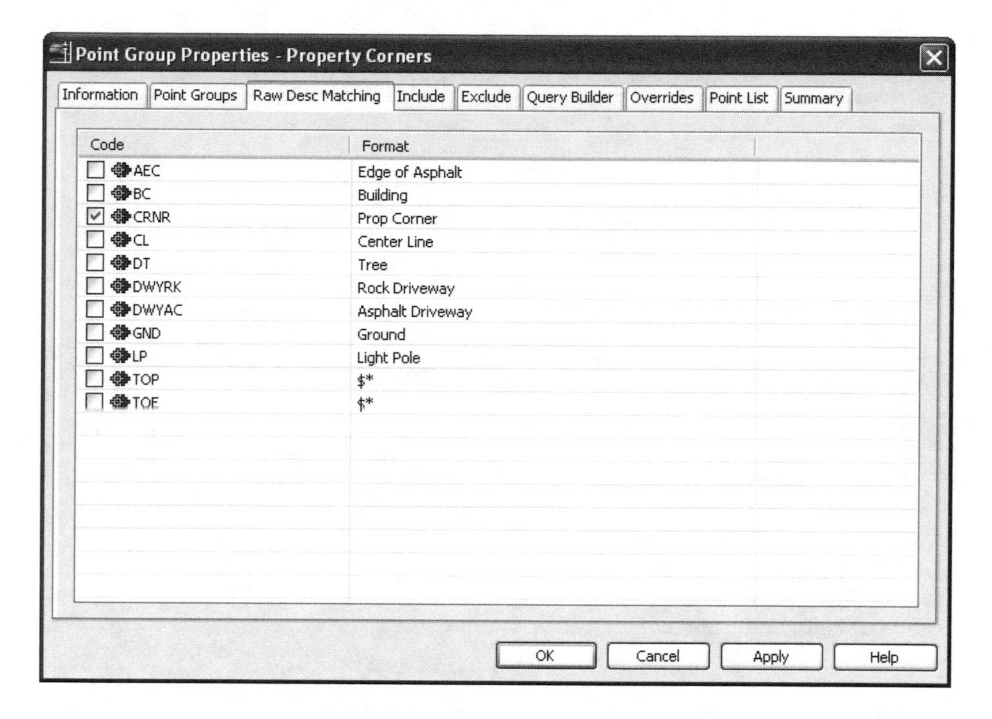

8. Select the Raw Description **CRNR**.

9. Click **<<Apply>>**.

10. Select the **Point List** tab in the *Point Group Properties* dialog box, and you will see the point numbers that match your filter.

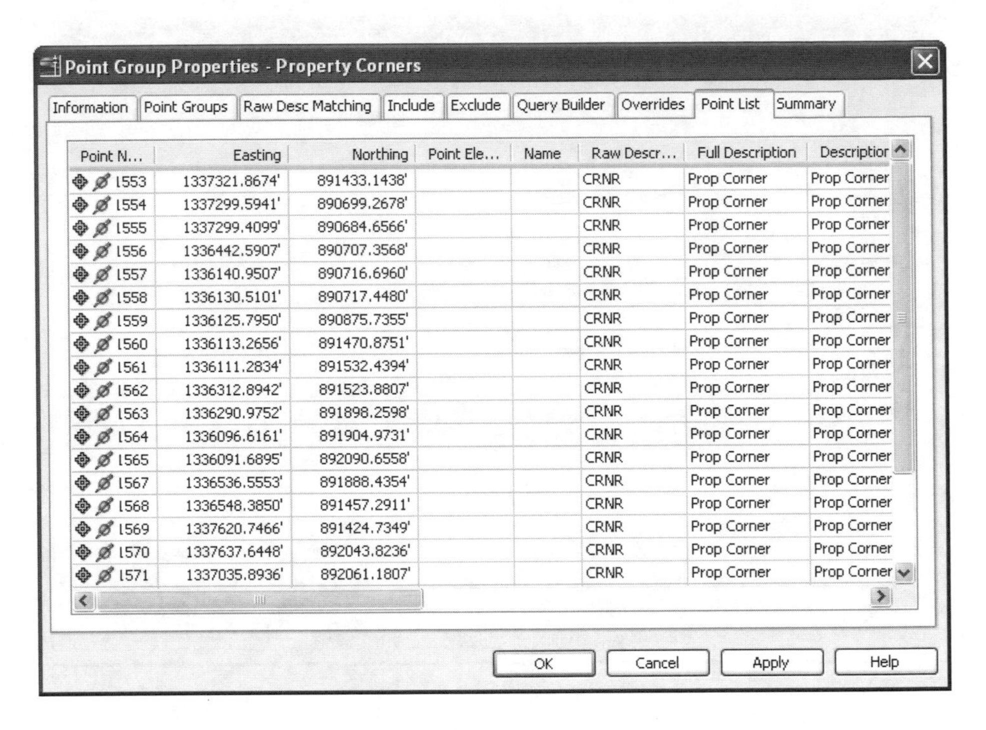

Point N...	Easting	Northing	Point Ele...	Name	Raw Descr...	Full Description	Descriptior
1553	1337321.8674'	891433.1438'			CRNR	Prop Corner	Prop Corner
1554	1337299.5941'	890699.2678'			CRNR	Prop Corner	Prop Corner
1555	1337299.4099'	890684.6566'			CRNR	Prop Corner	Prop Corner
1556	1336442.5907'	890707.3568'			CRNR	Prop Corner	Prop Corner
1557	1336140.9507'	890716.6960'			CRNR	Prop Corner	Prop Corner
1558	1336130.5101'	890717.4480'			CRNR	Prop Corner	Prop Corner
1559	1336125.7950'	890875.7355'			CRNR	Prop Corner	Prop Corner
1560	1336113.2656'	891470.8751'			CRNR	Prop Corner	Prop Corner
1561	1336111.2834'	891532.4394'			CRNR	Prop Corner	Prop Corner
1562	1336312.8942'	891523.8807'			CRNR	Prop Corner	Prop Corner
1563	1336290.9752'	891898.2598'			CRNR	Prop Corner	Prop Corner
1564	1336096.6161'	891904.9731'			CRNR	Prop Corner	Prop Corner
1565	1336091.6895'	892090.6558'			CRNR	Prop Corner	Prop Corner
1567	1336536.5553'	891888.4354'			CRNR	Prop Corner	Prop Corner
1568	1336548.3850'	891457.2911'			CRNR	Prop Corner	Prop Corner
1569	1337620.7466'	891424.7349'			CRNR	Prop Corner	Prop Corner
1570	1337637.6448'	892043.8236'			CRNR	Prop Corner	Prop Corner
1571	1337035.8936'	892061.1807'			CRNR	Prop Corner	Prop Corner

11. Click <<**OK**>> to save the group.

4.3.4 Creating a Point Group for Center Line Points

Repeat the procedure from the previous exercise to create a *Point Group* for the points that use the *CL* description key.

1. On the *Prospector* tab of the *Toolspace*, right-click on **Point Groups** and select ⇒ **New**.

2. Enter **"Center Line"** for the **Name**.

3. Click the **Point group Layer** button to open the *Object Layer* dialog box.

4. Select **Suffix** as the **Modifier**.

5. Enter "**-***" as the **Modifier value.**

6. Click **<<OK>>** to close the **Object Layer** dialog box.

7. Select the **Raw Desc Matching** tab in the *Point Group Properties* dialog box.

8. Select the Raw Description **CL**.

9. Click **<<OK>>** to save the group.

4.3.5 Creating a Point Group for Breakline Points

Repeat the procedure from the previous exercise to create a *Point Group* for the points that use the *AEC, DWYAC, DWYRK, Top,* and *TOE* description keys.

1. On the *Prospector* tab of the *Toolspace,* right-click on **Point Groups** and select ⇒ **New**.
2. Enter **"Breakline"** for the **Name**.
3. Click the **Point group Layer** button to open the *Object Layer* dialog box.
4. Select **Suffix** as the **Modifier**.
5. Enter **"-*"** as the **Modifier value.**
6. Click **<<OK>>** to close the **Object Layer** dialog box.
7. Select the **Raw Desc Matching** tab in the *Point Group Properties* **dialog box**.
8. Select the Raw Descriptions **AEC, DWYAC, DWYRK, Top, and TOE.**
9. Click **<<OK>>** to save the group.

4.3.6 Creating a Point Group for Tree Points

In this exercise you will repeat the procedure from the previous exercise to create a *Point Group* for the points that use the *DT* description key. You will also apply a *Point Style* to display a tree symbol as part of the point object.

1. On the *Prospector* tab of the *Toolspace,* right-click on **Point Groups** and select ⇒ **New**.

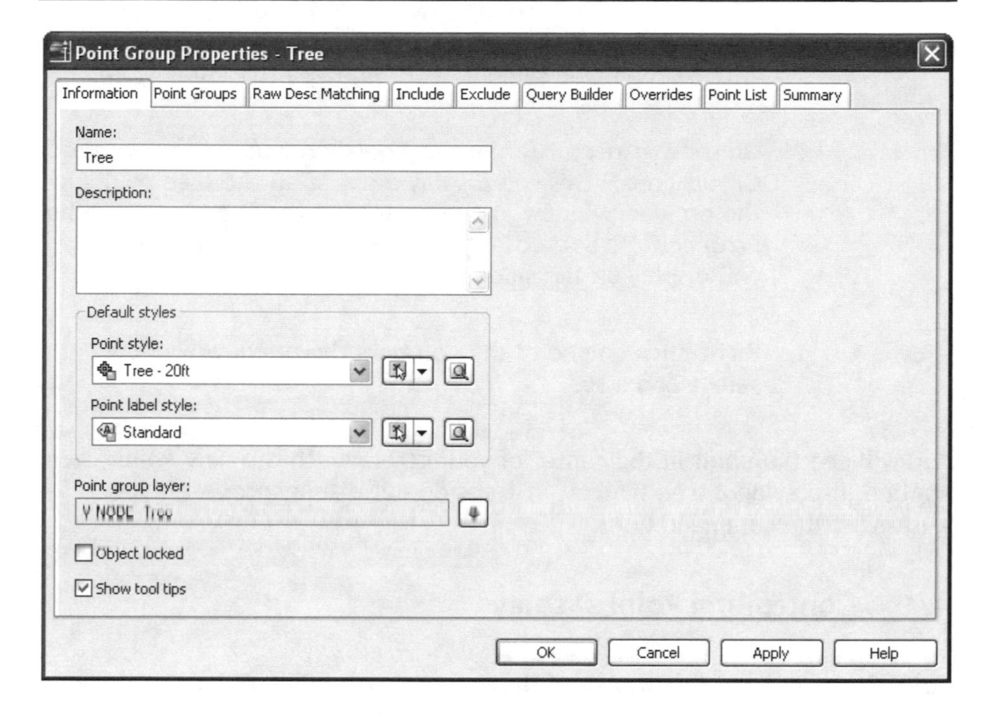

2. Enter **"Tree"** for the **Name**.

3. Set the **Point style** to **Tree - 20ft**.

4. Click the **Point group Layer** button to open the *Object Layer* dialog box.

5. Select **Suffix** as the **Modifier**.

6. Enter "**-***" as the **Modifier value.**

7. Click **<<OK>>** to close the **Object Layer** dialog box and return to the *Point Group Properties* dialog box.

8. Select the **Raw Desc Matching** tab in the *Point Group Properties* dialog box.

9. Select the *Raw Description* **DT**.

10. Click **<<OK>>** to save the group and display the tree points with a green tree symbol, as defined by the *Point Style*.

11. On the *Prospector* tab of the *Toolspace*, select the *Point Group* **Tree.** This will display a list of all the tree points in the preview window at the bottom of the *Prospector*, if the *Prospector* is docked. If the *Prospector* is not docked it will display on the side.

12. Right-click on one of the points in the preview window and select **Zoom to**.

You will see the point in the center of your screen with the *new Point Style* applied displaying a tree symbol. If trees do not display properly type **Regen** at the command line.

4.4 Controlling Point Display

Point Styles control the display of the point marker while *Point Label Styles* control the display of the *Point Label* text. In this section you will create a custom *Point Style* and *Point Label Style*. Then you will apply these new styles to a *Point Group* and explore controlling point display through Point Groups.

4.4.1 Creating Point Styles

Point Styles control the display of the point marker. They can use an *AutoCAD Point*, a custom marker similar to *Land Desktop*, or an *AutoCAD Block* for the marker. *Point Styles* are saved in the drawing and can be created and edited on the Settings tab of the *Toolspace*. *Point Styles* can be applied to individual points, point groups, or point group overrides.

1. On the *Settings* tab of the *Toolspace*, expand the **Point** node, and then expand the **Point Styles** node under it.

2. Under the *Point Styles* node, right-click on **Standard** and select ⇒ **Copy**.

You may need to scroll down to locate the *Point Style* called *Standard*.

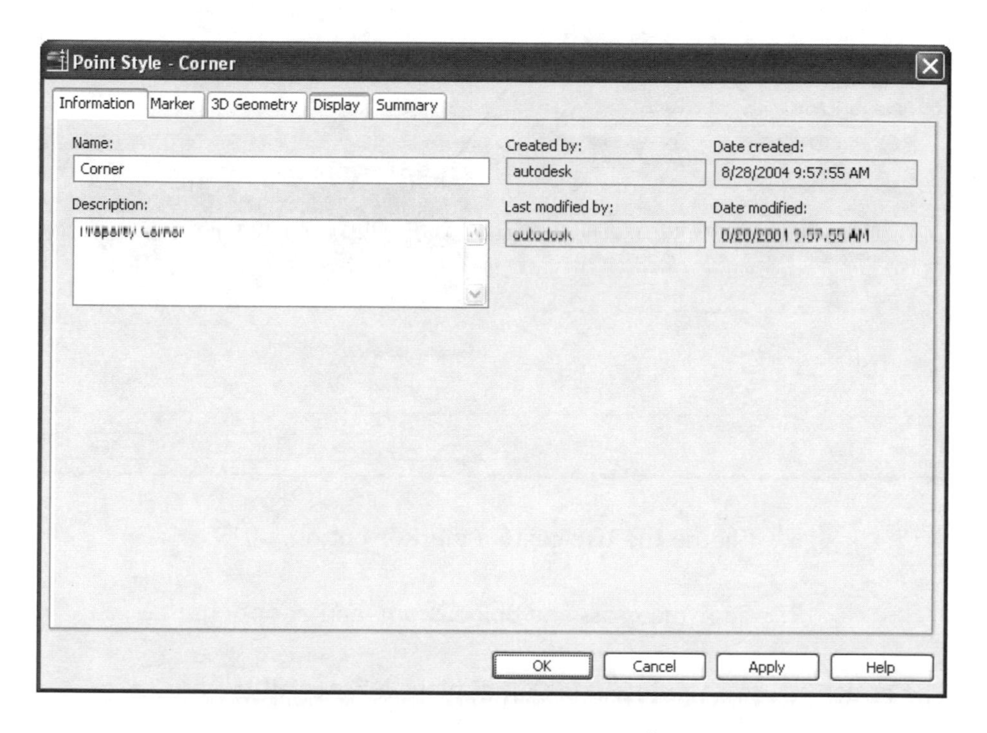

3. Change the **Name** of the new *Point Style to* **"Corner"**.

4. Enter **"Property Corner"** for the **Description** of the new *Point Style*.

5. Select the **Marker** tab of the *Point Style* dialog box.

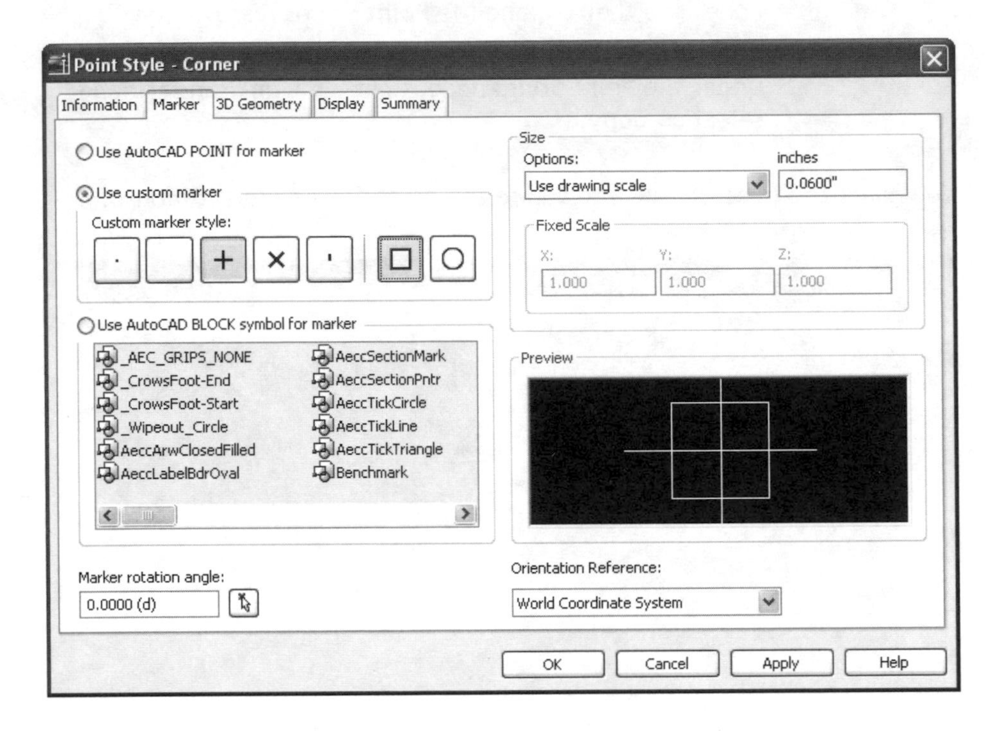

6. Choose the **Use custom marker** option.

7. Click the **cross** and **box** custom marker options.

The option to use an *AutoCAD BLOCK* displays a list of all the blocks currently defined in the drawing. You can also right-click on one of the blocks in the list to browse and insert any *AutoCAD BLOCK* on your system.

8. Select the **Display** tab of the *Point Style* dialog box.

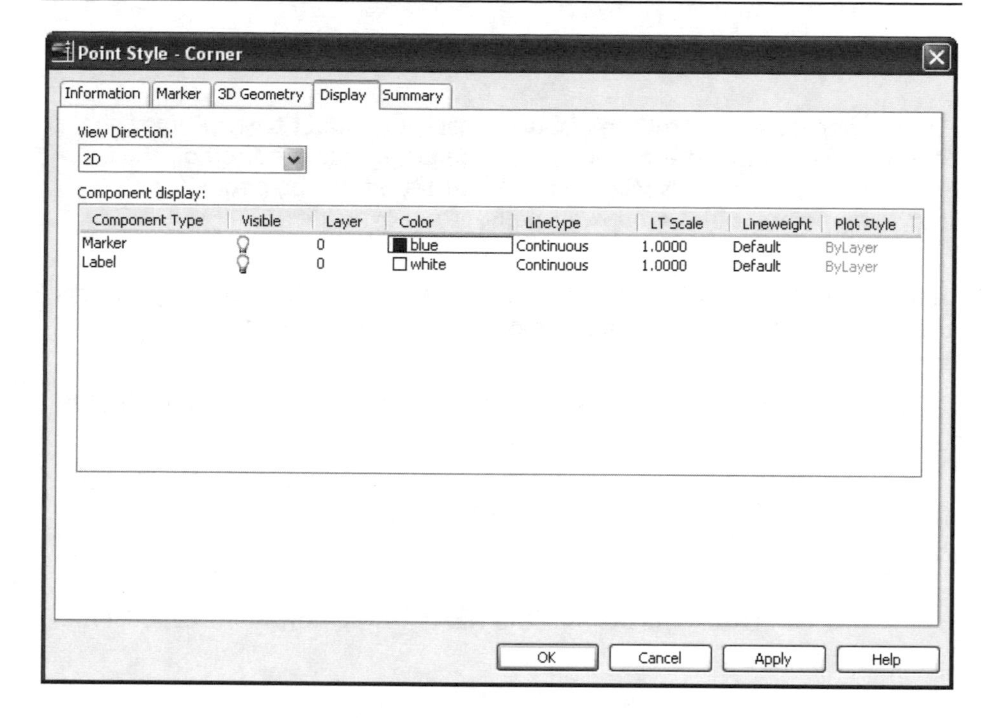

9. Set the **Marker** color to **Blue**.

10. Click <<**OK**>> to save the new *Point Style*.

4.4.2 Creating Point Label Styles

Point Label Styles control the display of the point label text. *Point Label Styles* are saved in the drawing and can be created and edited on the *Settings* tab of the *Toolspace*. *Point Label Styles* can be applied to individual points, point groups, or point group overrides.

In this exercise you will create *a Point Label Style* for the property corners that displays the *Point Number*, *Description*, *Northing*, and *Easting*.

1. On the *Settings* tab of the *Toolspace*, expand **Label Styles** under the *Point* node.

2. Under the *Label Styles* node, right-click on **Standard** and select ⇒ **Copy**.

3. Enter **"Corner"** for the **Name** of the new Point Label Style.

4. Enter **"Property Corner"** for the Description.

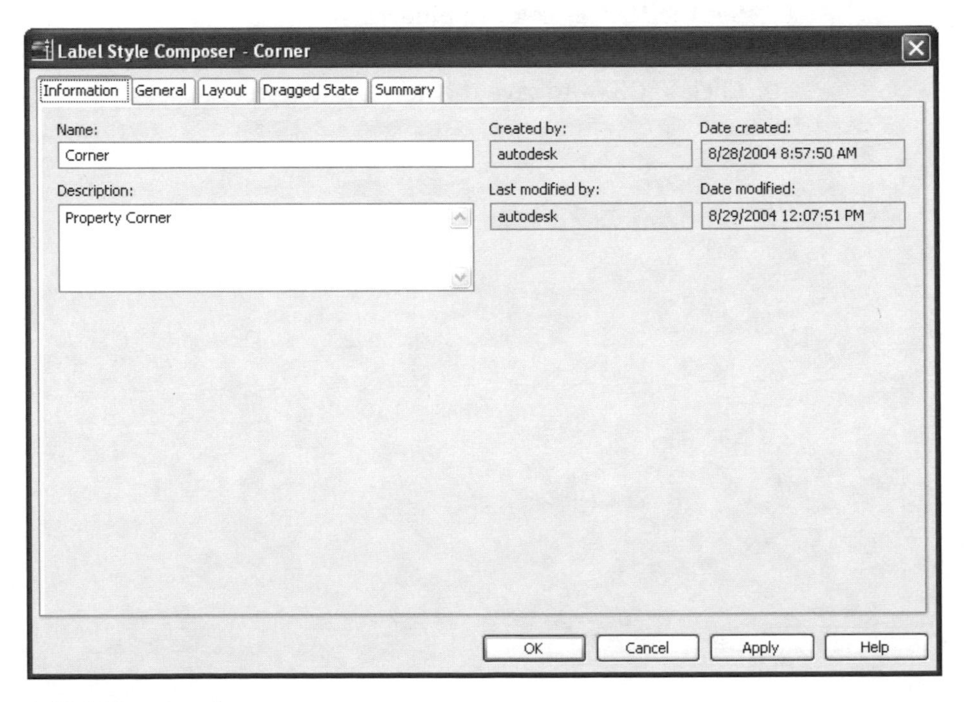

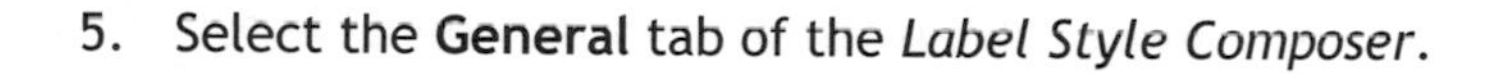

5. Select the **General** tab of the *Label Style Composer*.

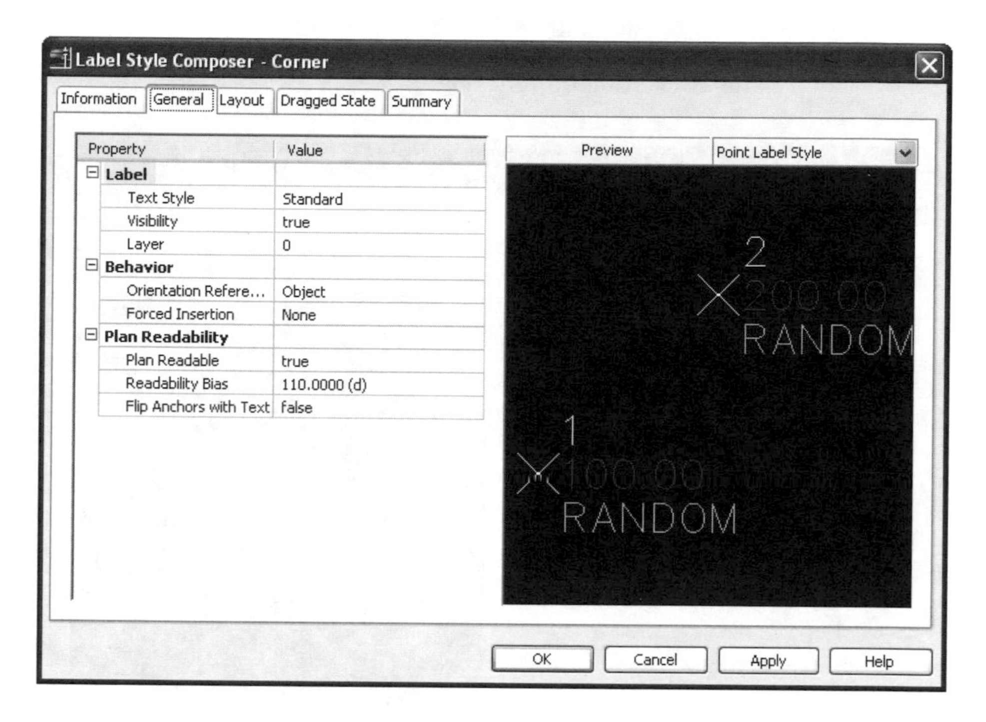

On the *General* tab you can set the Text Style for the label and the Layer that the label is created on. If you leave the layer set to 0 the label will be controlled by the Point layer. The *General* tab also allows you to set the Plan Readability for the label.

The *Label Style Composer* is very similar for all *Civil 3D* label styles. This consistency will make it easier to create label styles for any object type as you become familiar with the interface.

For this exercise you will not make any changes on the *General* tab.

6. Select the **Layout** tab of the *Label Style Composer*.

7. Select **Point Elevation** from the drop-down list to make it the active **Component name**.

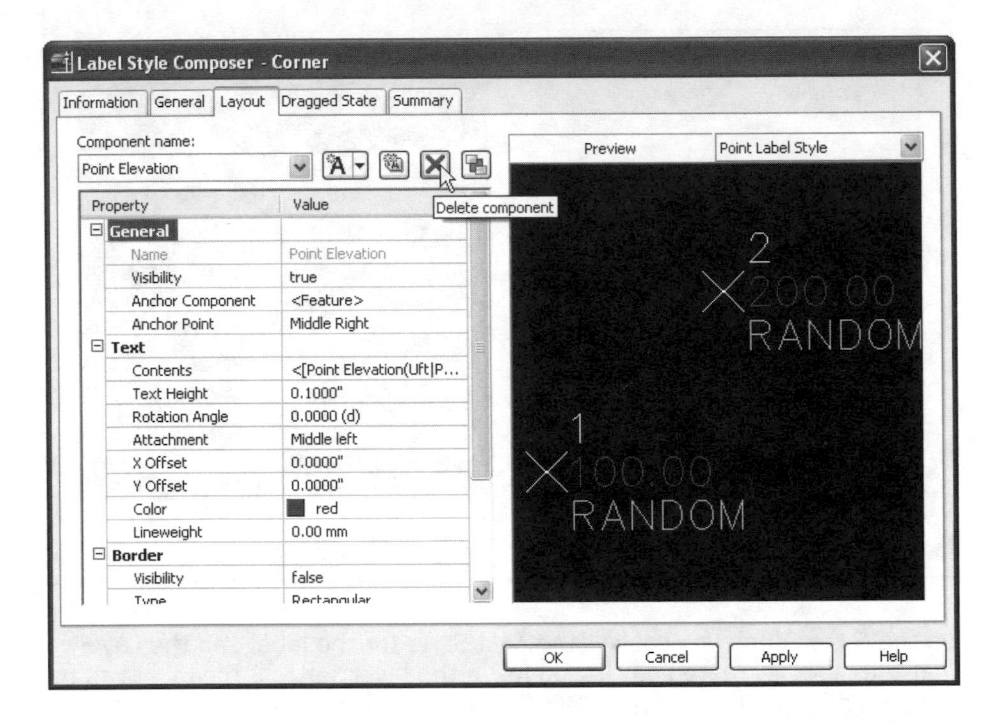

8. Click the Delete button to remove the **Point Elevation** component from this *Label Style*.

You will receive a warning that this label component is used as an anchor. Deleting this component will effect the position of the other text and will need to be fixed before the Label Style is complete.

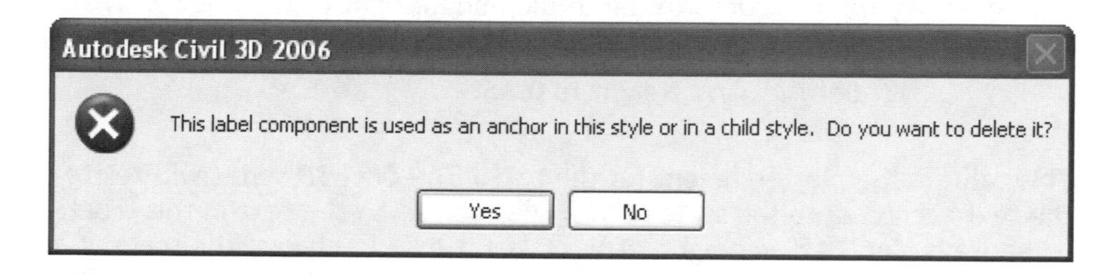

9. Click **<<Yes>>** to delete the component.

10. Select **Point Description** from the drop-down list to make it the active **Component name**.

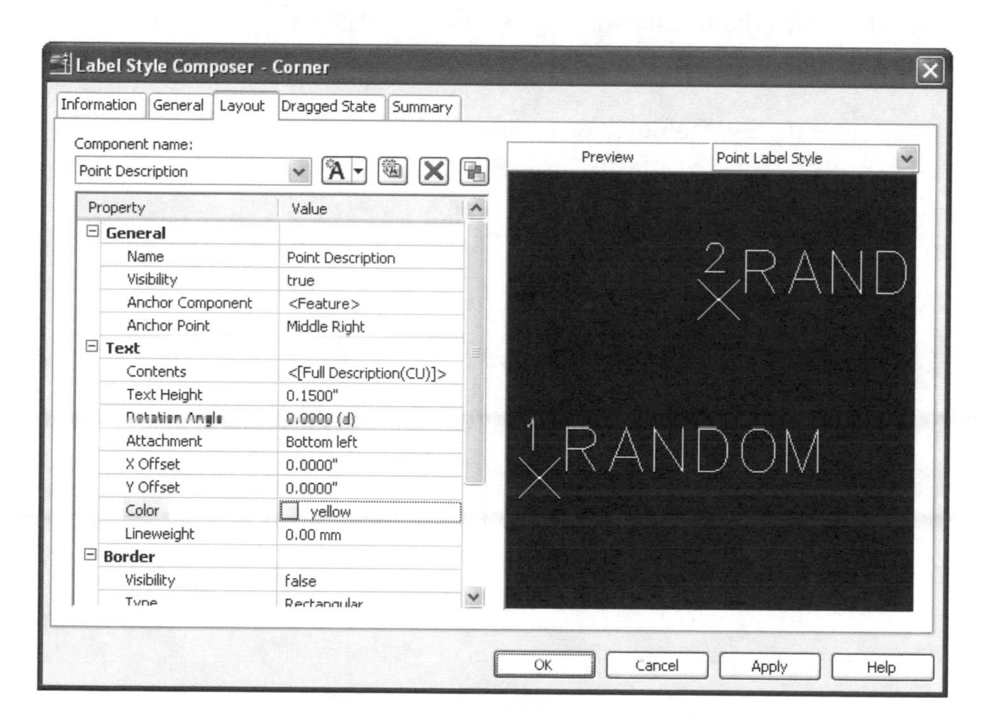

11. Confirm that the **Anchor Component** is set to **<Feature>**.

12. Click in the **Anchor Point** *Value* field to activate it, and then choose **Middle Right** from the drop-down list.

This attaches the Point Description to the middle right of the Point Marker.

13. Set the **Text Height** to **0.15**.

This value is the plotted height for the text. The *Civil 3D* styles will resize the text according to the scale of your viewports. So the text in this label will always plot 0.15" in height. This makes it easy to change the scale of a drawing or even have 2 viewports on the same sheet with different scaled viewports.

14. Set the **Attachment** to **Bottom left.**

This will attach the bottom left of the Point Description text to the middle right of the Point Marker that you set in a previous step.

15. Set the **Color** to **Yellow.**

16. Select **Point Number** from the drop-down list to make it the active **Component name.**

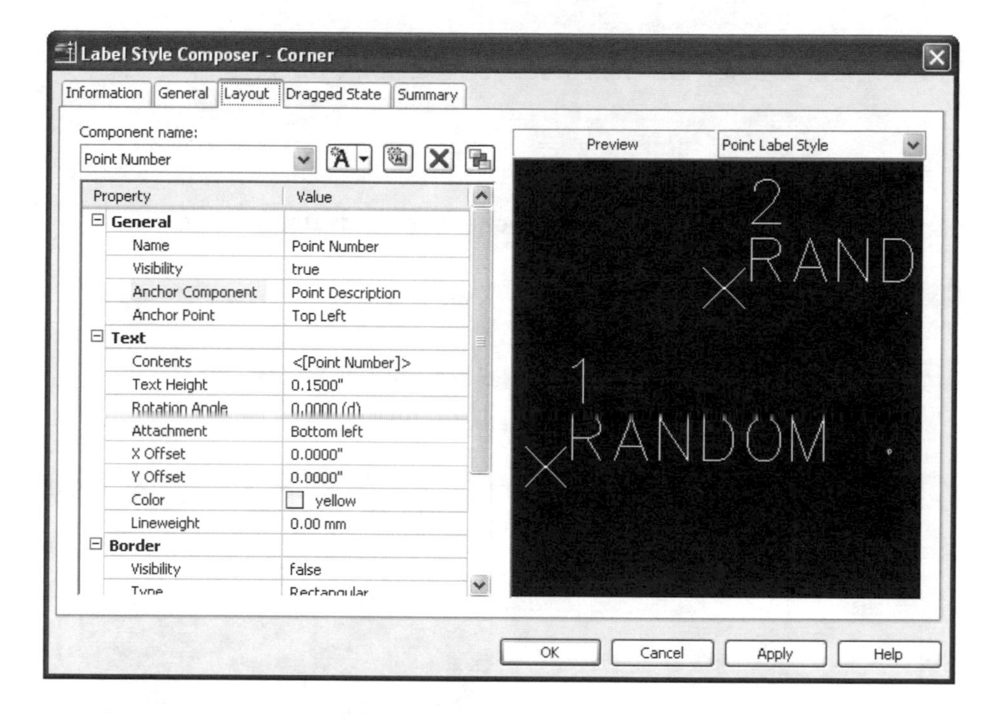

17. Set the **Anchor Component** to **Point Description**.

18. Confirm that the **Anchor Point** is set to **Top Left**.

This will attach the Point Number text to the top left of the Point Description text.

19. Set the **Text Height** to **0.15**.

20. Confirm that the **Attachment** is set to **Bottom left**.

This will attach the bottom left of the Point Number text to the top left of the Point Description text that you set in a previous step.

21. Confirm that the **Color** is set to **Yellow**.

22. Click the **Text button** to the right of the **Component name** drop down list.

This will create a new component for you to configure, called *Text.1*.

23. Enter **"Northing"** as the **Name** of the new component.

24. Set the **Anchor Component** to **Point Description.**

25. Set the **Anchor Point** to **Bottom left.**

This will attach the Northing text to the bottom left of the Point Description text.

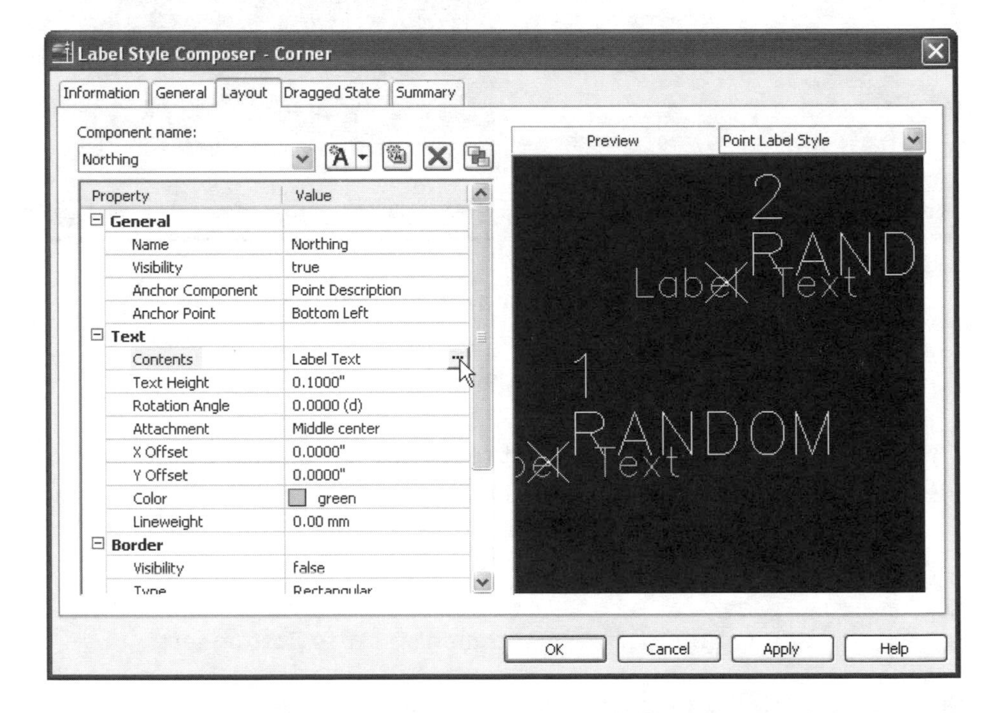

26. Click the **Text Contents** *Value* field to activate the ellipses <<...>> button.

27. Click the ellipses <<...>> button to open the *Text Component Editor*.

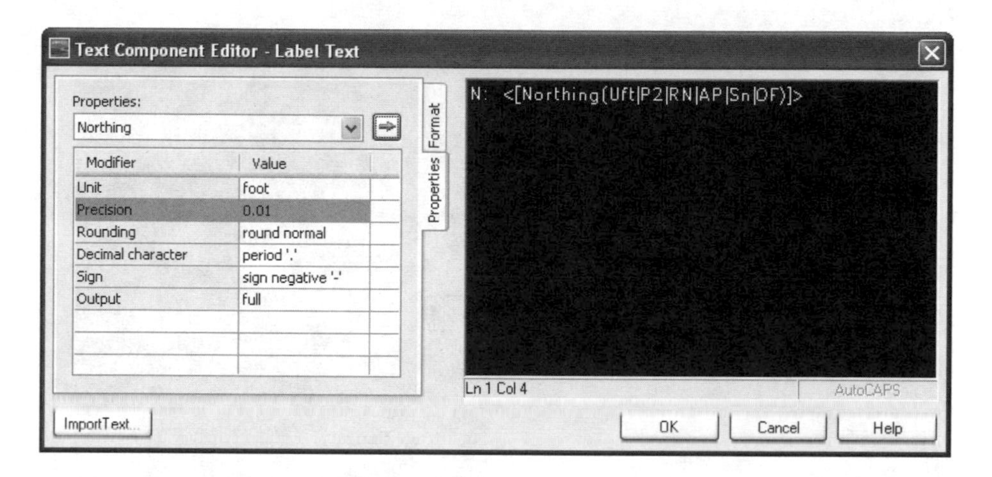

28. Delete the default **"Label Text"** displayed in the editor by dragging across it to highlight, then pressing **<<delete>>**.

29. Select **Northing** from the **Properties** drop down list.

Take a moment to explore all of the available properties that you can use to label points.

30. Set the **Precision** to **0.01** (two decimal places), by selecting it from the drop-down list.

31. Click the **right arrow** button to insert the code which will display the northing of the point.

32. Enter **"N: "** in the editor before the code to better describe the label.

33. Click **<<OK>>** to save the contents of the new component.

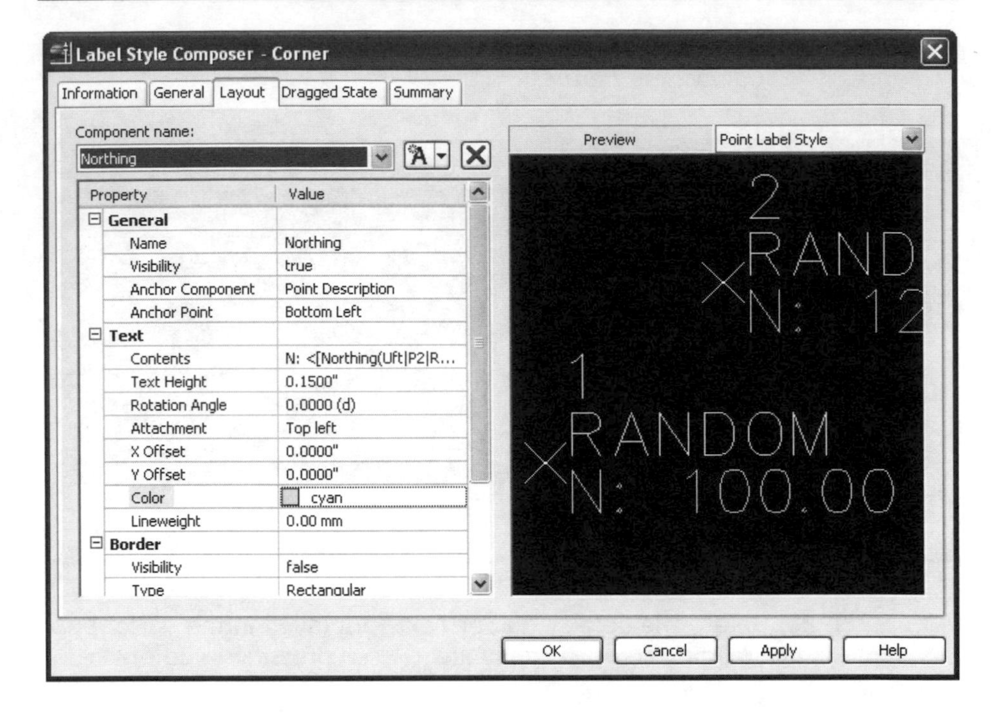

34. Back in the *Label Style Composer* set the **Text Height** to **0.15**.

35. Set the **Attachment** is set to **Top left**.

This will attach the top left of the Northing text to the bottom left of the Point Description text that you set in a previous step.

36. Set the **Color** to **Cyan**.

37. Click the **Text button** to the right of the *Component name* drop down list.

This will create a second new component for you to configure, called *Text.1*.

38. Enter **Easting** as the name of the new component.

39. Set the **Anchor Component** to **Northing.**

40. Set the **Anchor Point** to **Bottom Left.**

This will attach the Easting text to the bottom left of the Northing text.

41. Click in the **Text Contents** *Value* field to activate the ellipses <<...>> button.

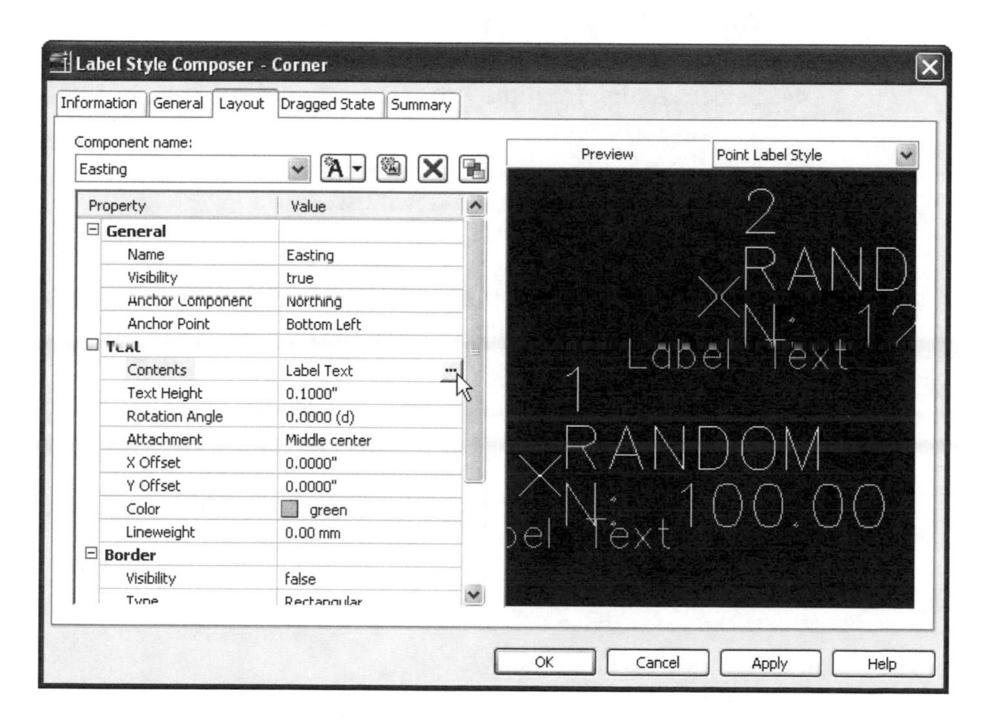

42. Click the ellipses <<...>> button to open the *Text Component Editor*.

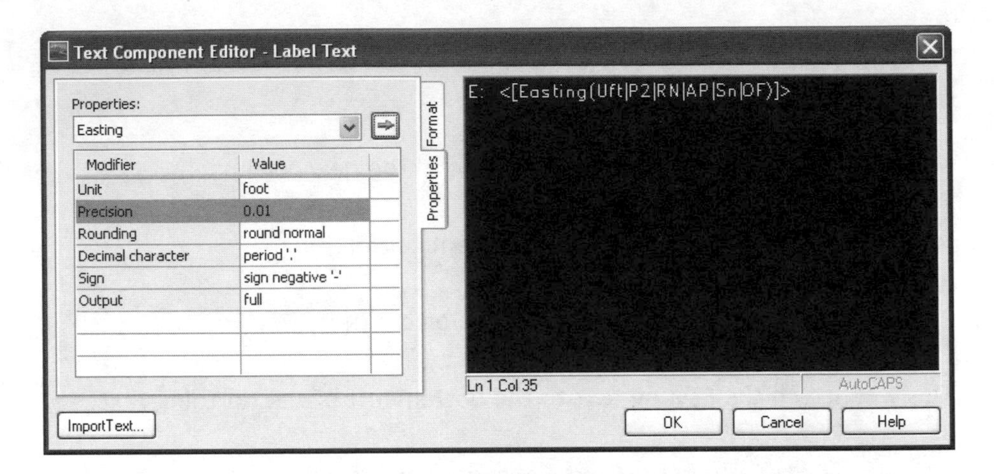

43. Delete the default **"Label Text"** displayed in the editor by dragging across it to highlight, then pressing **<<delete>>**.

44. Select **Easting** from the **Properties** drop down list.

45. Set the **Precision** to **0.01** (two decimal places), by selecting it from the drop-down list.

46. Click the **right arrow button** to insert the code which will display the Northing of the point.

47. Enter **"E: "** in the editor before the code to better describe the label.

48. Click <<**OK**>> to save the contents of the new component.

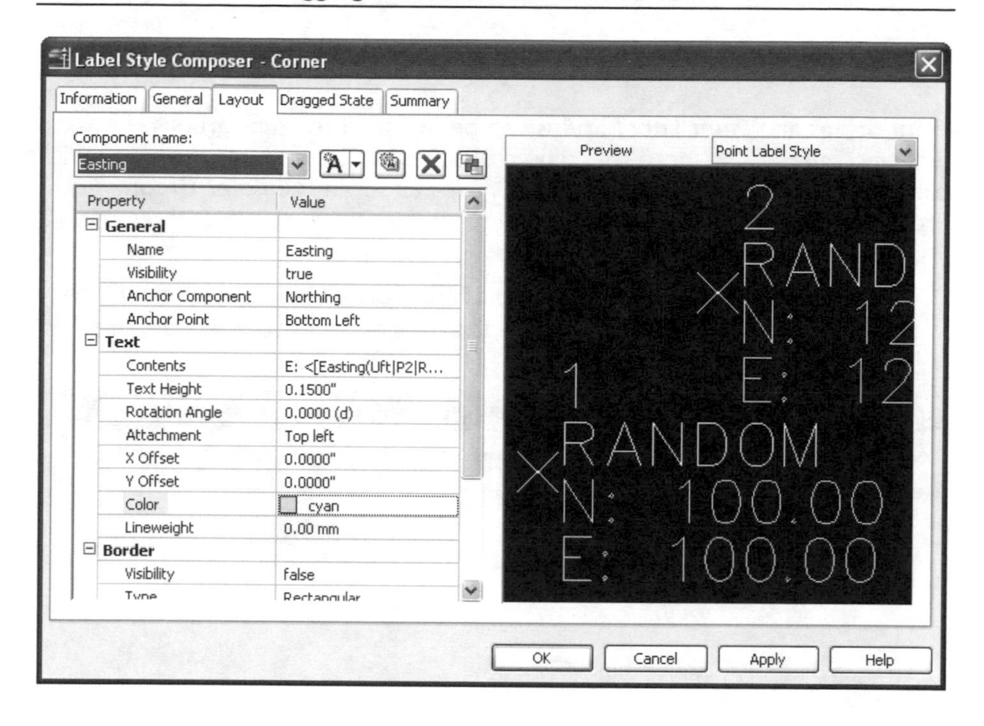

49. Back in the *Label Style Composer* set the **Text Height** to **0.15**.

50. Set the **Attachment** to **Top Left**.

This will attach the top left of the Easting text to the bottom left of the Northing text that you set in a previous step.

51. Set the **Color** to **Cyan**.

52. Click **<<OK>>** to save the new **Point Label Style**.

53. Save the drawing.

You have created and saved a new *Point Style* and *Point Label Style* in your drawing. However, neither style has been applied to any *Points*. They are only available and ready to be used. In the next exercise you will apply theses new styles to a *Point Group*.

4.4.3 Controlling Point Display with Point Groups

Point Styles and *Point Label Styles* can be assigned to point groups. If a point is contained in more than one group, which is common, the display is controlled by the order of the *Point Groups* as set in the *Point Group Properties*.

1. On the *Prospector* tab of the *Toolspace*, right-click on the *Point Group* **Property Corners** and select ⇒ **Properties**.

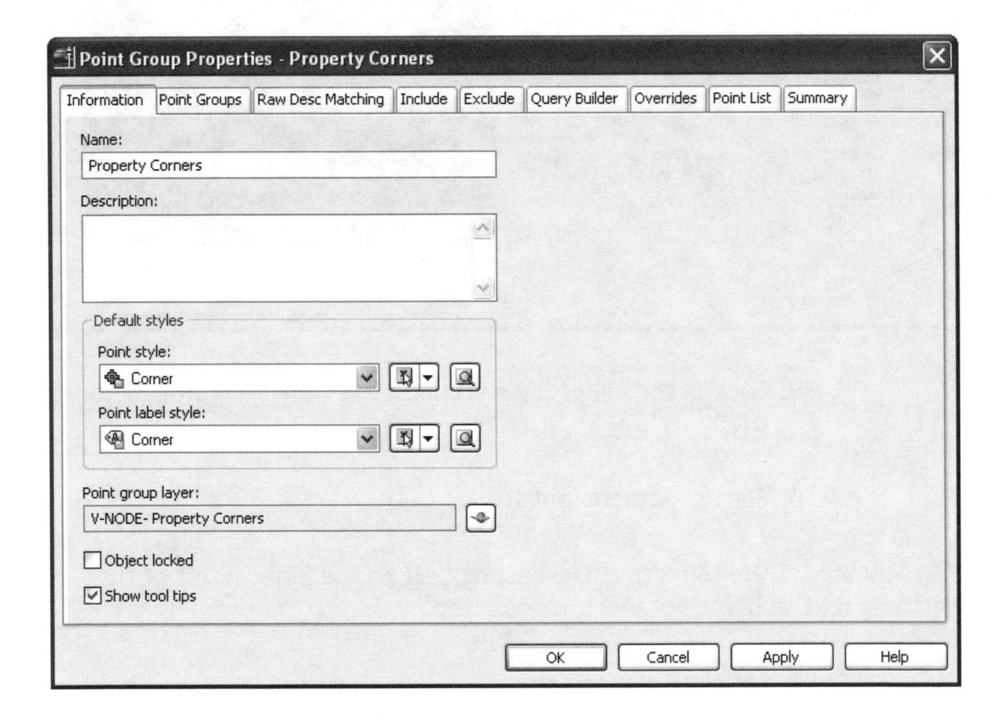

2. Set the **Point style** to **Corner** on the *Information* tab of the *Point Group Properties* dialog box.

3. Set the **Point label style** to **Corner.**

4. Click **<<OK>>** to close the **Point Group Properties** dialog box and apply the new *Point Style and Point Label Style* to the point group.

5. On the *Prospector* tab of the *Toolspace,* select the *Point Group* **Property Corners**. This will display a list of all the property corner points in the preview window at the bottom of the *Prospector*, if the *Prospector* is docked. If the *Prospector* is not docked it will display on the side.

6. Find point number **1556** in the preview window.

7. Right-click on point **1556** and select **Zoom to**.

You will see point 1556 in the center of your screen with the *new Point Style* and *Point Label Style* applied.

8. Select the label text for point 1556 to enable a circular and a square grip.

9. Select the square grip and drag the label to a new location where the text does not overlap other points.

You will see that the point label text changes to green, resizes, and draws a leader back to the point marker. This behavior is controlled by the *Dragged State* which is a component of the *Point Label Style*.

4.4.4 Controlling the Dragged State

The Dragged State is a component of the *Point Label Style* that controls the display of a label when it is moved, or dragged, away from its original location. In this exercise you will modify the *Point Label* Style named *Corner* to set the *Dragged State* to display blue text with rounded border and a red spline leader.

1. On the *Settings* tab of the *Toolspace*, expand **Label Styles** under the *Point* node.

2. Under the *Label Styles* node, right-click on **Corner** and select ⇒ **Edit.**

The *Point Style Corner* is displayed with an orange triangle beside the name on the *Settings* tab of the *Toolspace*. This indicates that the style is in use by an object(s) in the drawing.

3. Select the **Dragged State** tab of the *Label Style Composer*.

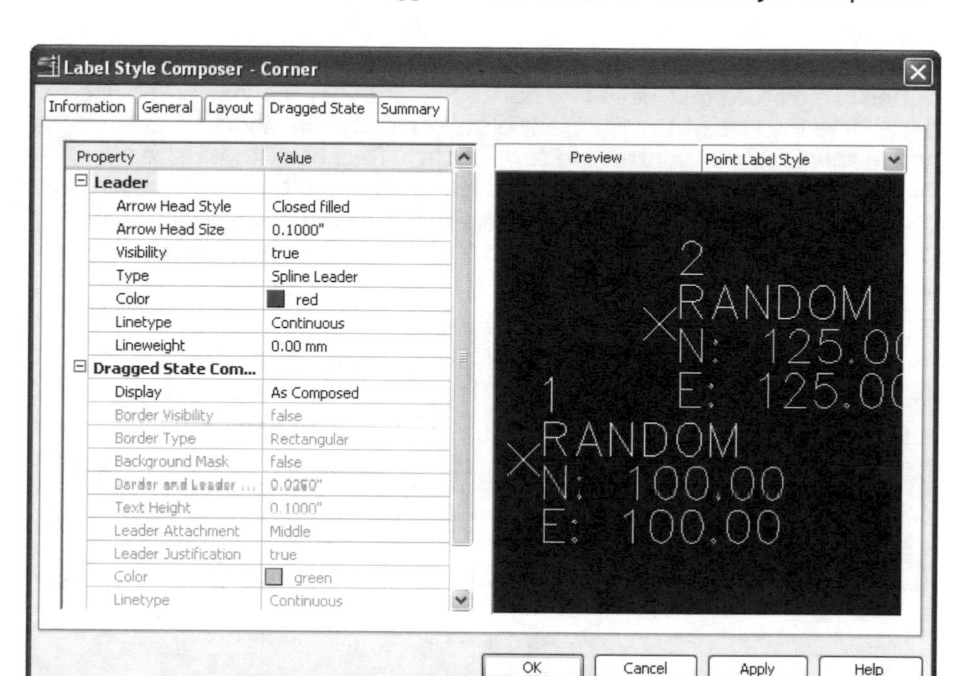

4. Set the **Leader Type** to **Spline Leader**.

5. Set the **Leader Color** to **Red**.

6. Set the **Display** to **As Composed**.

This option will keep the label text formatted exactly as it is when it is in the normal position on the point. If you want the text to be stacked you can use the *Stacked Text* option. This will align and stack the text while changing it all to the color and formatting defined below.

7. Click **<<OK>>** to save the changes to the *Point Label Style*.

Point 1556 now updates to show the point label that is in the dragged state with a red spline leader and the original text formatting as defined in the *Point Label Style*.

4.4.5 Controlling Point Label Size in Model Space

In the previous exercises you created *Label Styles* that included settings to define the plotted text height. This text height uses the viewport scale to determine the size of the text so it plots correctly. In Model Space there is no viewport scale to control the text height. To control these sizes in Model Space *Civil 3D* uses a *Drawing Scale* setting. In this exercise you will change the *Drawing Scale* to resize the *Point Labels*. This drawing scale controls the size of all *Civil 3D Label Styles* the same way, it is not limited to points.

1. On the *Settings* tab of the *Toolspace*, right-click on the drawing **Design** and select ⇒ **Edit Drawing Settings**.

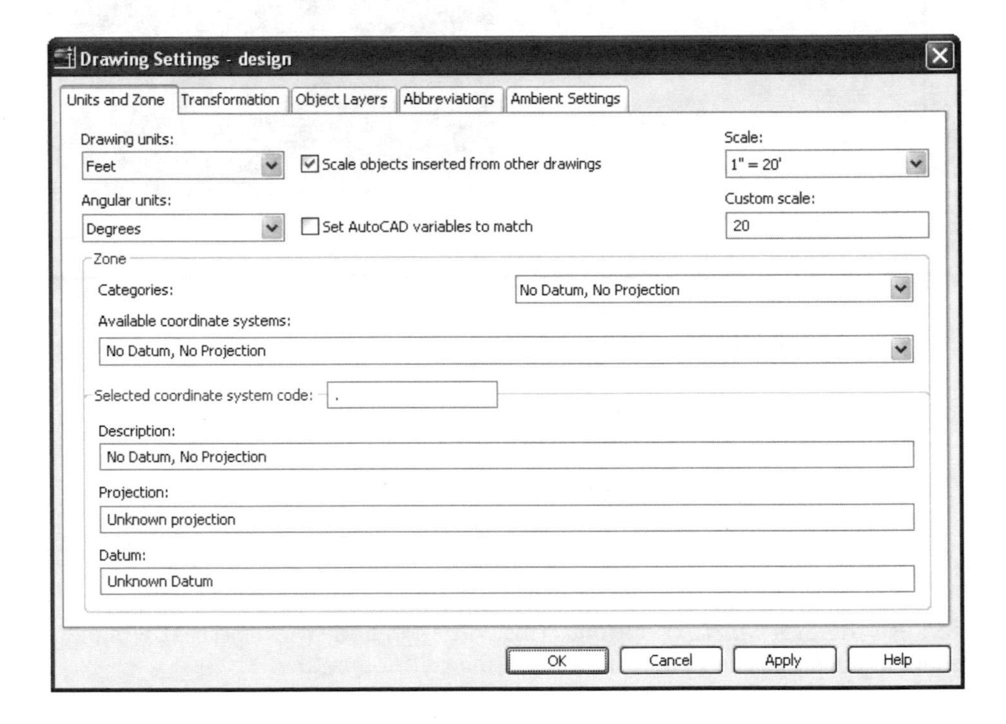

2. Set the **Scale** to **1"=20'**.

You can also set the drawing units and coordinate system in this dialog box along with setting the default object layers, abbreviations, and many other settings for the drawing.

3. Click **<<OK>>** to apply the new scale to the drawing.

It may take a moment to redisplay, but you will see all the labels resize to the new drawing scale.

4.4.6 Controlling Point Group Display Order

It is common for points to reside in more than one *Point Group*. However, only one *Point Group* can control the display of a point at one time. To give you control over which *Point Group* is used to display each point *Civil 3D* allows you to set the *Point Group Display Order*.

1. On the *Prospector* tab of the *Toolspace*, right-click on **Point Groups** and select ⇒ **Properties**.

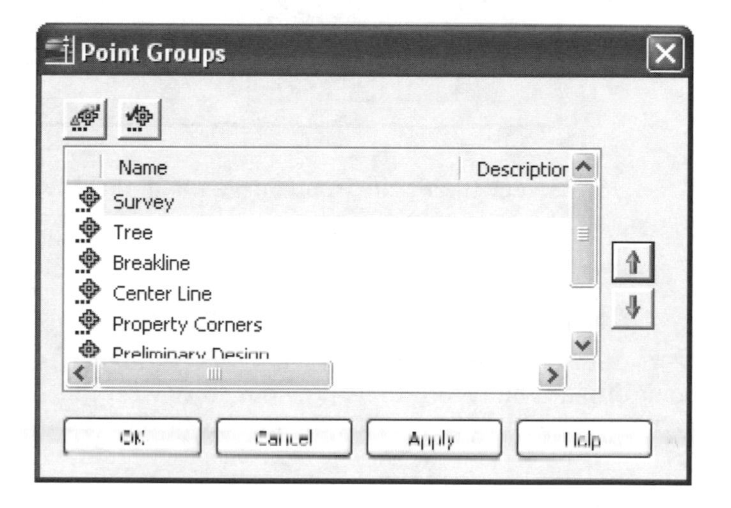

2. Select the *Point Group* **Survey** and move it to the **top** of the list using the **up arrow** button.

By moving the *Point Group Survey* to the top of the display order it will control the display of all the points in that group. Since the points in the groups *Tree, Breakline, Center Line,* and *Property Corners* are also included in the *Survey* group their display, and layers, will change to the definitions found in the *Survey* group.

3. Click <<**OK**>> to redisplay the points.

4. On the *Prospector* tab of the *Toolspace*, right-click on **Point Groups** and select ⇒ **Properties.**

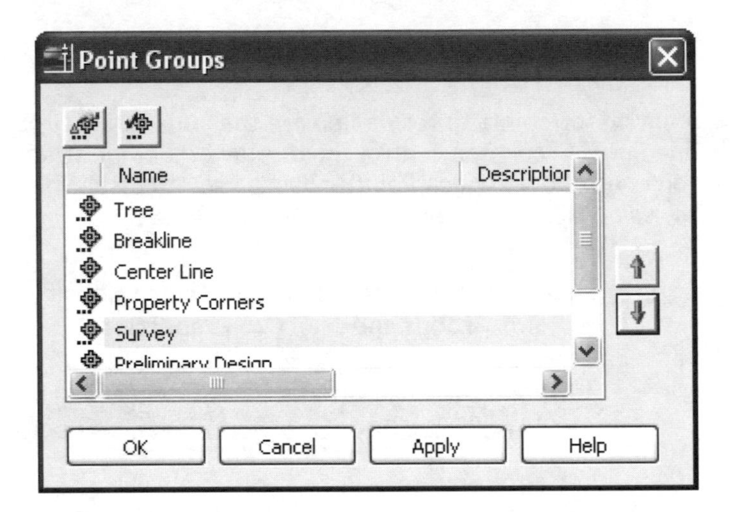

5. Select the *Point Group* **Survey** and move it below the group *Property Corners* using the **down arrow** button .

6. Click <<**OK**>> to redisplay the points.

The point display will return to its previous state with the *Tree* and *Property Corner* groups controlling the display of their points.

4.4.7 Managing Point Layers

Layer management can be very powerful, and complex, if you are using *Point Groups* and *Description Keys* together. By placing each *Point Group* on its own individual layer you will be able to turn the *Point Group* on and off with standard layer commands. As you have seen with the *Point Group Display Order*, the styles and layer of a point in multiple groups is determined by the first *Point Group* in the display order list that contains that point. This means that the point will jump from layer to layer as you change the *Point Group Display Order*.

In this chapter you have created several point groups that are all on specific layers that all start with *V-NODE-*.

1. Freeze the layer **V-NODE-Survey**.

All of the points in the *Survey* group are turned off with the exception of the points in the *Tree, Breakline, Center Line*, and *Property Corners* groups. If the points are not removed from the display, type `Regen` at the command line.

2. On the *Prospector* tab of the *Toolspace*, right-click on **Point Groups** and select ⇒ **Properties**.

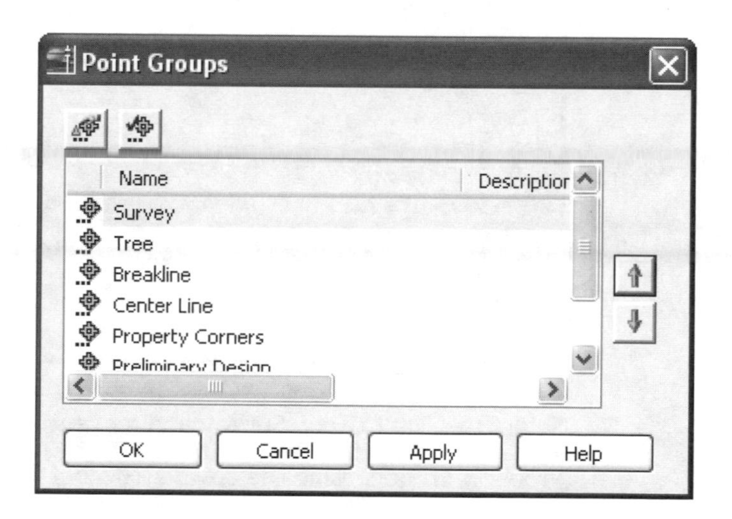

3. Select the *Point Group* **Survey** and move it to the **top** of the list using the **up arrow** button.

4. Click **<<OK>>** to redisplay the points.

All of the remaining points in the **Survey** group are turned off because the **Survey** group is now controlling the display of the points in the *Tree, Breakline, Center Line,* and *Property Corners* groups. If the points are not removed from the display type `Regen` at the command line.

5. Thaw the layer **V-NODE-Survey**.

All of the points in the *Survey* group are now displayed. However, they are all displayed using the styles defined in the *Survey Point Group*. If the points are not displayed type `Regen` at the command line.

6. On the *Prospector* tab of the *Toolspace,* right-click on **Point Groups** and select ⇒ **Properties.**

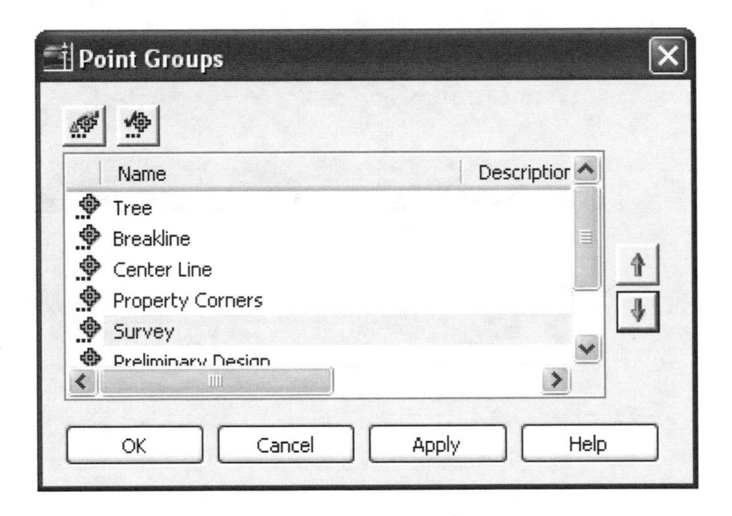

7. Select the *Point Group* **Survey** and move it below the group Property Corners using the down arrow button.

8. Click **<<OK>>** to redisplay the points.

The point display will return to its previous state with the *Tree* and *Property Corner* groups controlling the display of their points.

9. Save the drawing.

4.5 Drawing Linework Using Transparent Commands

The transparent commands in *Civil 3D* allow you to use *Point Numbers* and *Point* objects as locations during any *AutoCAD* command that asks you to specify one. There are many other transparent commands that do not involve points such as specifying a location by *Bearing* and *Distance* which you used in *Chapter 3*. You may want to experiment with some of the commands on the *Transparent Commands* toolbar. You can also enter them at the command line as you will in the following exercise.

4.5.1 Drawing Lines by Point Number

1. Create a new layer named **"EX-PROP-LINE"**, set it **Current** and color **Yellow.**

2. **Freeze** the layers **C-TOPO-Pre-EG, V-NODE-Breakline, V-NODE-Center Line, V-NODE-Survey,** and **V-NODE-Tree**.

You should only see the Property Corner Points displayed on your screen. If all the other points are not turned off type **`Regen`** at the command line.

3. Start the AutoCAD **Line** command.

4. Enter **`'PN`** to change the prompt to **`Point Number`**.

5. Enter points **1565, 1573, 1567, 1568, 1560, 1561, 1564, 1565**.

The point numbers can either be entered separated by commas without spaces, or with an <<ENTER>> after each number.

6. Press **<<ESCAPE>>** to end the `Point Number` prompt.

7. **<<ENTER>>** to end the line.

4.5.2 Drawing Lines by a Range of Point Numbers

1. Start the AutoCAD **Line** command.

2. Enter **`'PN`** to change the prompt to **`Point Number`**.

3. At the command line enter: **`1553-1560`**.

4. **<<ENTER>>** to draw the line.

5. Press **<<ESCAPE>>** to end the Point Number prompt.

6. **<<ENTER>>** to end the line.

7. Start the AutoCAD **Line** command.

8. Enter **`'PN`** to change the prompt to **`Point Number`**.

9. Now enter: **`1561-1564`**.

10. **<<ENTER>>** to draw the line.

11. Press **<<ESCAPE>>** to end the Point Number prompt.

12. **<<ENTER>>** to end the line.

13. Start the AutoCAD **Line** command.

14. Enter **`'PN`** to change the prompt to **`Point Number`**.

15. Now enter: `1568,1553,1569-1573`.

Enter the point numbers without using spaces between them.

16. **<<ENTER>>** to draw the line.

17. Press **<<ESCAPE>>** to end the Point Number prompt.

18. **<<ENTER>>** to end the line.

When finished the parcels should look like the graphic below.

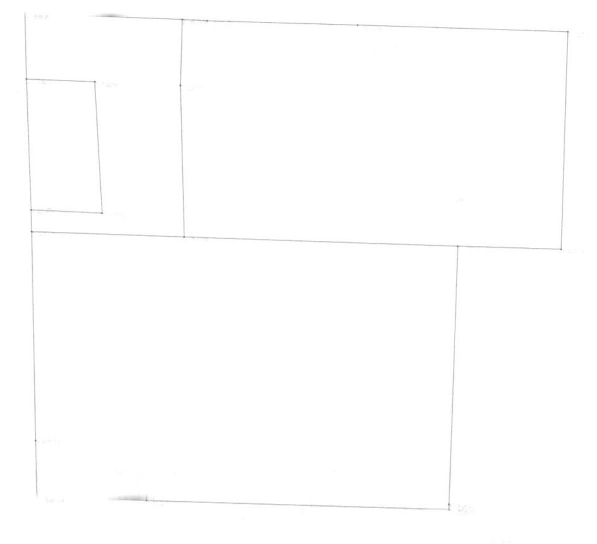

4.5.3 Drawing Lines by Point Object

1. On the *Prospector* tab of the *Toolspace,* select the *Point Group* **Property Corners**. This will display a list of all the property corner points in the preview window at the bottom of the Prospector, if the Prospector is docked. If the Prospector is not docked it will display on the side.

2. Find point numbers **1563** in the preview window.

3. Right-click on point **1563** and select **Zoom to**.

4. Zoom out until you can see points 1563 and 1567

5. Start the AutoCAD **Line** command.

6. Enter **`'PO`** to change the prompt to **`Point Object`**.

7. Select Points **`1563, 1567`** on the screen to draw a line between them.

This command does not show the rubber band line from the last point selected like you may be expecting to see. However, the command prompt will change from asking for the First Point to asking for the Next Point. If you select a point more than once you will create zero length lines at that location.

8. **<<ENTER>>** to end the Point Object prompt.

You can also use the Escape key.

9. **<<ENTER>>** a second time to end the line.

10. Start the AutoCAD **Line** command.

11. Enter **`'PO`** to change the prompt to **`Point Object`**.

12. **Zoom to** points **1556, 1574, 1575, 1576,** and **1558** at the southwestern corner of the site.

13. Select Points `1556, 1574, 1575, 1576, 1558.`

14. **<<ENTER>>** to end the Point Object prompt.

You can also use the Escape key.

15. **<<ENTER>>** a second time to end the line.

16. Save the drawing.

When finished the parcels should look like the graphic below.

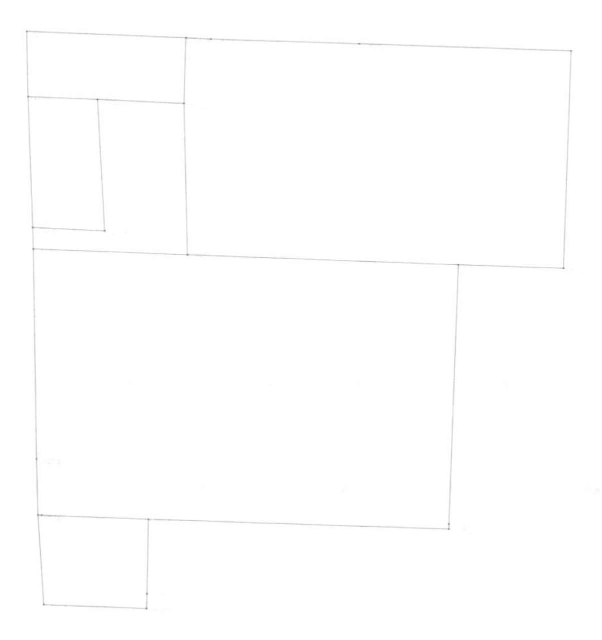

4.6 Working with Parcels

Parcels are closed polygons defined in *Civil 3D* and saved in the drawing as part of a *Site*. Parcels within a *Site* interact with each other and can not overlap. If parcels within the same *Site* overlap a third parcel will be created. If parcels share a common boundary line, editing that boundary line will cause one parcel to get larger and the other smaller.

4.6.1 Defining a Parcel from Existing Geometry

Parcels can be created from existing lines, curves, and polylines; or by using the *Parcel Layout Tools*. In this exercise you will define several parcels based on the lines that you created in the previous exercises.

1. **Freeze** the layer **V-NODE-Property Corners.**

You should only see the property lines on your screen. If all the points are not turned off type `Regen` at the command line.

2. Select **Parcels ⇒ Create From Objects.**

3. Select the 6 parcel lines that make up the parcel in the southwest corner of the site.

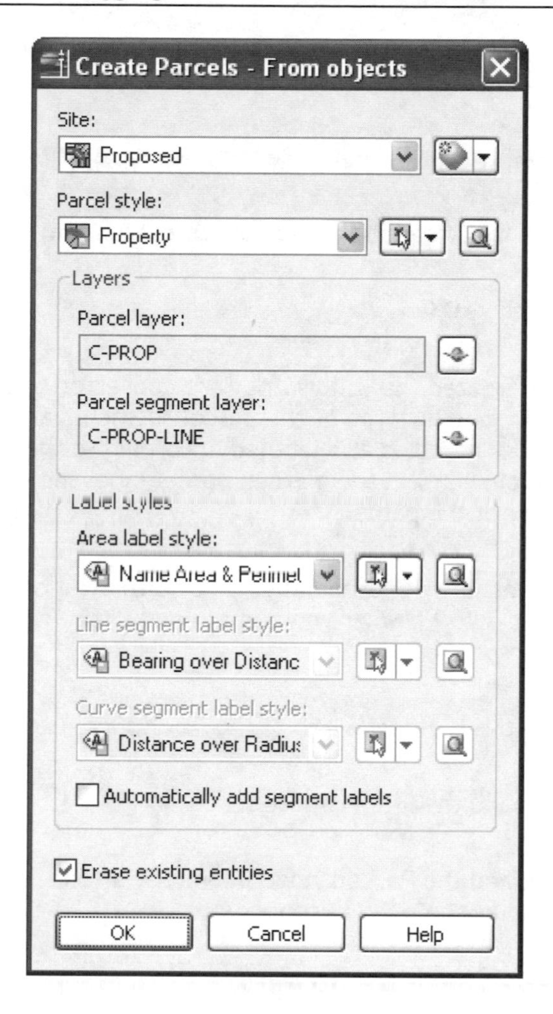

In *Civil 3D* a *Site* is a collection of objects that relate to and interact with each other. *Alignments, Parcels*, and *Grading Groups* can all be contained in a *Site*. By using multiple *Sites* in your drawing you can control which of these objects interact with each other. You may want to uses different sites for different phases of a project or for different design alternatives.

4. Click the **Create New Site** button to open the *Site Properties* dialog box.

5. **Name** the new Site "**Proposed**".

6. Click **<<OK>>** to create the new site and return to the *Create Parcels - From Objects* dialog box.

7. Confirm the **Parcel style** is set to **Property**.

8. Set the **Area label style** to **Name Area and Perimeter**.

9. Click **<<OK>>**.

When the parcel is created the yellow lines are converted to cyan parcel segments. The color and linetype of the parcel segments are controlled by the *Parcel Style*. The parcel is also labeled according to the parcel label style. In this example you will see a green parcel label showing the Name, Area, and Perimeter of the parcel.

If the parcel segments or the parcel label is not displayed confirm that the layers *C-PROP* and *C-PROP-LINE* are on and thawed. The Parcel will also be displayed in the Prospector.

10. On the *Prospector* tab of the *Toolspace*, expand **Sites**.

11. Expand the *Site* **Proposed**.

12. Expand the **Parcels** node under the *Site* **Proposed** to display the parcel **Property: 1**.

If there are no parcels displayed under the site then the parcel has not been defined. You may have converted the lines to parcel segments. However, if they are not closed they will not be defined as a parcel.

You can define additional objects as parcel segments or edit the existing parcel segments to close the area and the parcel will automatically be created. If you edit any of the parcel segments the parcel will update to display the new area and perimeter in the label. However, if you edit the parcel segments and the parcel is no longer closed, the parcel will be deleted and you will be left with only parcel segments. As you edit parcel segments you can use the Undo command to undo any changes.

13. Select **Parcels ⇒ Create From Objects.**

14. Select all the remaining parcel lines with a window.

15. Confirm the **Site** is set to **Proposed.**

16. Confirm the **Parcel style** is set to **Property.**

17. Set the **Area label style** to **Name Area and Perimeter.**

18. Click **<<OK>>.**

The remaining parcels will all be defined and labeled. Parcels created within the same site use a topology model and interact with each other. For example when two parcels share a boundary line there is only one line used, not overlapping duplicate lines. When that boundary line is moved or stretched both parcels changes size. If new parcel segments and parcels are created in the same location as existing parcels their area will be subtracted from the existing parcels where they overlap as long as all of the parcel segments are in the same Site. This is one way that you can subdivide parcels.

4.6.2 Creating a Parcel Area Report

1. Select **General** ⇒ **Reports Manager**.

This will open a new Tab called the *Toolbox* in the *Toolspace*.

2. On the *Toolbox* tab of the *Toolspace*, expand **Reports Manager**, and then expand **Parcel**.

3. Under the *Parcel* node, right-click on **Area_Report** and select ⇒ **Execute**.

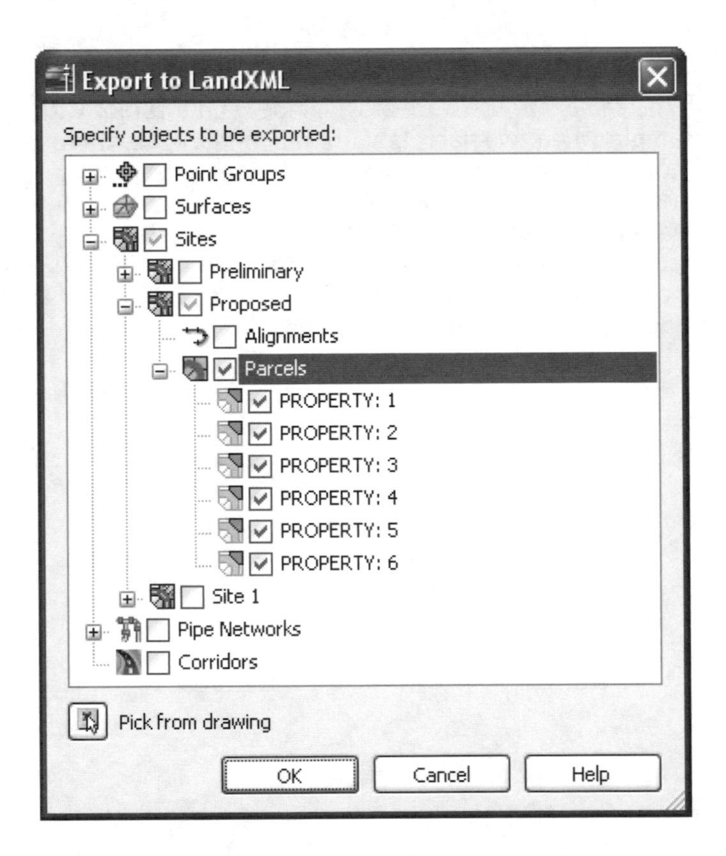

4. Confirm that all the Parcels are selected under the *Site* **Proposed** and click **<<OK>>**.

You have the option here to be selective about the parcels you want to include in the report. You can also improve performance by disabling options for other objects that will not be used in the report like surfaces and alignments.

The Parcel Area report is now generated and displayed in Internet Explorer. Here you can print the report, save the report, or copy from the report and paste into other programs like Word or Excel.

Your Company Name

123 Main Street

Suite #321

City, State 01234

Parcel Area Report
Project Name: C:\Cadapult Training Data\Civil 3D 2007\Level 1\Design.dwg
Report Date: 5/19/2006 11:18:18 AM

Client: Client Company
Project Description:
Prepared by: Preparer

Parcel Name	Square Feet	Acres	Perimeter (ft)
PROPERTY: 1	76146.988	1.748	1109.347
PROPERTY: 2	73799.060	1.694	1144.110
PROPERTY: 3	893954.810	20.522	3881.018
PROPERTY: 4	115319.040	2.647	1750.823
PROPERTY: 5	673061.014	15.451	3407.617
PROPERTY: 6	82694.981	1.898	1262.495

5. After reviewing the report, **close Internet Explorer** and return to *Civil 3D*.

4.6.3 Creating a Parcel Legal Description Report

1. If the *Toolbox* tab is not visible in the *Prospector* Select **General ⇒ Reports Manager**.

2. On the *Toolbox* tab of the Toolspace, **expand Reports Manager, and then** expand **Parcel**.

3. Review the different reports that are available.

You will notice that there are two different types of reports, XSL and VBA, each represented by a different icon. Custom reports can also be written using .NET and COM executables. The XSL reports are the reports that Civil 3D created in the 2006 and earlier versions. These same reports can also be created by using the *Autodesk LandXML Reporting 7* application that was installed with Civil 3D by opening a LandXML file that has been exported from Civil 3D or any other program that supports LandXML. The *Autodesk LandXML Reporting 7* application is used to change the report settings, like units and precision, for the XSL reports created in Civil 3D through the *Toolbox*.

In this exercise you will change some report settings in the *Autodesk LandXML Reporting 7* application to generate a Legal Description with the desired units and precision.

4. Minimize Civil 3D.

5. Launch the *Autodesk LandXML Reporting 7* application by double clicking it's shortcut on the desktop.

It can also be found through the Start menu in the *Autodesk* program group.

6. Select the **Settings** Tab

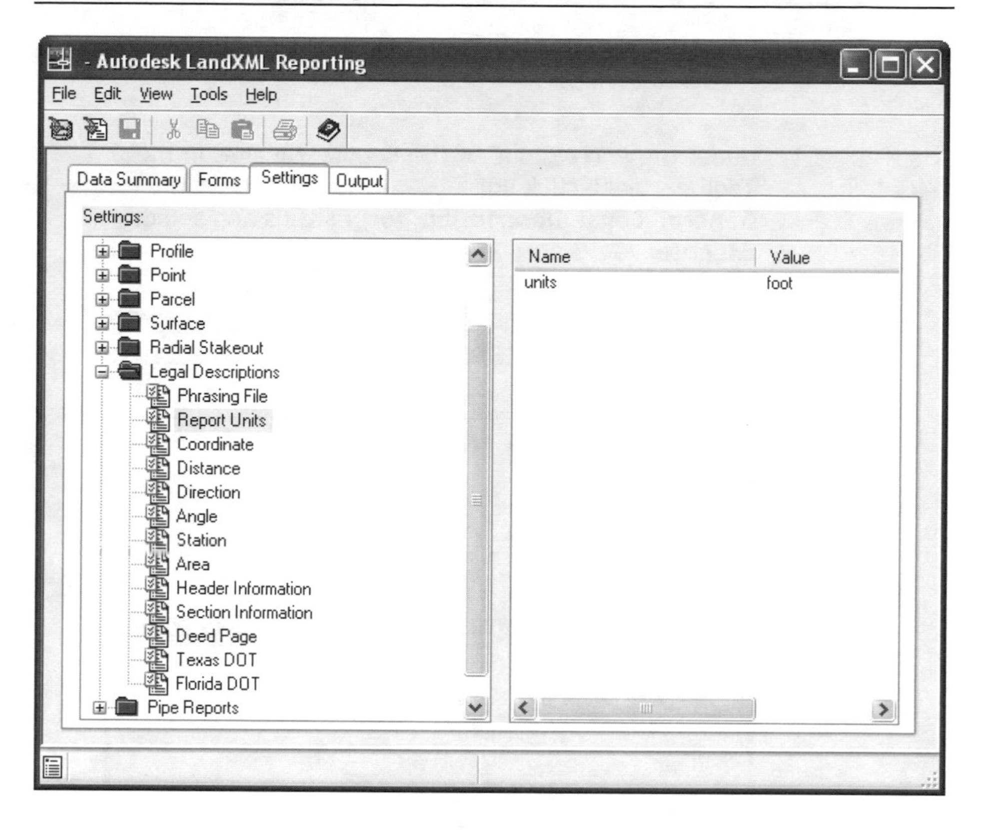

7. Expand the **Legal Description** settings and select the **Report Units** setting.

8. Set the **units** to **foot**.

9. Select the **Distance** settings.

10. Set the **Distance Precision** to **3** decimal places.

11. Close the *Autodesk LandXML Reporting 7* application and return to *Civil 3D*.

12. Under the *Parcel* node of the Report Manager in the Toolbox, right-click on **General_Legal_Description_for_Parcels** and select ⇒ **Execute**.

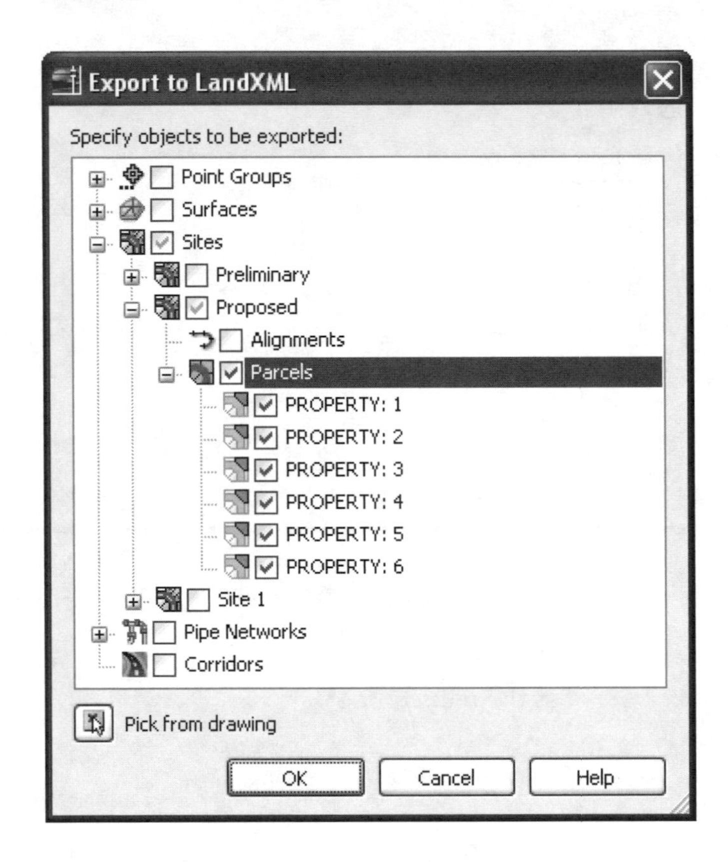

13. Confirm that all the Parcels are selected under the *Site* **Proposed** and click **<<OK>>**.

You have the option here to be selective about the parcels you want to include in the report. You can also improve performance by disabling options for other objects that will not be used in the report like surfaces and alignments.

Internet Explorer is now opened with options for you to select an individual or all parcels to include in the Legal Description report.

14. If you receive a warning about **blocked content** select the option to **allow** it. (This depends on your Internet Explorer security settings)

15. **Enable** the **Select all Parcels** option.

16. Click **<<Append to Report>>** near the bottom of the sidebar.

The *Legal Descriptions* of the parcels are now displayed. You can print, save, or copy from the report and paste into other programs like *Word* for further editing and formatting.

17. After reviewing the *Legal Descriptions*, **close Internet Explorer** and return to *Civil 3D*.

18. Save the drawing.

4.7 Labeling Linework

The *Add Labels* command in *Civil 3D* can be used to label lines and curves if they are either part of a parcel segment or if they are standard *AutoCAD* lines, arcs, or polylines. In this section you will first label the parcel segments that you created in the previous exercises, and then you will create a polyline representing a wetland boundary and label that. It is important to remember that the commands to label and edit parcel segments and *AutoCAD* objects are very similar. The major difference is that you will select a different feature in the *Add Labels* command.

4.7.1 Labeling Parcel Lines

Parcel Segment labels are defined by Parcel Line and Parcel Curve Label Styles. These styles are saved in the drawing and can be created and edited on the Settings tab of the *Toolspace* or on the fly in the *Add Labels* dialog box.

1. Select **Parcels ⇒ Add Labels**.

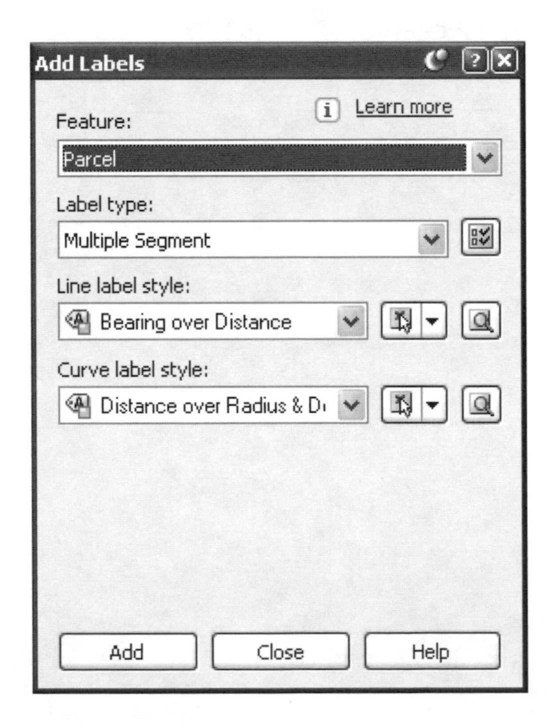

2. Set the **Label Type** to **Multiple Segment.**

3. Click **<<Add>>**.

When creating *Multiple Segment Labels* you must pick the *Parcel Label* at the center of the parcel rather than the lines that you want to label. This will select all of the *Parcel Segments* that make up the parcel and label them at their midpoints simultaneously.

4. Pick the **Parcel Label** at the center of one of the parcels.

Labels will be created at the midpoints of all the *Parcel Segments* around the selected parcel according to the *Label Style* selected in the *Add Labels* dialog box. These *Label Styles* can be modified similar to the way that you created and edited the *Point Label Styles* in previous exercises.

5. Set the **Label Type** to **Single Segment.**

6. Click **<<Add>>**.

When creating *Single Segment Labels* you must pick the *Parcel Segment* at the location that you want to create the label.

7. Select a **Parcel Segment** at the location that you want to create the line label.

A label will be created at the location you picked the *Parcel Segment* according to the *Label Style* selected in the *Add Labels* dialog box. These *Label Styles* can be modified similar to the way that you created and edited the *Point Label Styles* in previous exercises.

8. Click <<**Close**>> in the *Add Labels* dialog box when you are finished.

4.7.2 Working with Parcel Segment Labels

Parcel Segment Labels can be grip edited and erased with standard *AutoCAD* commands. Other commands specifically for labels can be found by right-clicking on them. In this exercise you will make several changes to the labels you just created to demonstrate how these editing commands work.

1. Select one of the **Parcel Segment Labels** and drag it to a new location.

If you move the label to a location along the *Parcel Segment* it will not change appearance. Otherwise, the label display will change to the *Dragged State* values defined in the *Label Style* and draw a leader back to the original location. You can use the Undo command to return the label to its previous location.

2. Select a label, right-click and select ⇒ **Reverse Label**.

This will reverse the direction of the label, for example, changing southeast to northwest.

3. Select a label, right-click and select ⇒ **Flip Label**.

This will flip the label text from one side of the line to the other.

4.7.3 Creating a Line Tag Style

To create Line and Curve Tables you must first tag the *Parcel Segments* with a Tag Label. In this exercise you will create a Tag Label Style.

1. Select **Parcels** ⇒ **Add Labels**.

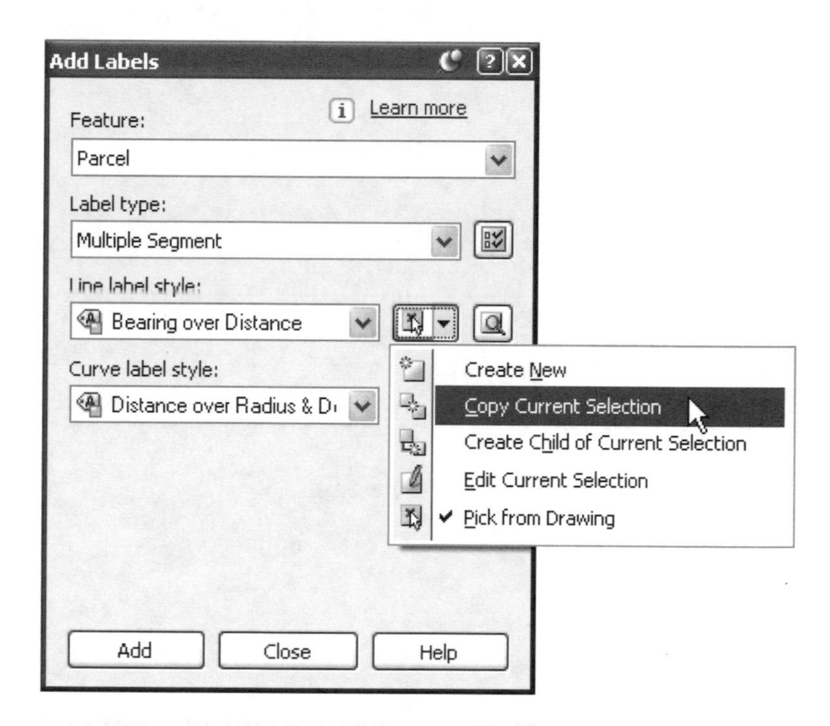

2. Set the **Label Type** to **Multiple Segment**.

3. Click the button to the right of the *Line label style* and select **Copy Current Selection** from the drop-down list.

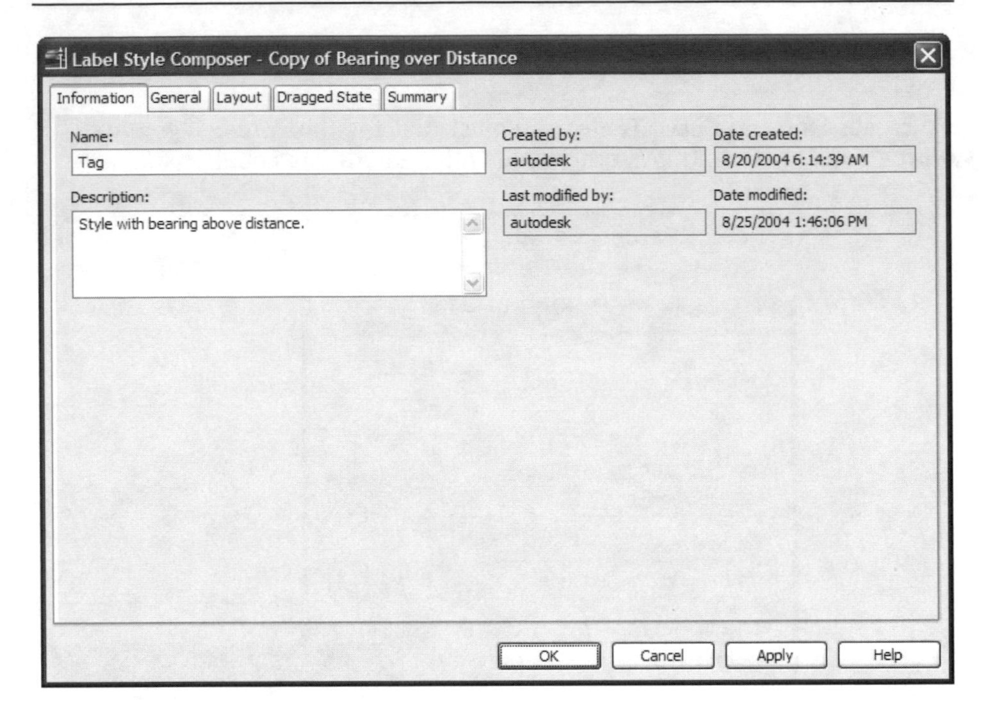

4. Name the new style **Tag**.

5. Select the **General** tab.

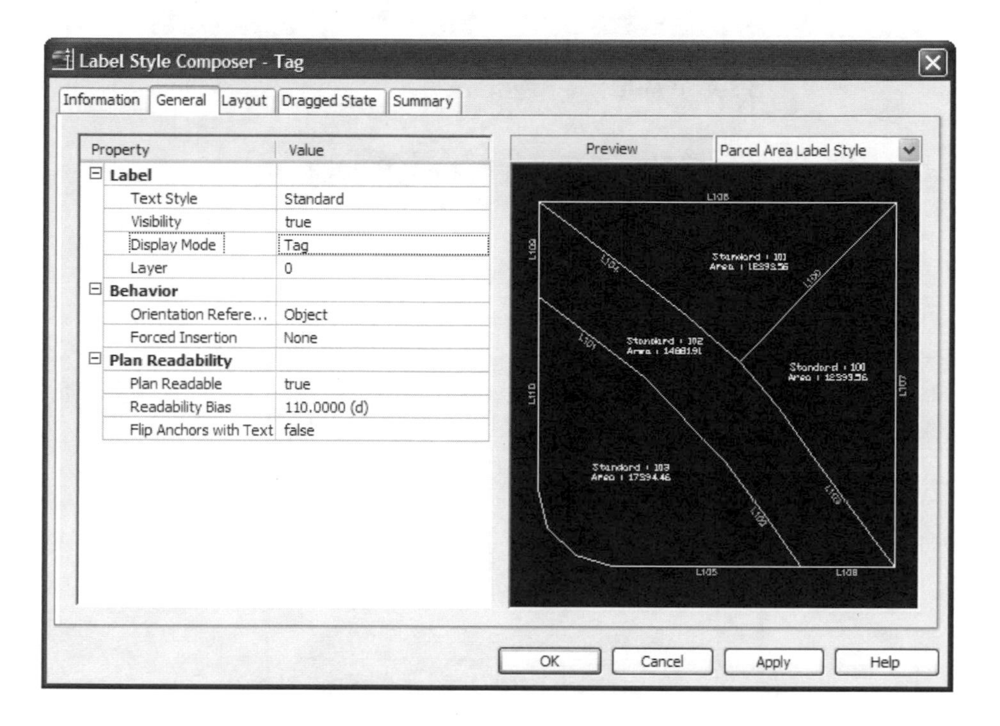

6. Set the **Display Mode** to **Tag**.

7. Click <<**OK**>> to save the new style.

4.7.4 Tagging Parcel Lines

1. Back in the *Add Labels* dialog box confirm the **Label Type** is set to **Multiple Segment**.

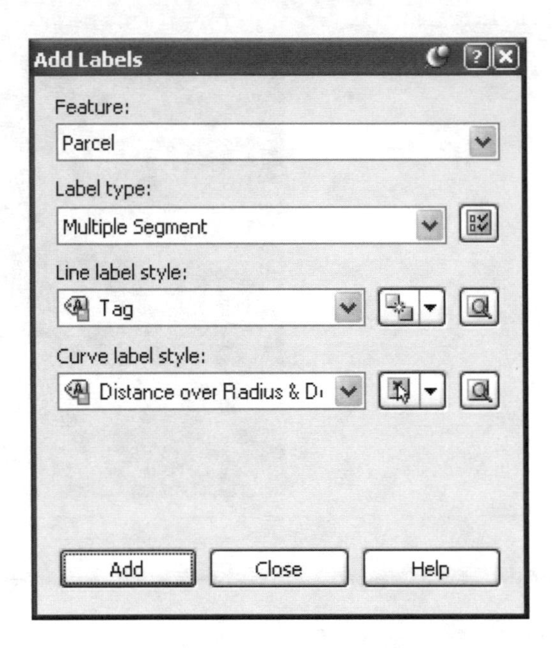

2. Confirm the **Line Label Style** is set to **Tag**.

3. Click <<**Add**>>.

4. Pick the **Parcel Label** at the center of one of the parcels that does not currently have segment labels.

Tag Labels will be created at the midpoints of all the *Parcel Segments* around the selected parcel according to the *Label Style* selected in the *Add Labels* dialog box.

5. Click <<**Close**>> in the *Add Labels* dialog box when you are finished.

4.7.5 Creating a Line Table

Once you have added *Tag* labels to the *Parcel Segments* you can create a *Line Table*.

1. Select **Parcels** ⇒ **Tables** ⇒ **Add Line.**

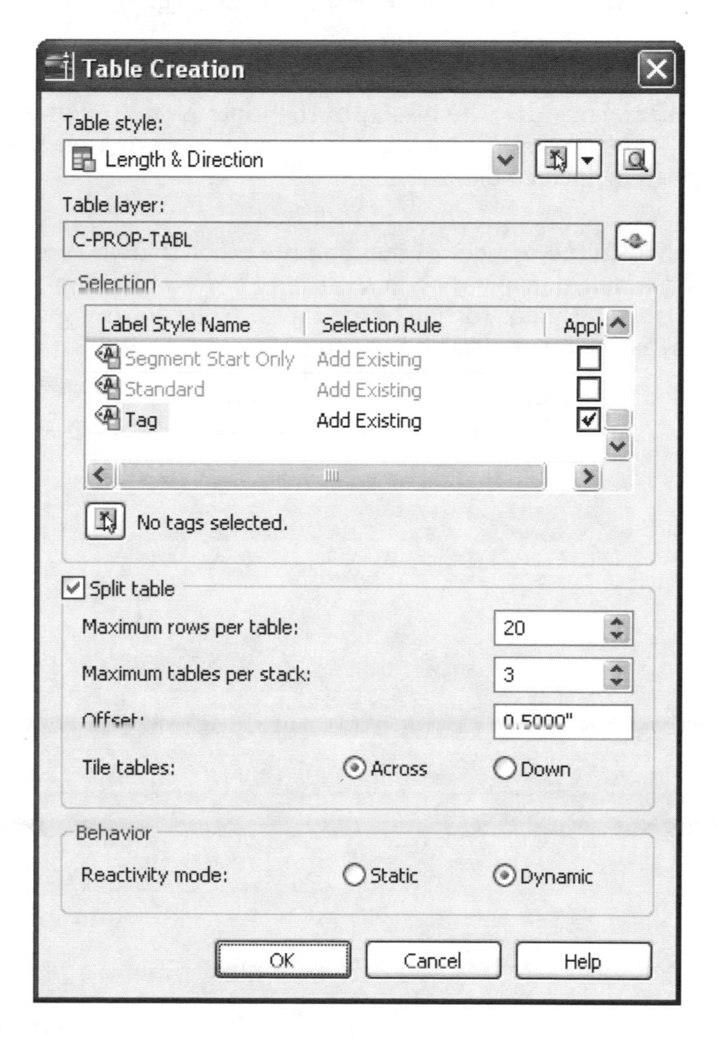

2. **Enable** the **Label Style** name **Tag.**

The display of the table, including the columns of data displayed is controlled by the *Table Style*. You can also set the layer the table will be created on as well as if the table is *Dynamic* or *Static*.

3. Click **<<OK>>.**

4. Select a location for the *Line Table*.

The table is a single object and can be moved or erased with standard *AutoCAD* commands. The display of the table is controlled by the *Table Style*. This includes the color of the lines and text as well as the data contained in each column.

Depending on the number of tags and the order that you created them, the line tag numbers may not be in sequential order in the table. The *Table Style* also determines sorting of the table. You can sort any column in the table in ascending or descending order.

5. Select the table, right-click and select ⇒ **Edit Table Style.**

6. Select the **Data Properties** tab of the *Table Style* dialog box.

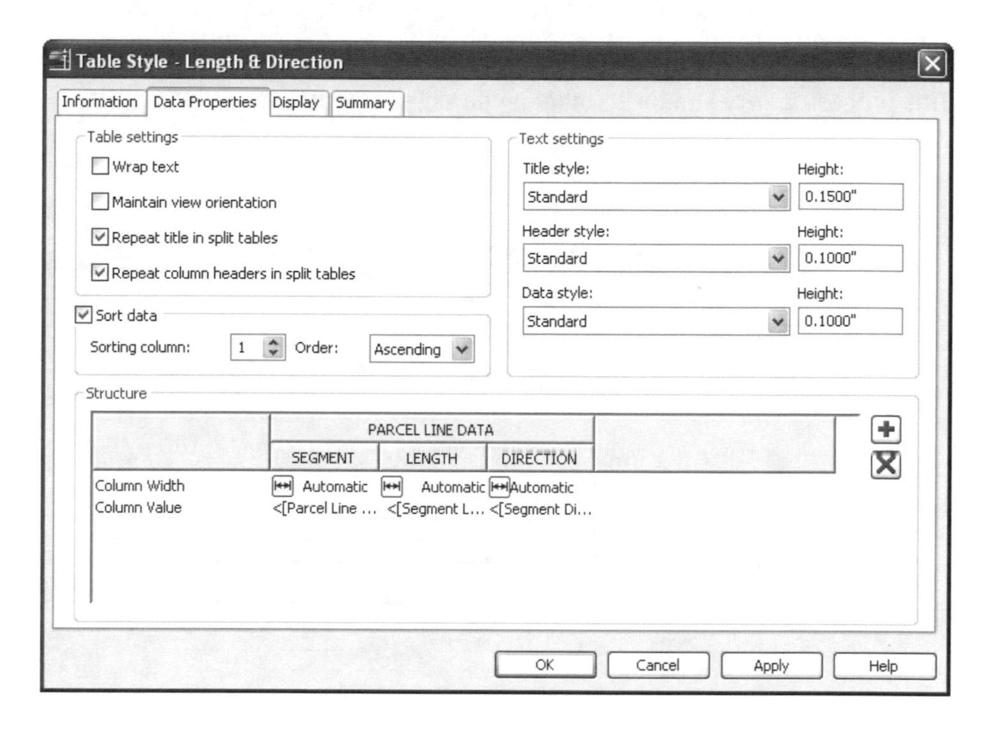

7. Enable the Sort Data option.

8. Confirm that the **Sorting Column** is set to **1**.

9. Confirm that the **Order** is set to **Ascending**.

10. Click **<<OK>>** to save the changes to the Table Style and sort the table.

4.7.6 Labeling AutoCAD Objects

In this exercise you will import a point file collected during a wetland delineation. You will then create the wetland boundary by connecting the points using the polyline command with the *Civil 3D* transparent point selection commands. Finally, you will label the bearing and distance of the polyline segments representing the wetland boundary.

This exercise brings together many of the procedures and concepts that you have covered earlier in this chapter as you import points and create the boundary of the wetland. For more detail you can refer back to those previous exercises. Once the wetland boundary is created you will learn to use the *Civil 3D* labeling commands to label standard *AutoCAD* objects. This process is very similar to labeling parcel segments.

1. Select **Points ⇒ Create Points.**

2. Click the **Import Points** button on the *Create Points* toolbar.

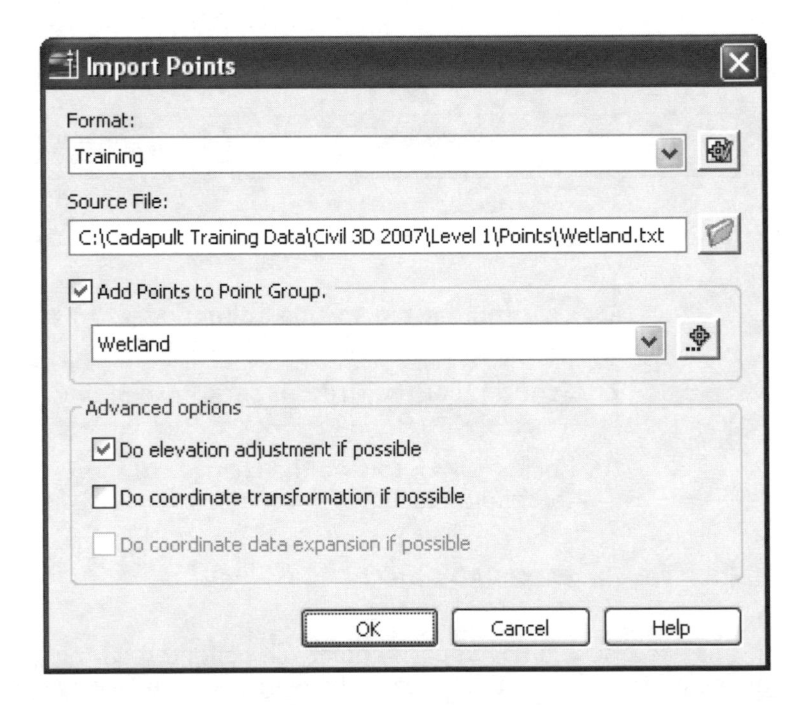

3. In the *Import Points* dialog box, select "**Training" as the Format.**

4. Click **the Browse** **button** to select *the Source File* you want to import:

C:\Cadapult Training Data\Civil 3D 2007\Level 1\Points\Wetland.txt

5. **Enable** the **Add Points to Point Group** option.

6. Click the Add Points button to the right of the drop down list, to enter a new point group called **"Wetland"**.

7. Click **<<OK>>** to import the points.

8. Close the **Create Points** toolbar.

9. Zoom in to the wetland points along the southern border of the largest parcel.

10. On the *Prospector* tab of the *Toolspace*, expand the **Point Groups node.**

11. Right-click on the *Point Group* **Wetland** and select ⇒ **Properties**.

12. Click the **Point group Layer** button to open the *Object Layer* dialog box.

13. Select **Suffix** as the **Modifier**.

14. Enter "**-***" as the **Modifier value.**

15. Click **<<OK>> to close the Object Layer** dialog box.

16. Click **<<OK>>** to save the changes to the group.

This will create a new layer called V-NODE-Wetland and move the *Point Group* to that layer.

17. On the *Prospector* tab of the *Toolspace*, right-click on the **Wetland** point group and select ⇒ **Lock Points.**

This locks all the points that are in the selected group so that they cannot be edited.

18. On the *Prospector* tab of the *Toolspace*, right-click on the **Wetland** point group and select ⇒ **Lock**.

This locks the *Properties* of the point group so that points cannot be added or removed from the group.

19. Create a new layer named **"EX-WETLAND-LINE"**, set it **Current** and color **Green.**

20. Start the AutoCAD **Polyline** command.

21. Enter **`'PN`** to change the prompt to **`Point Number`**.

22. At the command line enter: **`2001-2007,2001`**.

23. **<<ENTER>>** to draw the polyline.

24. Press **<<ESCAPE>>** to end the **`Point Number`** prompt.

25. **<<ENTER>>** to end the polyline.

26. Select **General** ⇒ **Add Labels.**

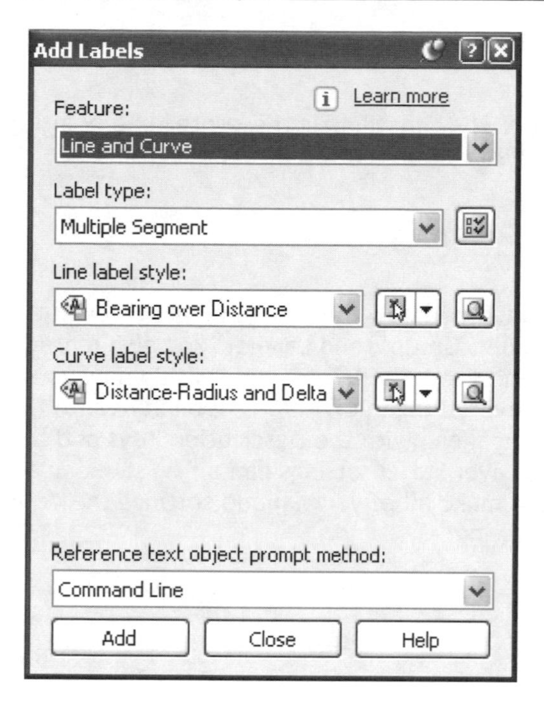

27. Set the **Feature** to **Line and Curve**.

28. Set the **Label type** to **Multiple Segment**.

29. Confirm that the **Line label style** is set to **Bearing over Distance**.

30. Click **<<Add>>**.

31. Select the wetland boundary polyline.

32. Click **<<Close>>** in the *Add Labels* dialog box when you are finished.

Labeling *AutoCAD* objects with the *Civil 3D* labeling commands is very similar to labeling parcel segments. Editing the labels and modifying label styles is also accomplished in the same way as it is with parcel segments.

4.8 Additional Exercises

Create a *Point Label Style* that labels the center line points with a blue point number and description rotated at a 45 degree angle.

4.9 Chapter Summary

In this chapter you learned how to manage points through the use of Description Keys, Point Groups, and Layers. You also learned how to create and label Parcels and generate Parcel Reports. By now you can see how dependant almost everything in *Civil 3D* is upon Styles. All of the Styles that you are creating along with the Description Keys and the default Object Layers and Layer Suffix settings can all be saved in a drawing template. This can make all of your unique settings the defaults each time you start a new drawing.

Chapter 5

Building a Survey Quality Surface

In this chapter you will use points and breaklines from the survey data to create a survey quality existing ground surface. You will learn ways to leverage the use of *Point Groups* to efficiently build and edit a *Surface*. You will also explore various ways of editing and analyzing surfaces including the use of the preliminary surface to add extra data beyond the limits of the survey.

- **Building a Surface from Survey Data**
- **Editing Surfaces**
- **Surface Analysis**
- **Working with Contours**
- **Labeling Contours**

Dataset:

To start this chapter you will continue working in the drawing named **Design.dwg**. You can continue with the drawing that you currently have from the end of the previous chapter or, if you are starting in the middle of the book, you can open the drawing **CH-05.dwg** located in the folder **C:\Cadapult Training Data\Civil 3D 2007\Level 1\Chapter Drawings.** Opening the drawing from the dataset provided will ensure that you have the drawing set up correctly for the exercises in the following chapter and overwrite any mistakes that you may have made in previous exercises.

5.1 Overview

In this chapter you will use points and breaklines from the survey data to create a survey quality existing ground surface. You will learn ways to leverage the use of *Point Groups* to efficiently build and edit a *Surface* by editing the source data and also editing the *Surface* itself. You will also explore various ways of editing and analyzing surfaces including the use of the preliminary surface to add extra data beyond the limits of the survey. Finally, you will learn to display and label contours working with *Surface Styles* and *Contour Label Styles*.

5.1.1 Concepts

Creating an accurate Surface is one of the most important parts of any *Civil 3D* project. The *Profiles, Sections, Corridor Models*, and *Grading* as well as *Volume Calculations* that you create in future chapters are all based on this *Surface*. This chapter will explore ways to create an existing ground surface from survey data as well as ways to check, display, analyze, and edit the surface.

5.1.2 Styles

Styles are saved in the drawing and can be created and edited on the Settings tab of the *Toolspace or on the fly during many object creation and editing commands.*

Surface Styles control the display of surfaces. You have individual control over many components of the surface including contours, triangles, points, borders, slope analysis, elevation bands, watersheds, and flow arrows.

Contour Label Styles control the display of contour labels including text style, size, color, and border.

5.2 Building a Surface from Survey Data

Any time you build a surface the most important step is to understand what data you have available to work with. In this chapter you will work with points that will be managed with a *Point Group* and breaklines that you will create based on some of those same survey points.

5.2.1 Creating a Point Group to Be Used As Surface Data

Before you create the surface you need to create a *Point Group* that will be used to select only the points that you want to use for the surface data. Points that should not be included in the surface should not be included in the point group. Points for utility potholes or points that are part of the project for horizontal control and do not have accurate surface elevations are examples of points that should not be included in this group.

1. Continue working in the drawing **Design.dwg**.

This drawing contains the *Points, Alignment, Parcels,* and *Surface* from the previous chapters. Currently only the parcel lines and labels are displayed.

2. On the *Prospector* tab of the *Toolspace,* right-click on **Point Groups** and select ⇒ **New.**
3. Enter **"Topo"** for the **Name.**
4. Click the **Point group Layer** button to open the *Object Layer* dialog box.
5. Set the **Modifier** to **Suffix.**
6. Enter **-*** as the **Modifier** *value.*
7. Click **<<OK>>** to close the *Object Layer* dialog box.

8. Select the **Raw Desc Matching** tab in the *Point Group Properties* dialog box.

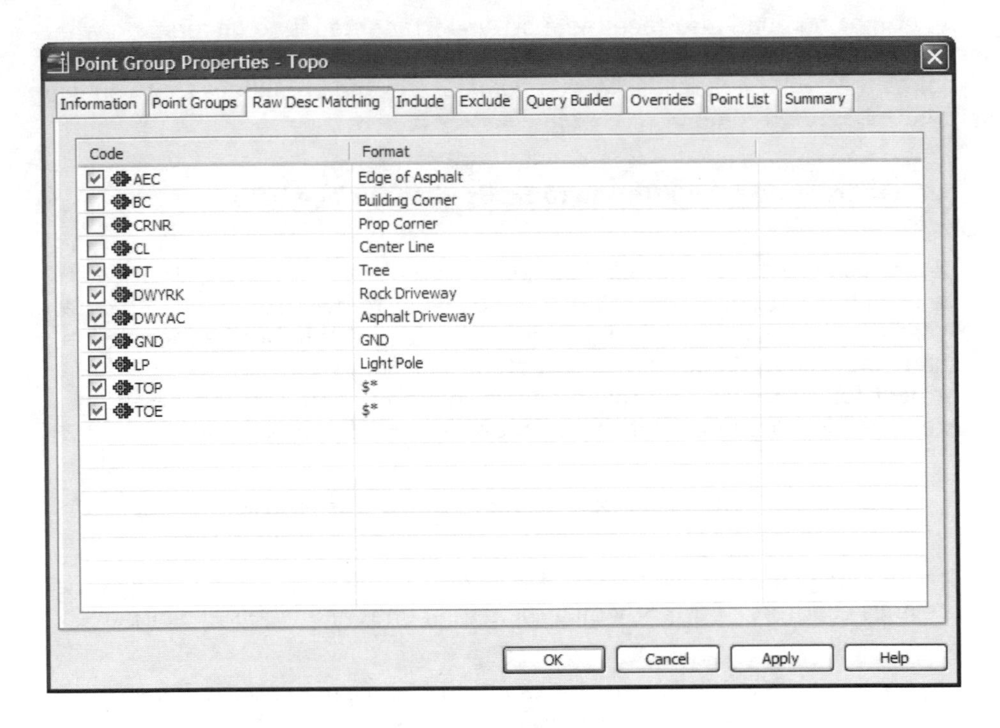

9. Select the description keys **"AEC, DT, DWYRK, DWYAC, GND, LP, TOP,** and **TOE"**.

10. Click **<<OK>>** to create the *Point Group*.

The points in the *Topo* group are now displayed because the new group is at the top of the point group display order and a new layer *V-NODE-Topo* was created for the group. If the points are not displayed turn on and thaw the layer *V-NODE-Topo*.

5.2.2 Creating the Survey Surface

1. On the *Prospector* tab of the *Toolspace*, right-click on **Surfaces** and select ⇒ **New**.

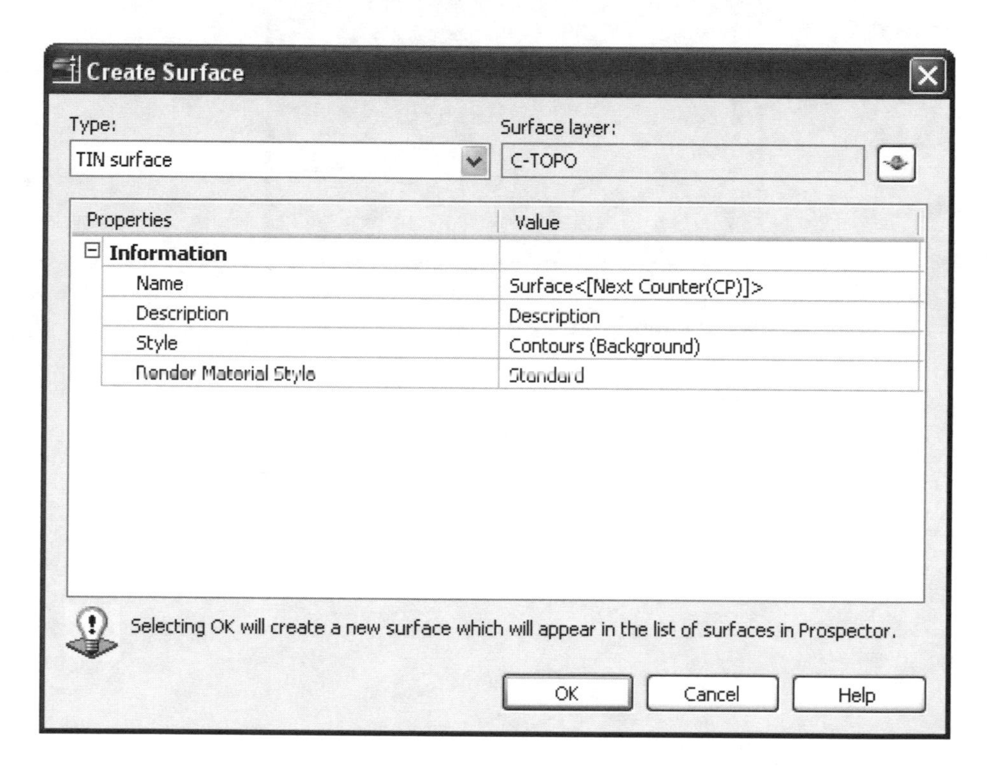

2. Confirm that **TIN surface** is selected as **Type.**

3. Click the **Surface Layer** button to open the *Object Layer* dialog box.

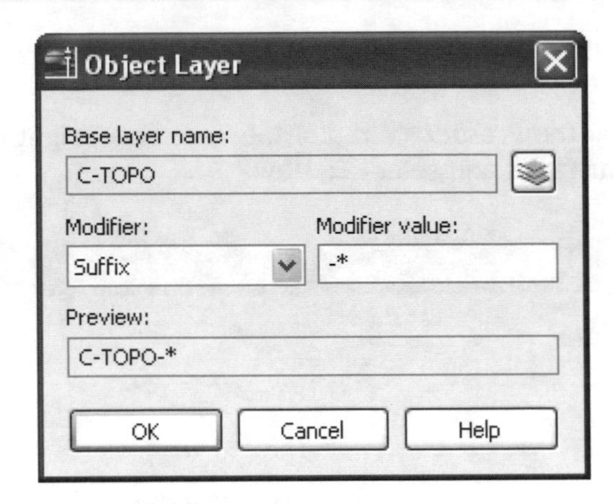

4. Set the **Modifier** to **Suffix**.

5. Enter **-*** as the **Modifier** value.

6. Click <<**OK**>> to close the *Object Layer* dialog box.

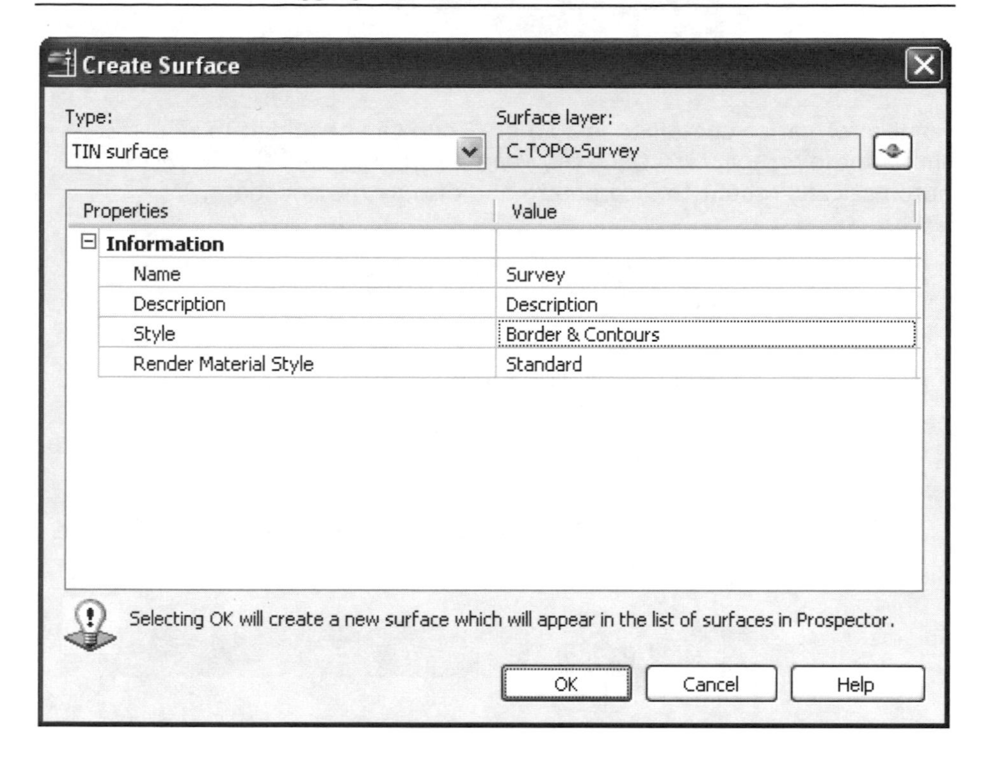

7. Enter **"Survey"** for the **Name**.

8. Set the **Style** to **Border & Contours**.

9. Click **<<OK>>** to close the *Create Surface* dialog box and create the surface.

At this time the surface has not been given any data so it is not displayed. However, it has been created and you will see it in the *Prospector*. This is where you will access the surface definition commands and add data to the surface.

5.2.3 Adding Point Group Data to a Surface

Point information contained in a *Point Group* can be added to a *Surface* through the *Prospector*. Once the *Point Group* is added the *Surface* is automatically rebuilt to incorporate and display the new data.

1. On the *Prospector* tab of the *Toolspace*, expand the **Surfaces** node.

2. Expand the *Surface* **Survey**.

3. Expand the **Definition** node under *Survey*.

4. Right-click on **Point Groups** under the *Definition* node and select ⇒ **Add**.

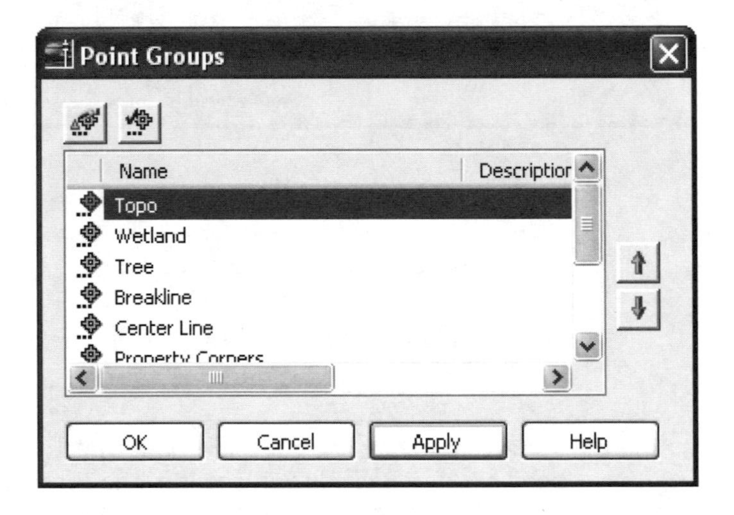

5. Select the *Point Group* **Topo**.

6. Click <<**OK**>> to add the point group data to the surface.

The surface is rebuilt with the point group data and displays 5 foot contours colored brown and green with a yellow border. This display is controlled by the surface style you selected when you created the surface. If the surface is not visible turn on and thaw the layer *C-TOPO-Survey*.

5.2.4 Creating Breaklines by Point Number

Civil 3D does not use special commands for drawing and defining breaklines the way that *Land Desktop* and many other programs do. Instead, you draw the breaklines with standard *AutoCAD* commands, like the *3D Polyline* command, and then define these objects as breaklines after they have been drawn.

1. Create a new *Layer* named **Breaklines-Survey** and set it **Current.**

2. Thaw the layer **V-NODE-Breakline**.

3. Freeze the layers **C-ANNO, C-PROP, C-PROP-LINE, C-PROP-TABL, EX-WETLAND-LINE, V-NODE-TOPO,** and **V-NODE-WETLAND**.

4. On the *Prospector* tab of the *Toolspace,* right-click on **Point Groups** and select ⇒ **Properties.**

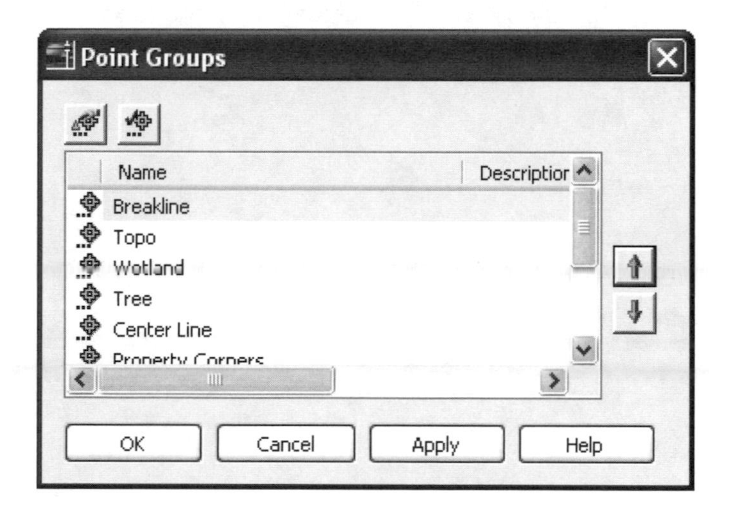

5. Select the *Point Group* **Breakline** and move it to the **top** of the list using the **up arrow** button.

6. Click **<<OK>>** to redisplay the points.

The drawing will now display the surface as contours and points that you will use for breaklines.

7. Enter `"3P"` at the command line to start the **3D Polyline** command.

8. Enter `'PN` to change the prompt to **Point Number**.

9. At the command line enter: `1408-1447`.

10. **<<ENTER>>** to draw the line.

11. Press **<<ESCAPE>>** to end the Point Number prompt.

12. **<<ENTER>>** to end the line.

13. Enter `"3P"` at the command line to start the **3D Polyline** command.

Use the points in the following list of points to draw the breaklines the same way that you drew the previous line. Be sure to use the **'PN** transparent command to change the prompt to **Point Number** and to end the command completely after drawing each line. Also be sure not to use spaces when entering the list.

Point Numbers

1448-1486
1008-1021
1191-1209
1226-1257
1258-1278
1281-1324
1295,1661-1710
1622-1660,1294
1286,1348-1398,1287
1022-1074
1075-1105
1155-1158
1159-1160
1153-1154
1143-1151
1130-1142
1121-1129

5.2.5 Creating Breaklines by Point Selection

1. On the *Prospector* tab of the *Toolspace*, select the *Point Group* **Breakline**.

This will display a list of all the points used in the surface in the preview window at the bottom of the *Prospector*, if the *Prospector* is docked. If the *Prospector* is not docked it will display on the side.

2. Find point number **1110** in the preview window.

3. Right-click on point **1110** and select **Zoom to**. You may want to zoom out some to see the surrounding points.

4. Enter `"3P"` at the command line to start the **3D Polyline** command.

5. Enter **'PO** to change the prompt to **Point Object**.

6. Pick point **1110** from the screen.

7. Then pick points **1109, 1108, 1107, and 1106** to draw a breakline between the TOP points toward the northeast corner of the site.

When using the `'PO` transparent command to draw lines between point objects you will not see the "rubber band" line that you normally see with the line command. Also, you may notice that when you pan or zoom in the middle of the command the last segment you have drawn disappears. Don't worry, it is not gone and will reappear if you draw the next segment.

8. **<<ENTER>>** to end **the Point Object** prompt.

9. **<<ENTER>>** again to end the line.

10. Starting at point **1116,** define a second breakline along the bottom of the ditch using the 'PO transparent command and points **1116, 1117, 1118, 1119**, and **1120.**

11. **<<ENTER>>** to end the `Point Object` prompt.

12. **<<ENTER>>** again to end the line.

13. Starting at point **1111,** define a third breakline along the bank of the ditch using points **1111, 1112, 1113, 1114,** and **1115.**

14. **<<ENTER>>** to end the `Point Object` prompt.

15. **<<ENTER>>** again to end the line.

16. Save the drawing.

5.2.6 Adding Breaklines to the Surface

1. Enter **LAYISO at the command line.**

This command can also be found in the Format menu if it is visible in your workspace. You can also manually turn off all the layers with the exception of the layer Breaklines-Survey.

2. Pick one of the breaklines to isolate the **Breaklines-Survey** layer.

3. Confirm that the **Definition** node under the *Surface* **Survey** is expanded on the *Prospector* tab of the *Toolspace*.

4. Right-click on **Breaklines** under the **Definition** node and select ⇒ **Add.**

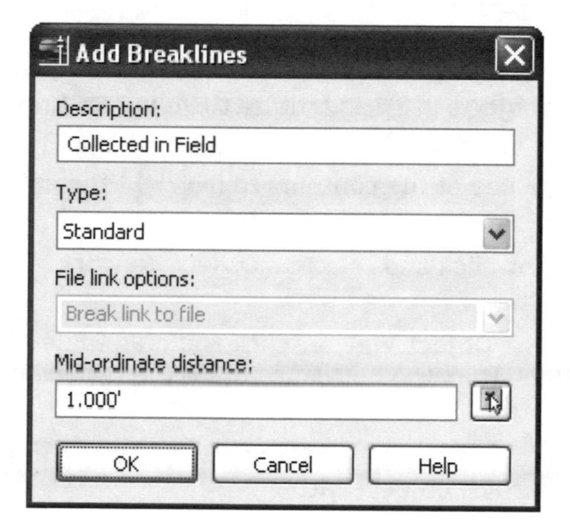

5. Enter a **Description** for the breakline set of **Collected in Field.**

6. Confirm that the **Type** is set to **Standard.**

7. Click **<<OK>>**.

8. Select the **Breaklines** with a crossing window.

9. **<<ENTER>>** to add the breaklines to the surface.

10. Click the Layer Previous button to restore the previous layer state.

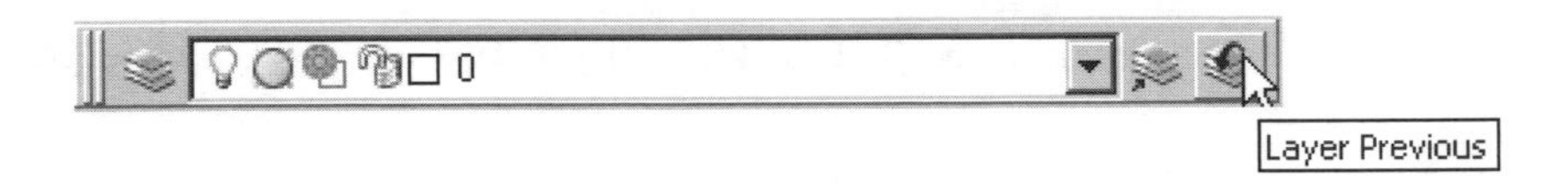

11. **Regen** to display the *Surface* with the new breakline data.

12. Save the drawing.

5.2.7 Viewing the Surface

The Object Viewer is a separate window that will allow you to view a selected object or objects in 3D and rotate them in real-time.

1. Pick one of the **contours** to highlight the entire surface.

2. **Right-click** and select ⇒ **Object Viewer**.

3. In the **Object Viewer**, click and drag while holding down the left mouse button to rotate the surface in 3D.

Once you rotate to a 3D view the contours will change to 3D faces. This is controlled by the surface object style.

4. If the surface is not shaded right-click and select **Visual Styles ⇒ Conceptual**.

5. Continue to rotate the surface to examine it from different angles. You will notice a large hole, or spike, in the surface.

6. When you are finished viewing the surface close the object viewer window to return to the drawing editor.

5.3 Editing the Surface

Civil 3D surfaces can be edited by either modifying the data used to build the surface or by using the surface editing commands found in the *Prospector* to directly edit the triangulation. In this section you will learn to use both methods and gain exposure to several different ways to edit a surface.

5.3.1 Editing Point Data

The surface is displayed showing contours and you will see the large hole that you identified earlier in the object viewer in the southwest quarter of the surface.

If a point is creating a hole in your surface there are a number of ways to correct it and the best way depends on the data in your project. If the point causing the hole is completely incorrect and has no value to the project you can erase the point from the drawing and rebuild the surface. Another alternative is that this point may not have a good surface elevation but it still might be valuable to your project. This can be common with control points or the invert of a manhole. In this situation you should edit the point to remove it from the group and rebuild the surface. Finally, you might have survey notes giving you information to edit the point and correct the elevation error. In this case you will edit the point and rebuild the surface. This final example is what you will do in this exercise. You will locate and edit the point that is creating the hole in the surface to give it the correct elevation.

1. Thaw the layer **V-NODE-TOPO**.

You may need to `Regen` if the points do not display.

2. Zoom to point **1221**. This is the point with an incorrect elevation that is creating the hole.

3. Select point **1221** and right-click to select **Edit Points.**

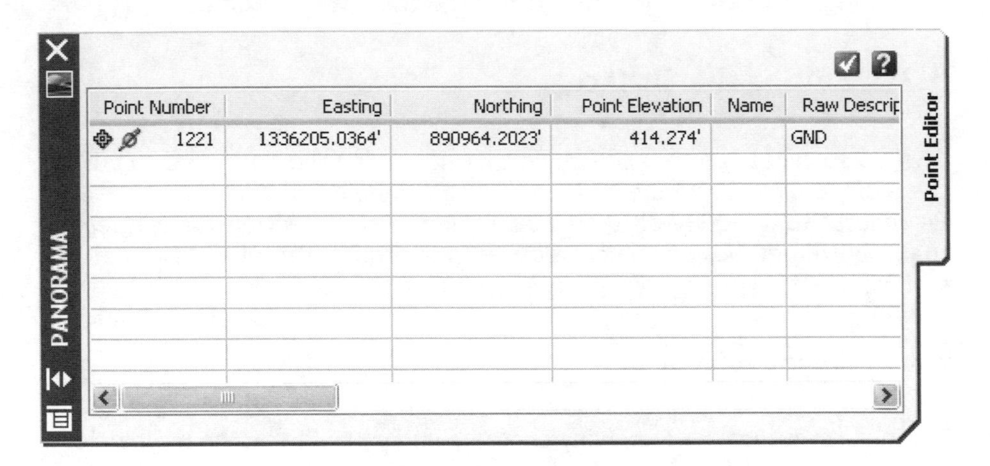

4. In the *Point Editor* right-click on point **1221** and select **Unlock.**

5. Change the elevation of point **1221** to **434.5.**

6. Right-click on point **1221** and select **Lock.**

7. Close the *Point Editor* by clicking the **green check button.**

8. On the *Prospector* tab of the *Toolspace,* right-click on the *Surface* **Survey** and select ⇒ **Rebuild.**

Once the surface is rebuilt the contours will update to reflect the change. You can also review the surface in the Object Viewer to confirm that the hole created by the incorrect elevation for point 1221 is now gone.

5.3.2 Editing Breaklines

The breaklines that you have defined in your surface are standard *AutoCAD 3D polylines*. This means that they can be edited with the **Polyedit** command or any other standard *AutoCAD* editing command like **Properties**, **Trim**, **Extend**, **Copy**, and **Erase**. Once you edit a breakline the surface and the breakline definition will display as out of date in the *Prospector*. You will then rebuild the surface to update it to display the changes made to the breakline.

If you set the surface option to *Rebuild Automatic*, the surface will automatically be rebuilt after each edit. This is a nice feature but may be inconvenient with larger surfaces because you will have to wait for them to rebuild and redisplay after each edit. If this option is disabled you can make several changes and rebuild when you are ready to review them.

1. Zoom to point **1151**.
2. Enter "**PE**" to edit the polyline.
3. Select the breakline at point **1151**.
4. Enter "**E**" to edit the polyline vertices.
5. Use the **Next** option to move to point **1151**.
6. **Insert** a vertex at point **1152** to extend the breakline. You can use the **'PN** transparent command to enter the point number.
7. Exit the **Poly Edit** command.

The breaklines and the surface will now show an icon that indicates they are out of date.

8. Right-click on the surface **Survey** and select ⇒ **Rebuild** to incorporate the edits to the breakline.

The contours around the additional segment that you added to the breakline will update reflecting the change to the breakline. If you make a mistake you can use the **Undo** command to backup and edit the breakline again.

5.3.3 Deleting Lines

TIN lines can only be deleted from a surface if you display the surface using a style that makes the lines visible. When deleting lines from a surface it is important to use the ⇒ *Delete Line* command found in the *Edits* node under the definition of the surface in the Prospector. This command will allow you to select and delete individual TIN lines from a surface. If you use the *AutoCAD* **Erase** command it will erase the entire surface.

The ⇒ *Delete Line* command is best suited to cleaning up the edges of a surface. If you use it to delete a line on the interior of a surface it will create a hole, or a void area, in the surface where the line was removed. If you need to delete triangles on the interior of a surface use the *Delete Point* command instead. This will remove the selected point and retriangulate over the area.

1. On the *Prospector* tab of the *Toolspace,* right-click on the *Surface* **Survey** and select ⇒ **Properties**.

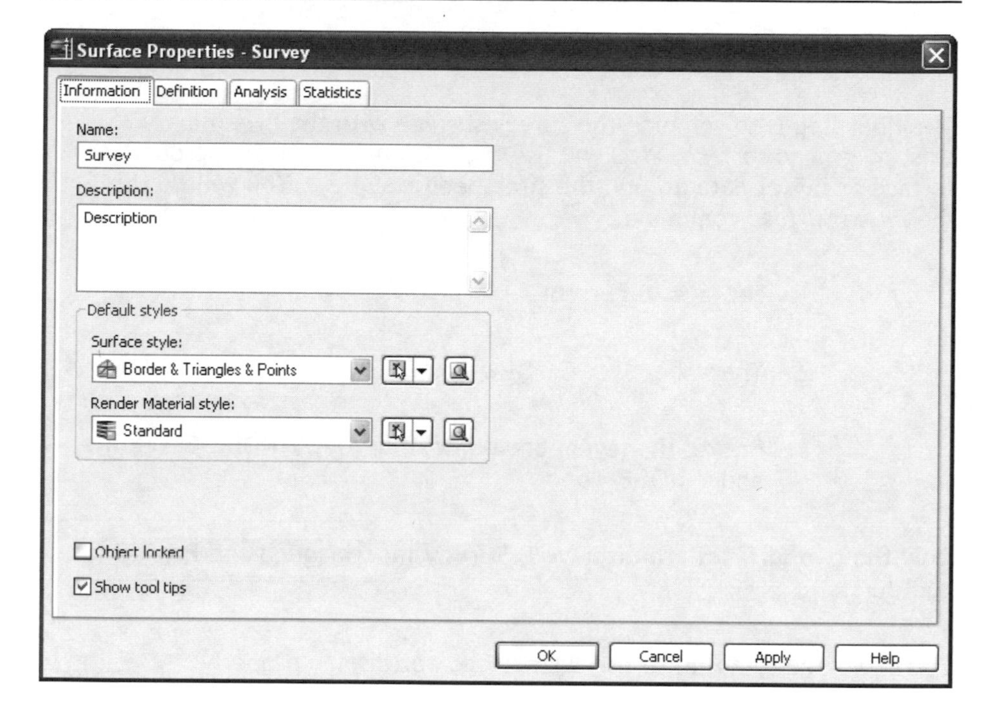

2. Set the **Surface Style** to **Border & Triangles & Points.**

3. Click **<<OK>>** to display the surface triangles.

4. On the *Prospector* tab of the *Toolspace*, confirm that the **Definition** node of the *Surface* **Survey** is expanded.

5. Right-click on **Edits** under the **Definition** node, and select ⇒ **Delete Line.**

6. Pick the long lines along the edge of our surface that have triangulated outside of our survey data.

Long skinny triangles are typically incorrect and need to be deleted. As you look at the TIN lines that connect each point ask yourself, "Should these points be associated with each other in the surface?" A line between two points will create a straight slope three dimensionally between them.

7. Save the drawing when you are finished deleting lines.

5.3.4 Pasting Surfaces

The final step is to combine the survey surface with the preliminary existing ground surface. You will use the preliminary existing ground surface as buffer data around the site-specific survey. You will do this with the *Paste Surface* command.

1. Set layer **0** current.

2. Thaw the layer **C-TOPO-Pre-EG**.

3. Freeze the layers **Breaklines-Survey, V-NODE-Breakline,** and **V-NODE-Topo**.

Only the two surfaces are displayed, **Survey** (as triangles) and **Pre-EG** (as contours).

You may need to **`Regen`** if the surfaces do not display properly.

4. At the command line enter `"CP"` to start the *AutoCAD* **`Copy`** command.

5. Pick one of the contours from the surface **Pre-EG** and press **<<Enter>>**.

6. Type `"D"` for **displacement**.

7. **<<Enter>>** to accept the default displacement of `0,0,0`.

This will create a copy of the *Surface Pre-EG* at the same coordinates. You will also see the new surface displayed in the *Prospector* with the name *Pre-EG (1)*. The original surface and the copy are on the same layer and use the same surface style, so they will look exactly the same in the drawing editor. To separate them onto different layers you can first rename the new surface and change its style so that it is displayed differently.

8. On the *Prospector* tab of the *Toolspace*, right-click on the *Surface* **Pre-EG (1)** and select ⇒ **Properties.**

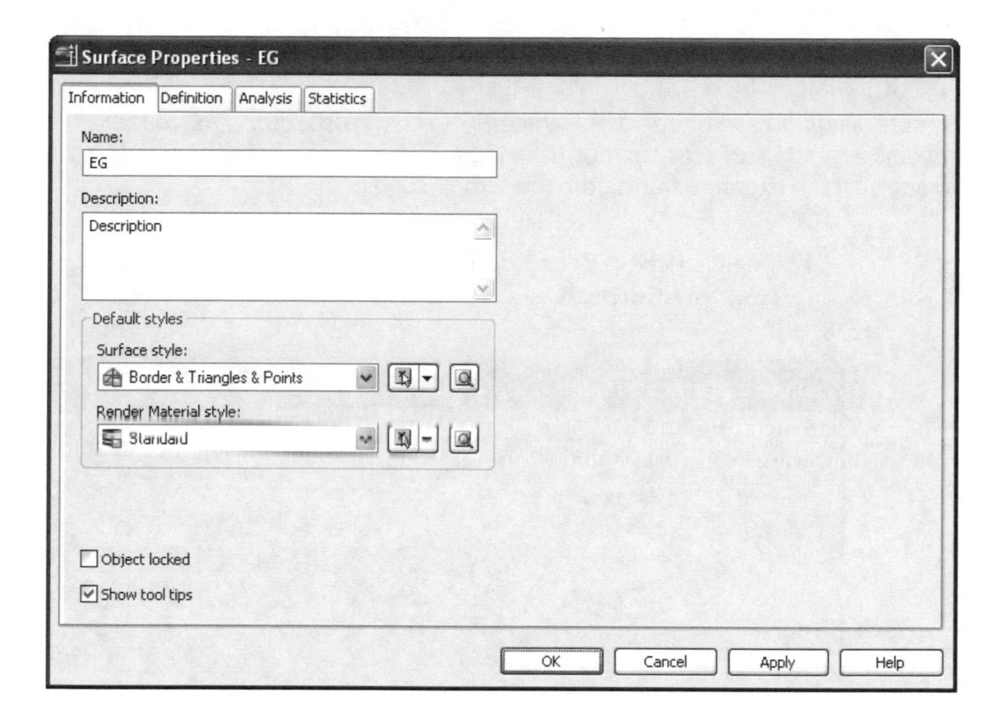

9. On the *Information* tab of the *Surface Properties* dialog box, enter a new **Name** of **EG.**

10. Select **Border & Triangles & Points** from the **Surface style** list.

11. Click **<<OK>>** to rename the surface and display it as triangles.

12. Create a new layer named **"C-TOPO-EG"**.

13. Use the *AutoCAD* **Properties** command to move the surface **EG** (the large surface displayed as triangles) to the layer **C-TOPO-EG.**

14. Freeze the layers **C-TOPO-Pre-EG** and **C-TOPO-Survey.**

15. Expand the *Surface* **EG** on the *Prospector* tab of the *Toolspace*.

16. Expand the **Definition** node under *Surface* **EG.**

Be sure that you are under the surface *EG* in the *Prospector*. As you continue to add surfaces to your drawing it can become easy to have the wrong surface expanded and edit the wrong surface.

17. Right-click on **Edits** under the *Definition* node, and select ⇒ **Paste Surface.**

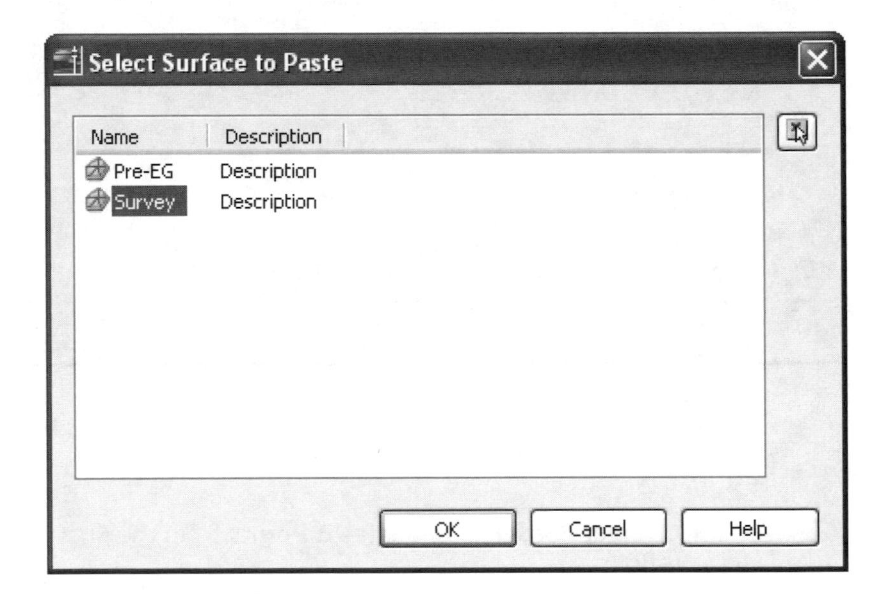

18. Select **Surface Survey** to paste into **EG.**

19. Save the drawing.

5.4 Surface Analysis

A surface can be analyzed and displayed in many different ways. The display of the surface is controlled by the *Surface Style*. Depending on the information that you want to display you may need to create a new surface style for your analysis. In this section you will display slope arrows and elevation bands. However, the process is similar for other types of surface analysis like slope analysis and watersheds.

5.4.1 Displaying Slope Arrows

Slope arrows can be displayed by editing the surface properties to display the surface using one of the styles predefined in the template you used to start the drawing.

1. On the *Prospector* tab of the *Toolspace*, right-click on **Surface EG** and select ⇒ **Properties.**

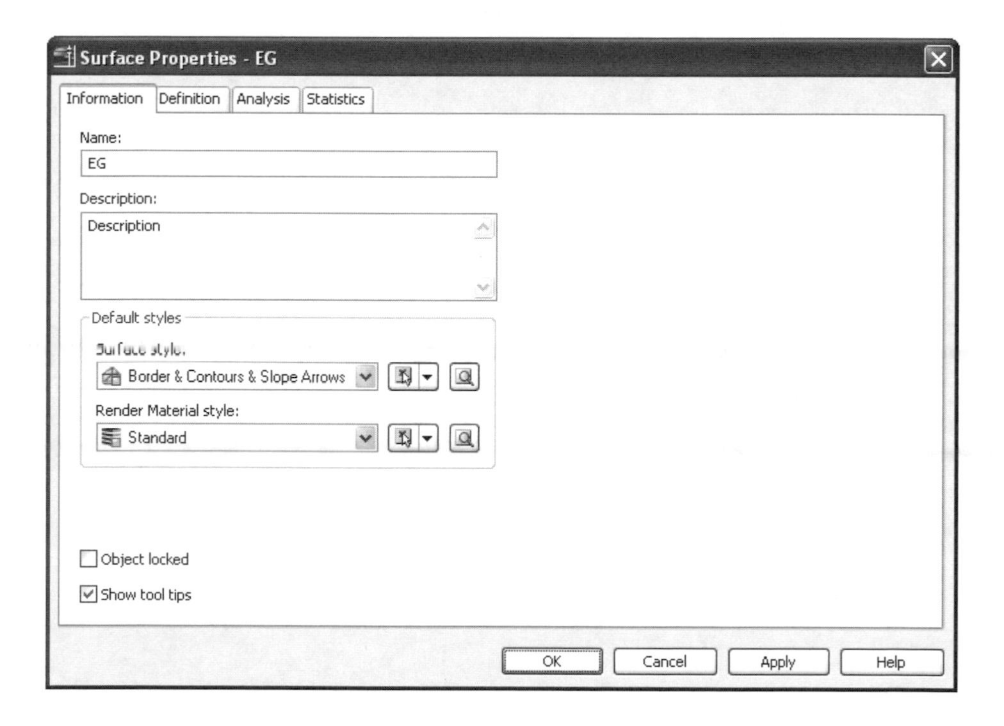

2. Select **Border & Contours & Slope Arrows** from the **Surface style** list on the *Information* tab of the *Surface Properties* dialog box.

3. Click **<<OK>>** to redisplay the surface with contours and an arrow showing the direction of slope on each triangle in the surface.

5.4.2 Elevation Banding

1. On the *Prospector* tab of the *Toolspace*, right-click on **Surface EG** and select ⇒ **Properties**.

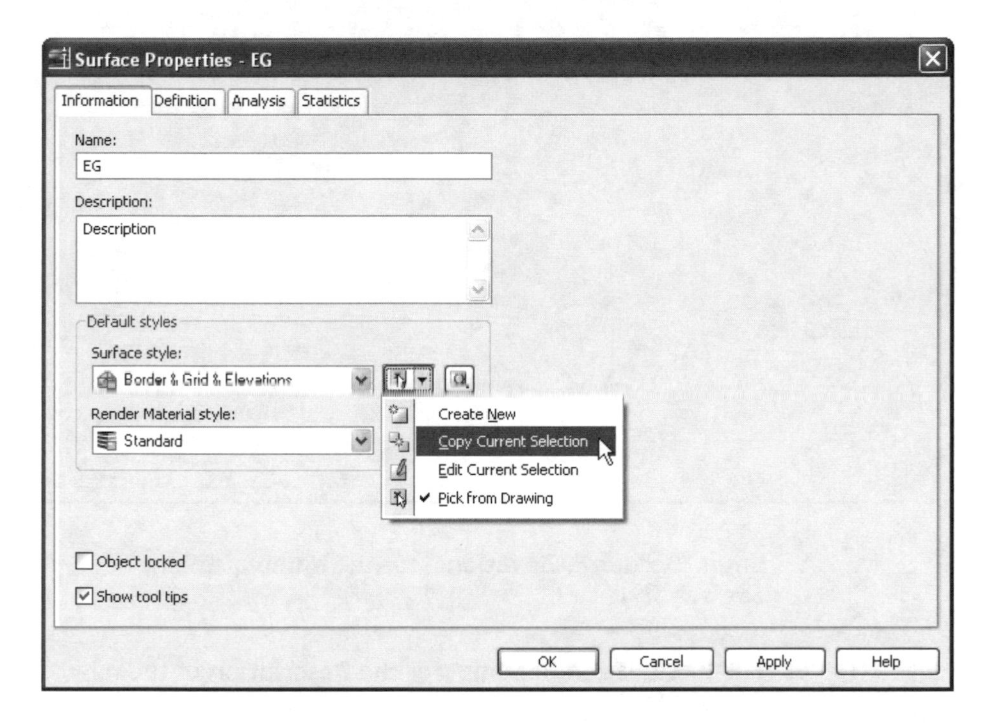

2. Select **Border & Grid & Elevations** from the **Surface style** list on the *Information* tab of the *Surface Properties* dialog box.

3. Click the **down arrow** button next to the **Surface style** on the *Information* tab of the *Surface Properties* dialog box to display the other style editing commands.

4. Click the **Copy Current Selection** button.

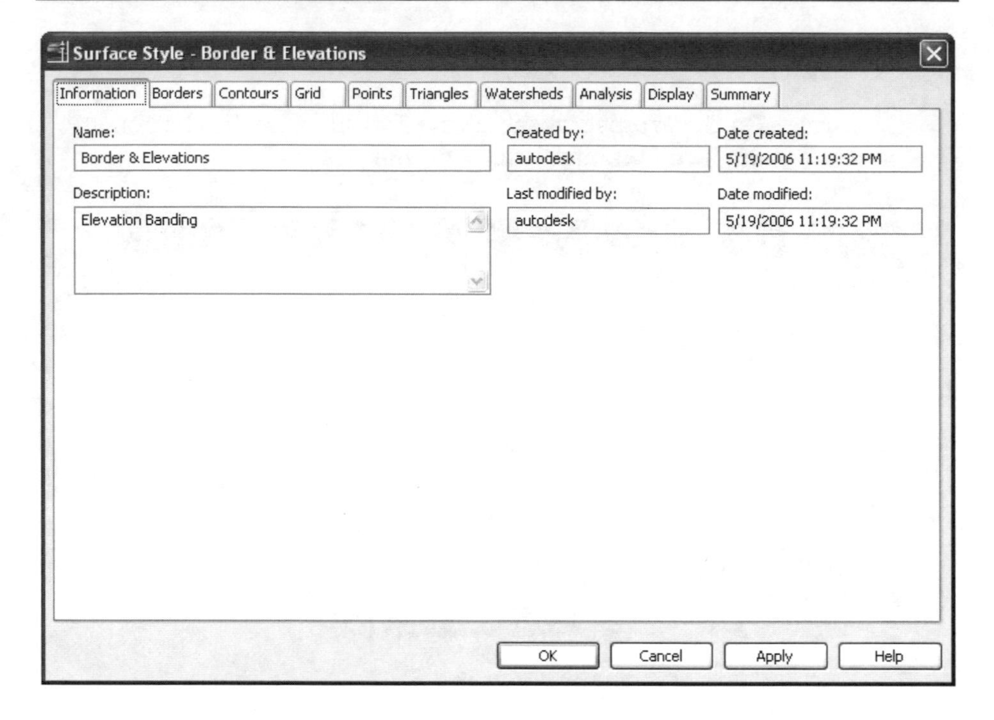

5. Enter "**Border & Elevations**" for the **Name** of the new *Surface Style*.

6. Enter "**Elevation Banding**" for the **Description** of the new *Surface Style*.

7. Select the **Analysis** tab of the *Surface Style* dialog box.

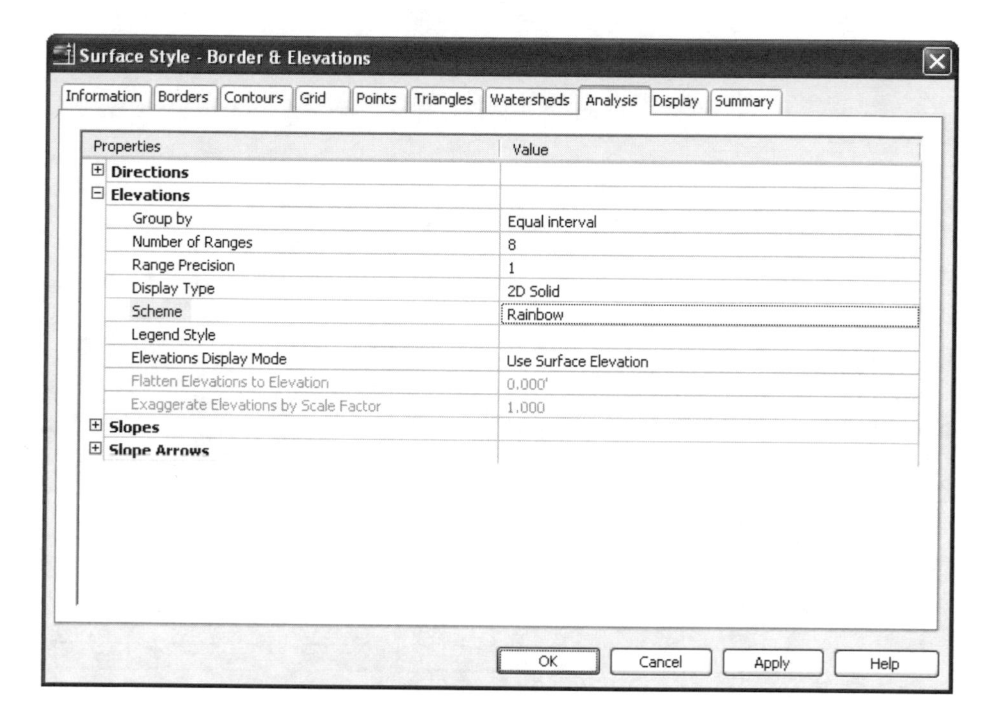

8. Expand the **Elevations** node.

9. Set the **Range Precision** to **1**.

10. Set the **Scheme** to **Rainbow**.

11. Select the **Display** tab of the *Surface Style* dialog box.

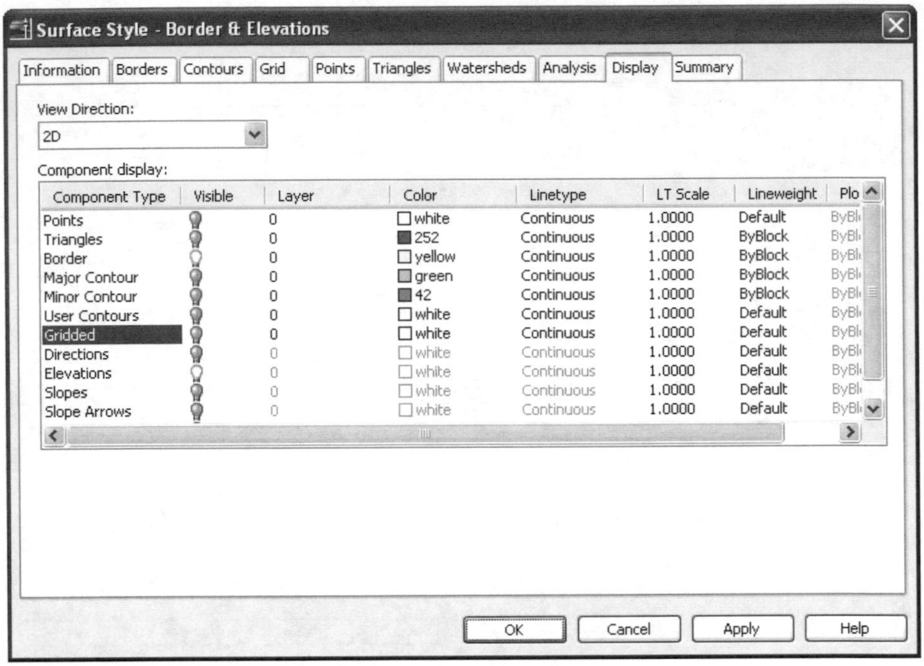

Surface Style - Border & Elevations

Information | Borders | Contours | Grid | Points | Triangles | Watersheds | Analysis | Display | Summary

View Direction:

2D

Component display:

Component Type	Visible	Layer	Color	Linetype	LT Scale	Lineweight	Plo
Points		0	white	Continuous	1.0000	Default	ByBl
Triangles		0	252	Continuous	1.0000	ByBlock	ByBl
Border		0	yellow	Continuous	1.0000	ByBlock	ByBl
Major Contour		0	green	Continuous	1.0000	ByBlock	ByBl
Minor Contour		0	42	Continuous	1.0000	ByBlock	ByBl
User Contours		0	white	Continuous	1.0000	Default	ByBl
Gridded		0	white	Continuous	1.0000	Default	ByBl
Directions		0	white	Continuous	1.0000	Default	ByBl
Elevations		0	white	Continuous	1.0000	Default	ByBl
Slopes		0	white	Continuous	1.0000	Default	ByBl
Slope Arrows		0	white	Continuous	1.0000	Default	ByBl

OK | Cancel | Apply | Help

12. Confirm that the **View Direction** is set to **2D**.

On the display tab you will see the list of components that are available to display for a surface. By turning the visibility of the components on and off you can control how the style displays the surface. There are also two view directions available, 2D and 3D. Changing the visibility of components between the 2D and 3D view directions is how the style can change the display of the surface from contours in plan view to triangles in a rotated 3D view.

13. Turn **Off** the **Gridded** *component*.

14. Confirm that the **Elevations** component is turned **On**.

15. Click **<<OK>>** to save the new **Surface Style** and return to the *Surface Properties* dialog box.

16. Select the **Analysis** tab of the *Surface Properties* dialog box.

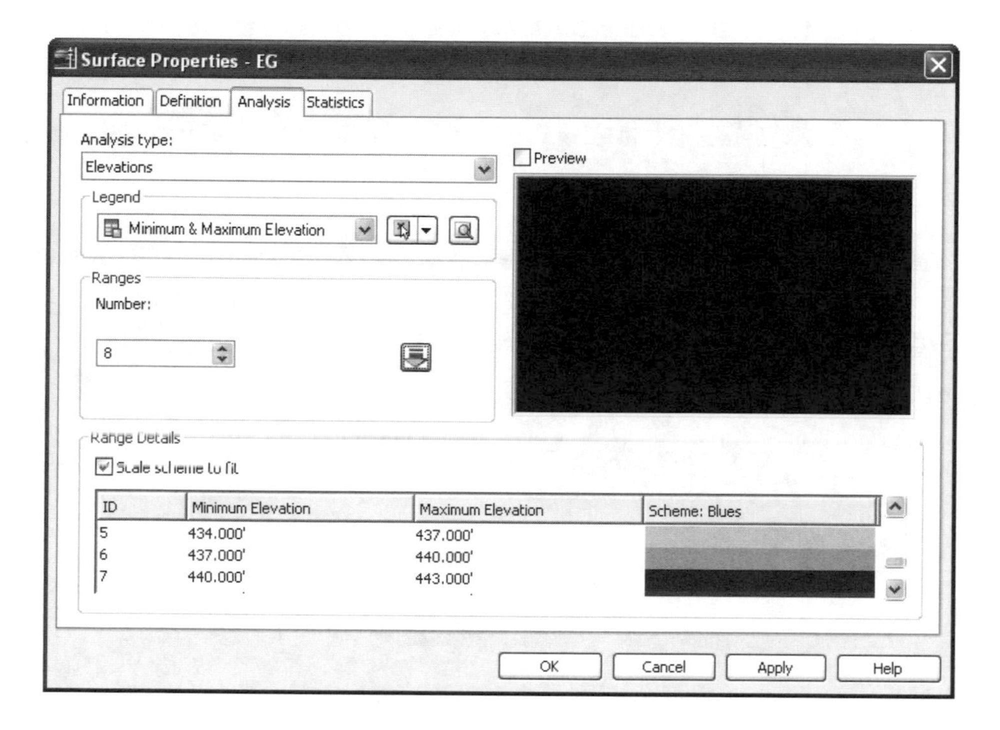

17. Confirm that the **Analysis type** is set to **Elevations.**

18. Click the **Run Analysis** button to analyze the surface.

The minimum and maximum elevations as well as the colors can all be changed by double clicking on the fields in the table.

19. Click **<<OK>>** to close the *Surface Properties* dialog box and apply the new *Surface Style*.

The surface should now be displayed with 2D solids representing the elevation bands. You may need to **Regen** if the display of the surface does not update.

20. Save the drawing.

5.4.3 Slope Analysis

1. On the *Prospector* tab of the *Toolspace*, right-click on **Surface EG** and select ⇒ **Properties.**

2. Select the **Information** tab of the *Surface Properties* dialog box.

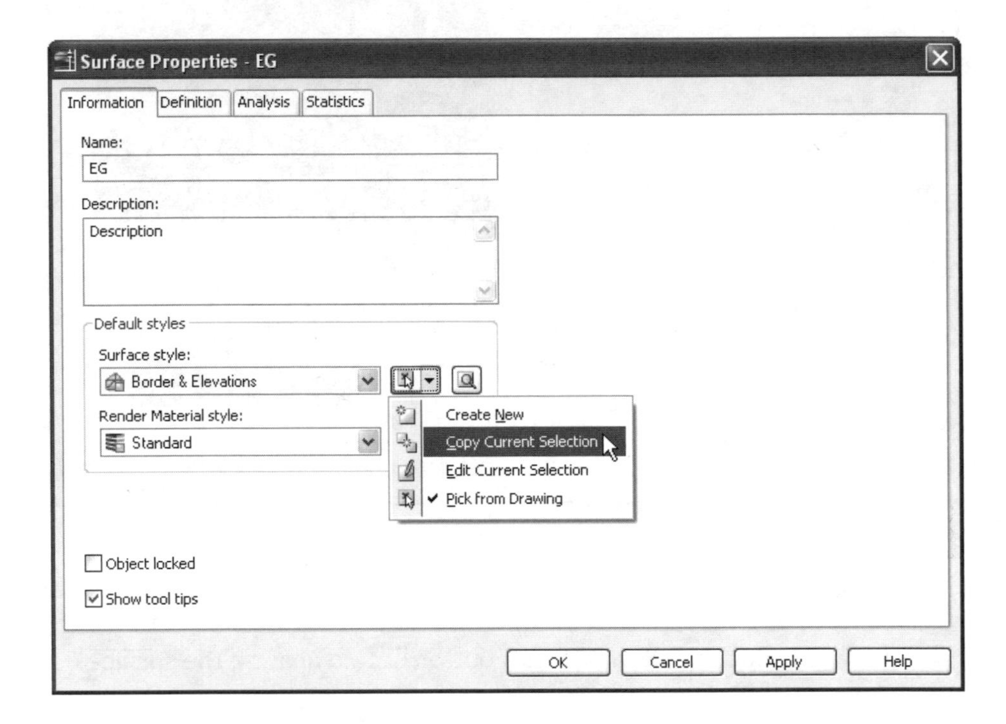

3. Confirm that **Border & Elevations** is the current **Surface style.**

4. Click the **down arrow** button next to the **Surface style** on the *Information* tab of the *Surface Properties* dialog box to display the other style editing commands.

5. Click the **Copy Current Selection** button.

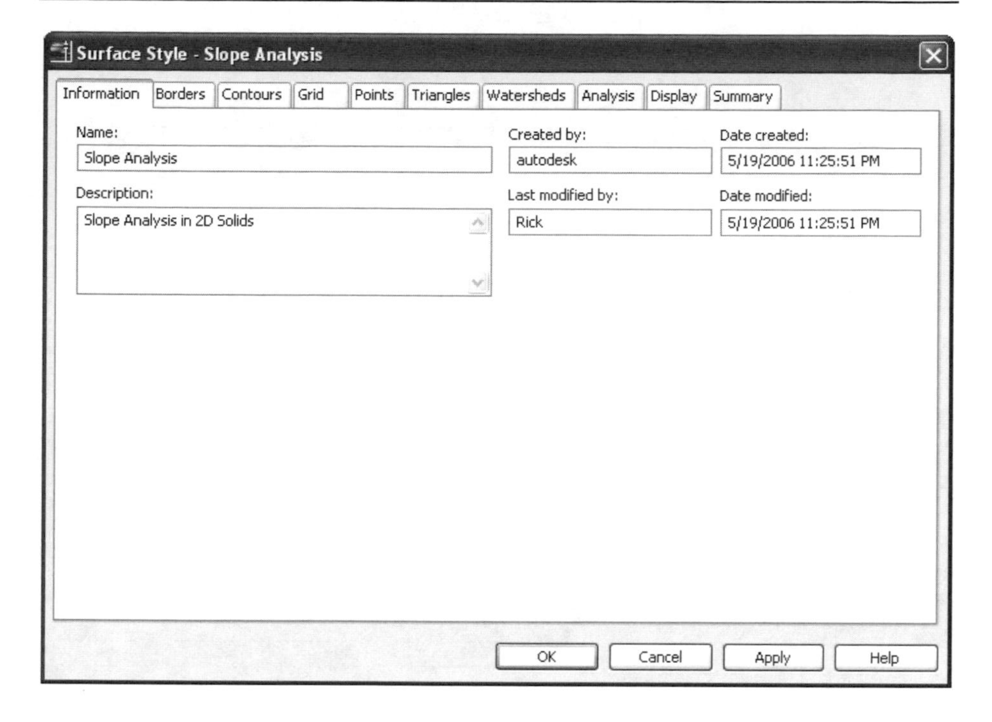

6. Enter "**Slope Analysis**" for the **Name** of the new *Surface Style*.

7. Enter "**Slope Analysis in 2D Solids**" for the **Description** of the new *Surface Style*.

8. Select the **Analysis** tab of the *Surface Style* dialog box.

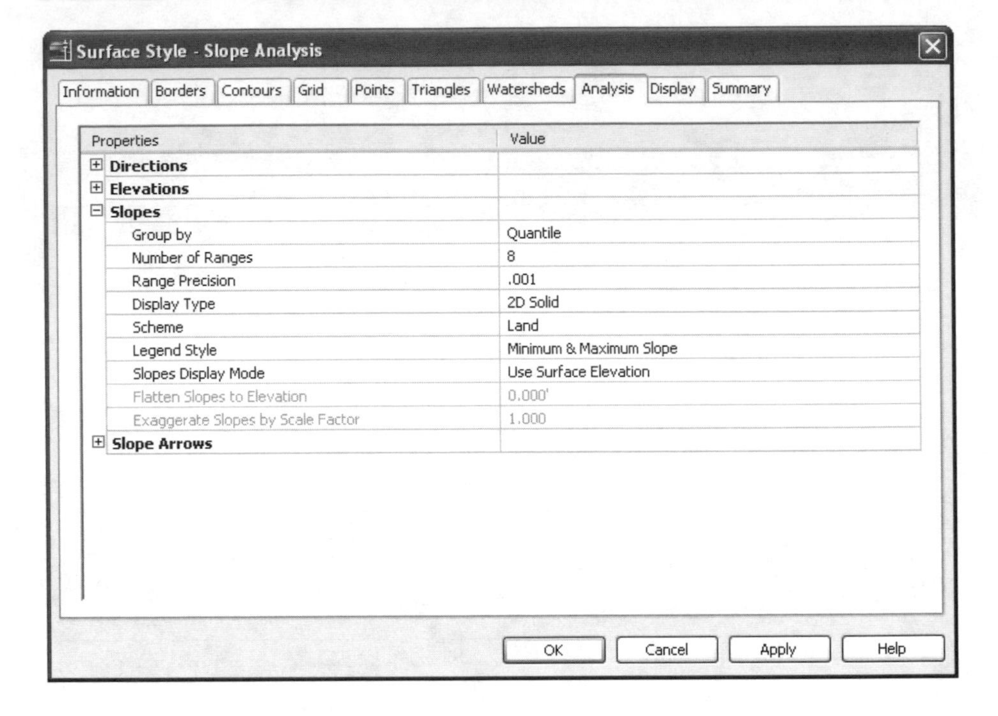

9. Expand the **Slopes** node.

10. Set the **Group by** value to **Quantile**.

11. Set the **Range Precision** to **0.01**.

12. Set the **Scheme** to **Land**.

13. Select the **Display** tab of the *Surface Style* dialog box.

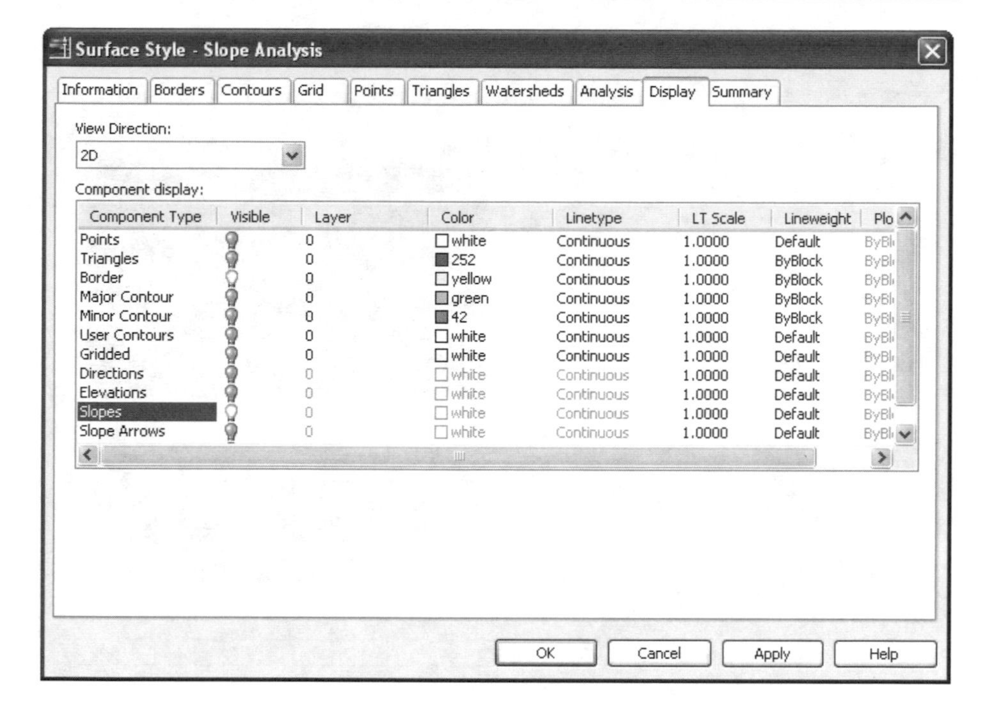

14. Confirm that the **View Direction** is set to **2D**.

15. Turn **Off** the **Elevation** *component*.

16. Turn **On** the **Slopes** *component*.

17. Click <<**OK**>> to save the new **Surface Style** and return to the *Surface Properties* dialog box.

18. Select the **Analysis** tab of the *Surface Properties* dialog box.

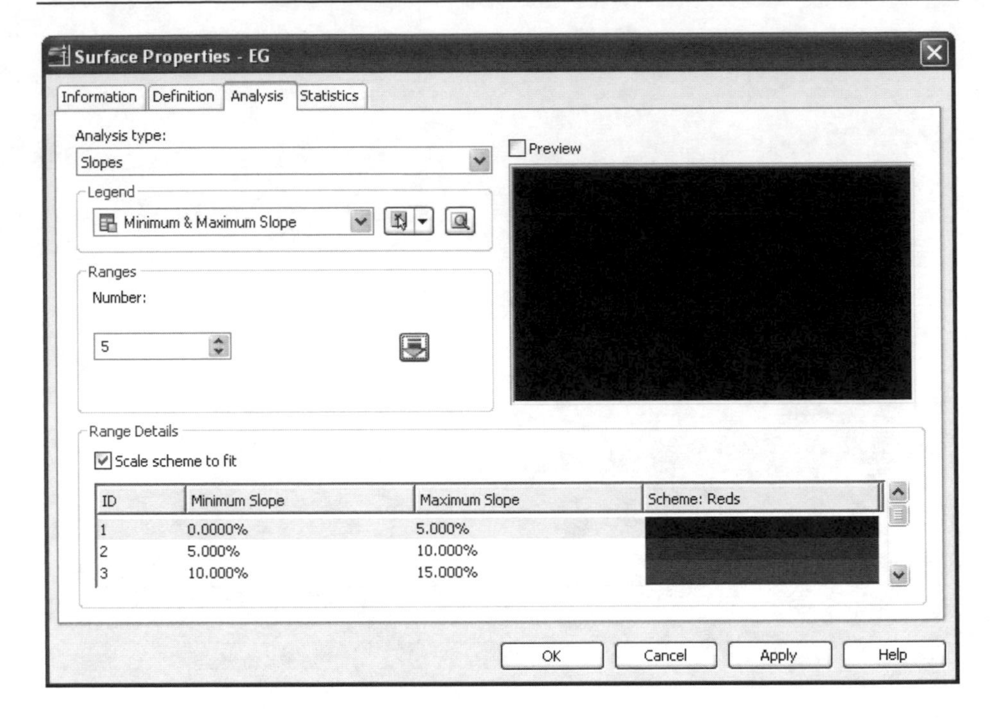

19. Set the **Analysis type** to **Slopes**.

20. Set the Number of Ranges to **5**.

21. Click the **Run Analysis** button to analyze the surface.

22. In the Range Details table enter slope ranges in 5% increments.

Minimum Slope	Maximum Slope
0%	5%
5%	10%
10%	15%
15%	20%
20%	100%

23. Click **<<OK>>** to close the *Surface Properties* dialog box and apply the new *Surface Style*.

The surface should now be displayed with 2D solids representing the different slope ranges. You may need to **Regen** if the display of the surface does not update.

24. **Save** the drawing.

5.5 Working with Contours

Displaying a surface as contours is one of the most common types of surface display. In this section you will create a new surface style to learn how styles control all the aspects of contours. You will also label the contours and work with contour label styles.

5.5.1 Displaying a Surface as Contours

There is no *Create Contour* command in Civil 3D. Contours are a display of a *Surface* that is controlled by the surface style. To display contours you simply change the surface style.

1. On the *Prospector* tab of the *Toolspace*, right-click on **Surface EG** and select ⇒ **Properties.**

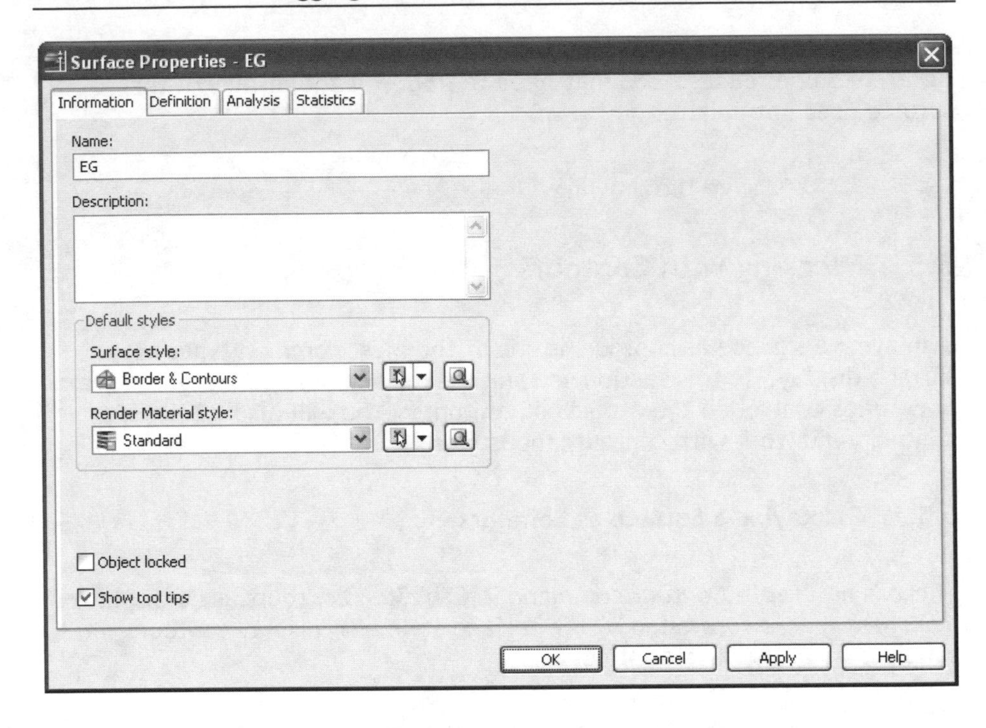

2. Select **Border & Contours** from the **Surface style** list on the *Information* tab of the *Surface Properties* dialog box.

3. Click **<<OK>>**.

The surface is now displayed showing 5 foot contours.

5.5.2 Controlling Contour Display

The current surface style displays the surface with contours at an interval of 5 and 25 feet. Since the site is relatively flat you may want to decrease the contour interval to show more detail. You might also want to smooth the contours to improve their appearance and change the color of the contour lines to match your organizations standards. All of these properties are controlled by the surface style. The style is saved in the drawing and should be saved to a drawing template if it is your standard and will be used in future projects.

1. On the *Prospector* tab of the *Toolspace*, right-click on **Surface EG** and select ⇒ **Properties**.

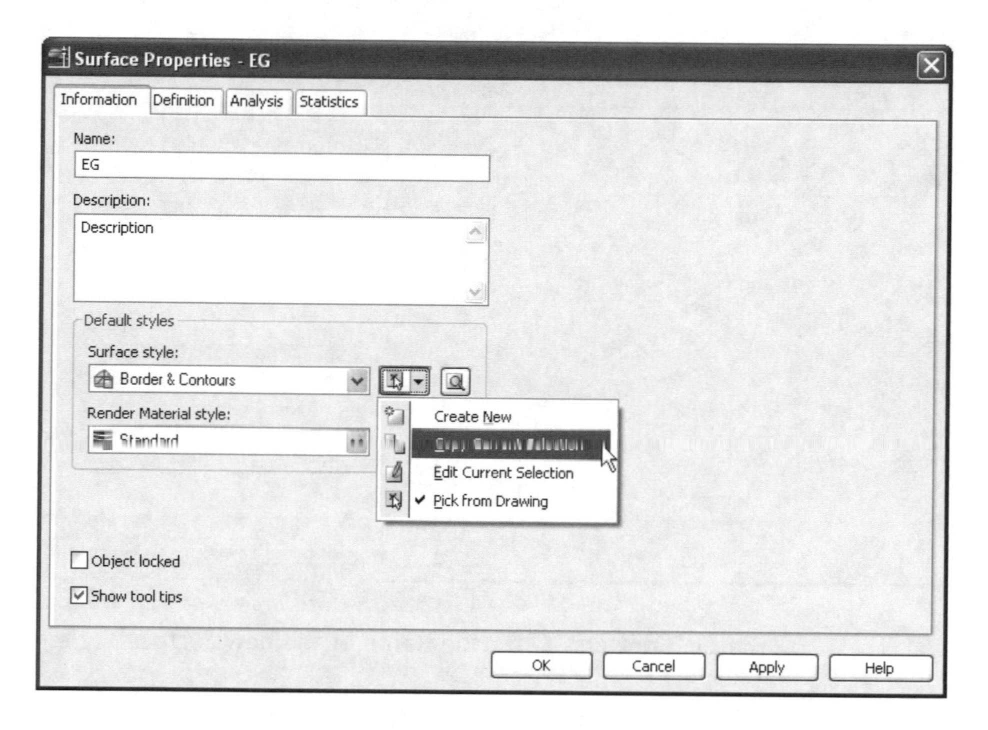

2. Confirm that **Border & Contours** is the current **Surface style**.

3. Click the **down arrow** button next to the **Surface style** on the *Information* tab of the *Surface Properties* dialog box to display the other style editing commands.

4. Click the **Copy Current Selection** button.

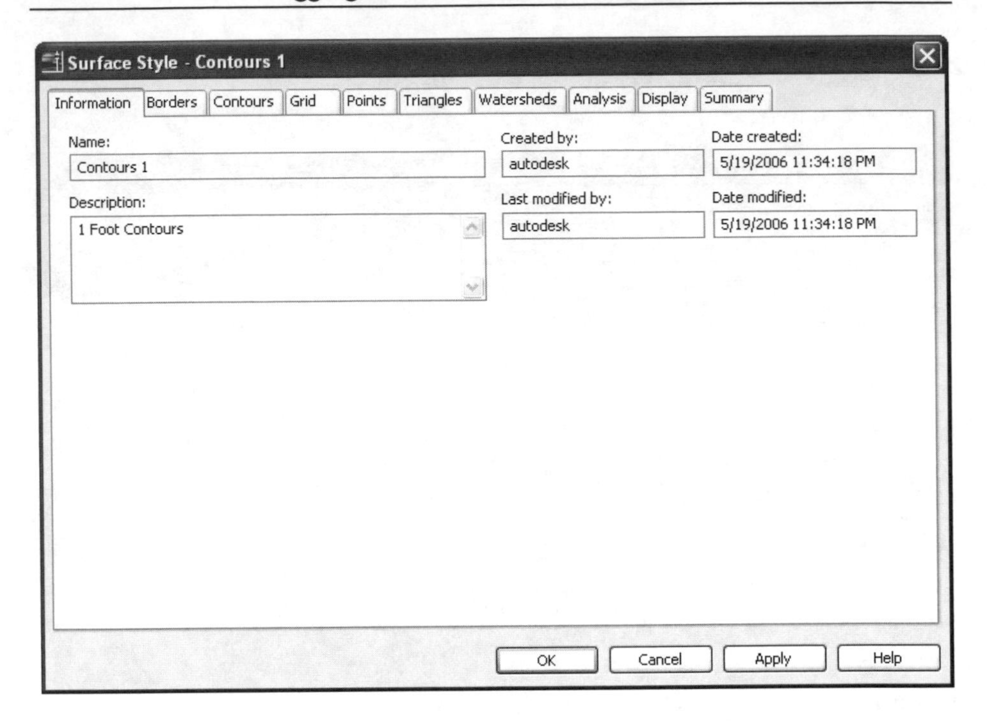

5. Enter **"Contours 1"** for the **Name** of the new *Surface Style*.

6. Enter **"1 Foot Contours"** for the **Description** of the new *Surface Style*.

7. Select the **Contours** tab of the *Surface Style* dialog box.

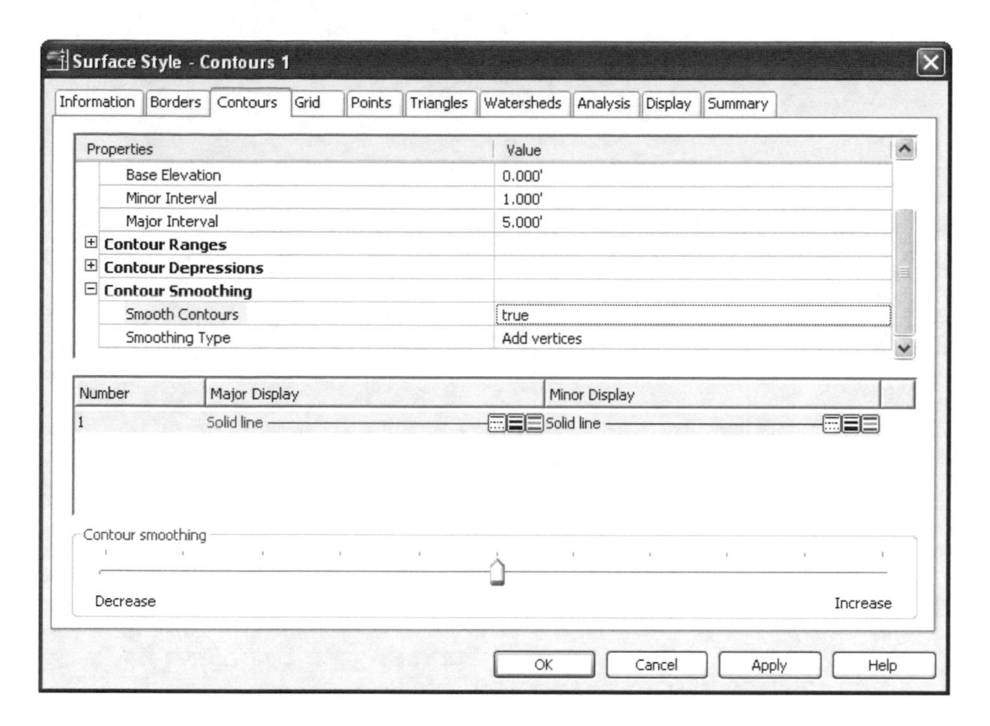

8. Expand the **Contour Intervals** node.

9. Set the **Minor Interval** to **1**.

10. Click in the value field of the **Major Interval** and it should automatically change to **5**.

11. Expand the **Contour Smoothing** node.

12. Set the **Smooth Contours** value to **true**.

13. Confirm that the **Smoothing Type** is set to **Add vertices**.

14. Increase the **Contour smoothing** slider to about **50%**.

You can also control the display of depression contours on this tab.

15. Select the **Display** tab of the *Surface Style* dialog box.

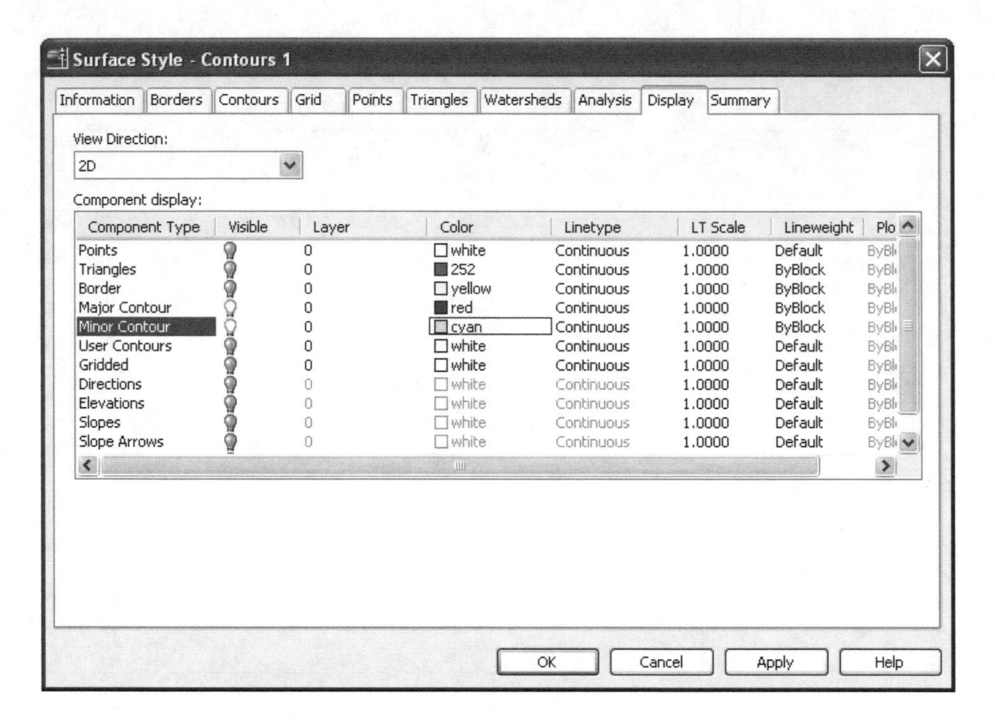

16. Confirm that the **View Direction** is set to **2D**.

17. Turn **off** the **Border** component.

18. Set the *Color* of the **Major Contour** component to **Red**.

19. Set the *Color* of the **Minor Contour** component to **Cyan**.

20. Click **<<OK>>** to save the new *Surface Style* and return to the *Surface Properties* dialog box.

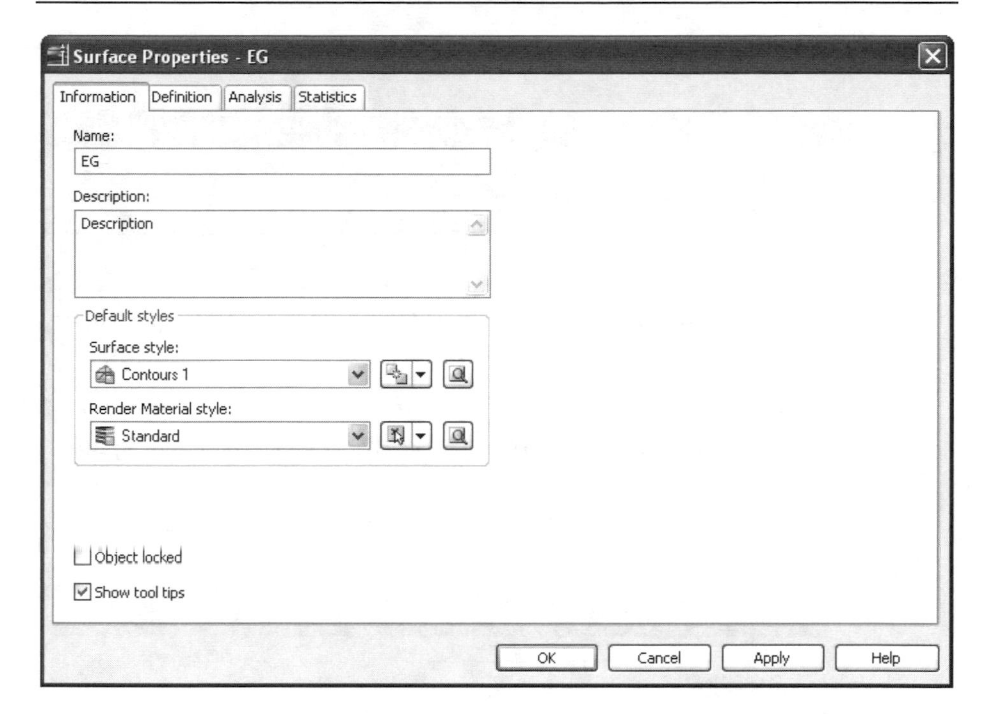

21. Click <<**OK**>> to close the *Surface Properties* dialog box and apply the new *Surface Style*.

The surface should now be displayed with red and cyan, one foot, smoothed contours. You may need to **Regen** if the display of the surface does not update.

5.5.3 Labeling Contours

Contours are labeled by drawing or selecting a line across the contours that determines the contour label position. This will only label contours and will not impact any other lines that may currently be displayed on the screen.

1. Select **Surfaces** ⇒ **Labels** ⇒ **Add Contour Labels**.

2. Select the *Surface* **EG**.

The Create Contour Labels toolbar will now be displayed.

3. Click the **Select Layer button** on the *Create Contour Labels* toolbar.

4. Click the **Layer button** in the *Select Layer* dialog box.

5. Click the **New button** in the *Layer Selection* dialog box.

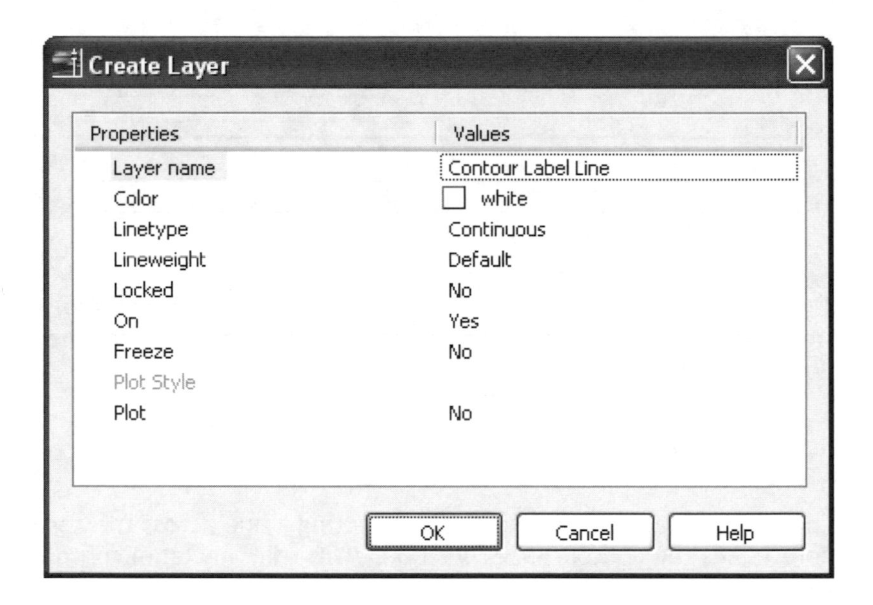

6. Enter a **Layer Name** of "**Contour Label Line**".

7. Set the **Plot** option to **No**.

This will allow you to have the contour label line visible for editing purposes while insuring that it will not plot at any time.

8. Click **<<OK>>** to create the new layer and close the *Create Layer* dialog box.

9. Click **<<OK>>** to set the layer and close the *Layer Selection* dialog box.

10. Click **<<OK>>** to close the *Select Layer* dialog box and return to the *Create Contour Labels* toolbar.

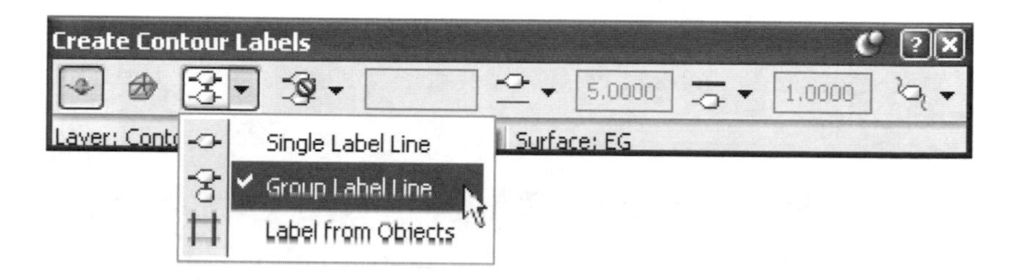

11. Click the **down arrow** button next to the third button from the left on the *Create Contour Labels* toolbar to confirm that the **Group Label Line** option is selected.

12. Click the **down arrow** button next to the fourth button from the left on the *Create Contour Labels* toolbar and select the **Label Multiple Group Interior** option.

This will add multiple labels at a specified interval along each contour. Once this option is selected it will activate the field to the right of the button for you to enter the desired interval.

13. Enter **300** in the field on the *Create Contour Labels* toolbar.

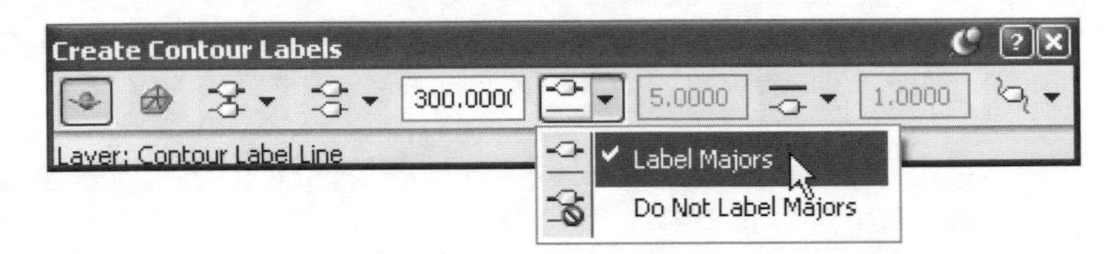

14. Click the **down arrow** button next to the fifth button from the left on the *Create Contour Labels* toolbar to confirm that the **Label Majors** option is selected.

If you select Label Majors you will get the option to select the contour label style.

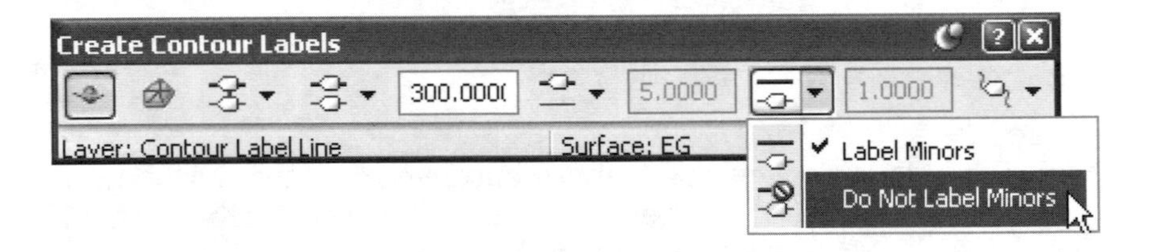

15. Click the **down arrow** button next to the sixth button from the left on the *Create Contour Labels* toolbar and select the **Do Not Label Minors** option.

You will be labeling only the 5 foot contours.

16. Draw a line across the surface at the location you would like the contours labeled.

17. Close the *Create Contour Labels* toolbar.

18. Zoom in to a location that the new contour label line crosses one of the red, major contour lines and examine the label.

5.5.4 Editing Contour Labels

The position of the contour labels can be changed by moving, or grip editing, the contour label line. To change the appearance of the contour labels like the font or color of the text you must change the contour label style.

1. Select **Surfaces ⇒ Labels ⇒ Contour Label Line Properties.**

2. Select the **Contour Label Line.**

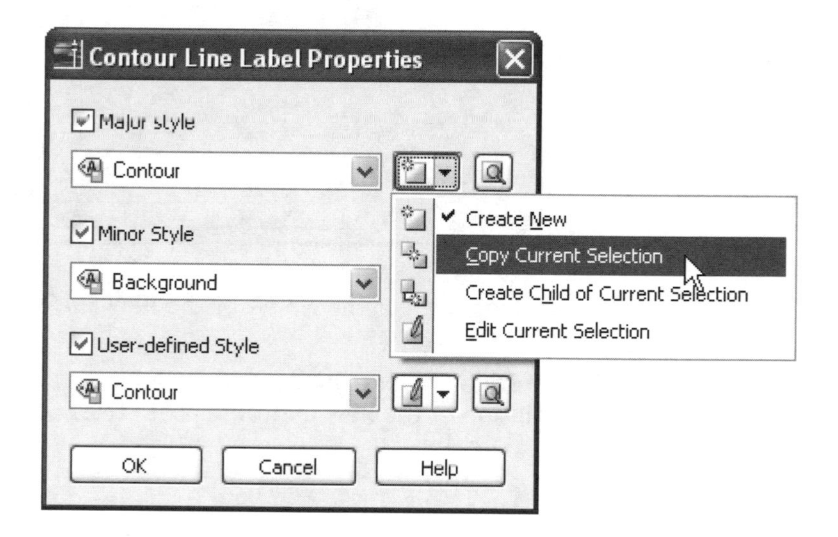

3. Confirm that the **Major style** is set to **Contour** in the *Contour Line Properties* dialog box.

4. Click the **down arrow** button next to the **Major style** to display the other style editing commands.

5. Click the **Copy Current Selection** button.

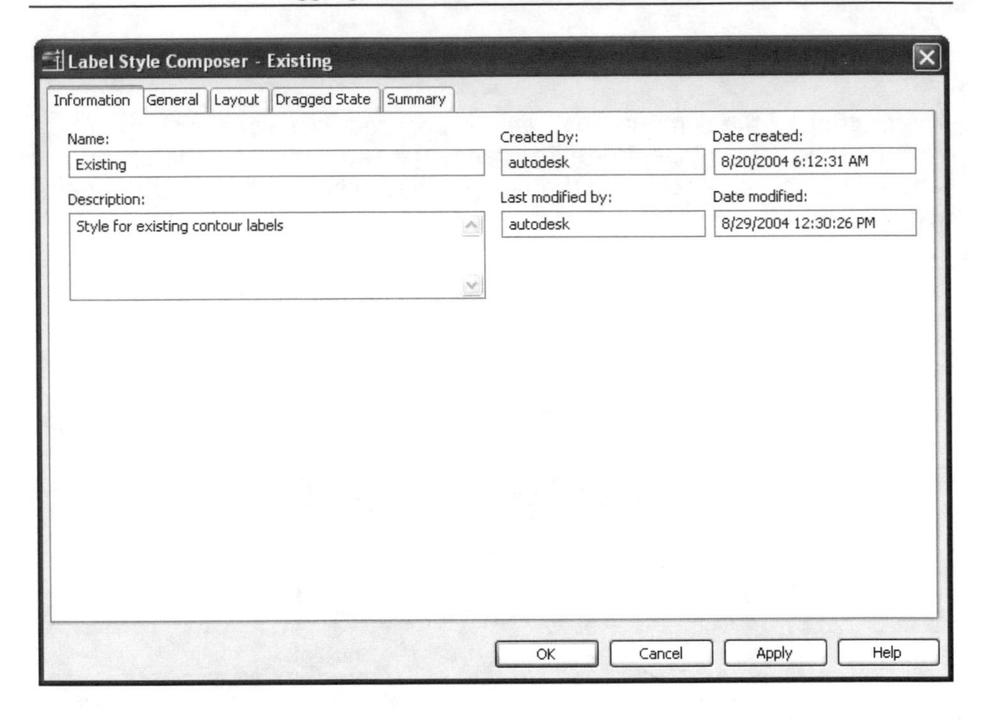

6. Enter **"Existing"** for the **Name** of the new *Surface Style*.

7. Enter **"Style for existing contour labels"** for the **Description** of the new Contour Label Style.

8. Select the **Layout** tab of the *Label Style Composer* dialog box.

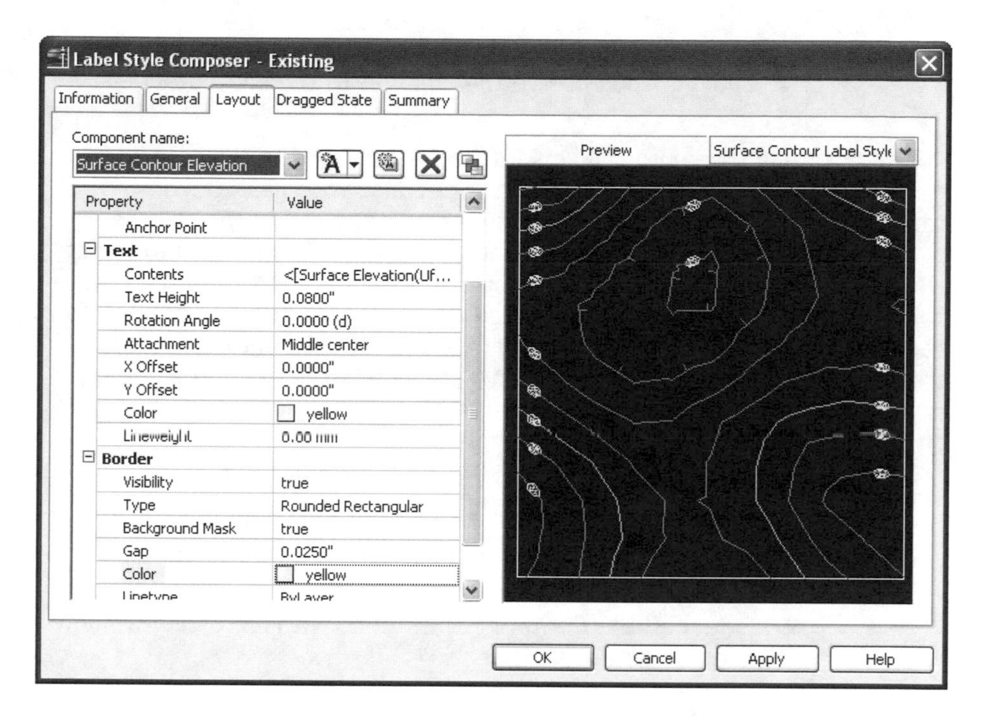

9. Confirm that **Surface Contour Elevation** is set as the active **Component name.**

10. Set the **Text Height** to **0.08**.

11. Set the **Color** to **Yellow.**

12. Under the *Border* node, set the **Visibility** to **true.**

13. Confirm that the **Type** is set to **Rounded Rectangular**.

14. Set the **Color** to **Yellow.**

15. Click **<<OK>>** to save the new **Contour Label Style** and return to the *Contour Line Properties* dialog box.

16. Click <<**OK**>> to close the *Contour Line Properties* dialog box and apply the new Contour Style to the contour label line.

17. Zoom in and observe the changes to the contour labels.

Notice that only the contour labels that were attached to the selected contour label line were updated. This is because you created a new contour label style and that new style was only applied to the selected contour label line. To update all of the contour labels you need to select all of the contour label lines and change their properties to use the new contour label style. Notice that a small contour label line was created for each contour label at the distance specified when the contours were created with the group interior option. These small lines control the style of the label and its position. You can move or erase these labels by moving or erasing the individual contour label lines.

18. Select **Surfaces ⇒ Labels ⇒ Contour Label Line Properties.**

19. Select all the **Contour Label Lines** with a crossing window over the entire surface.

This command will automatically filter the selected object so that only contour label lines are selected.

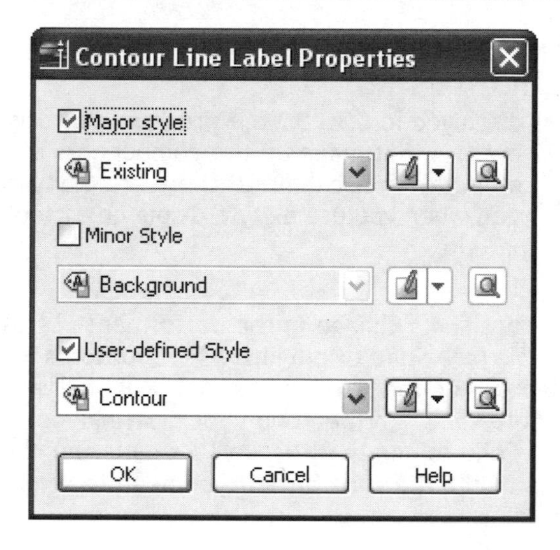

20. Set the **Major style** to **Existing** in the *Contour Line Properties* dialog box.

21. Click **<<OK>>** to close the *Contour Line Properties* dialog box and apply the new **Contour Label Style** to the contour label lines.

All of the contour labels should now display as yellow text with a rounded border according to the new label style.

5.5.5 Controlling Surface Display for Performance

When any object is displayed in *Civil 3D* the program must display all of the individual *AutoCAD* entities that make up the components that the object style is configured to display. For example, there are many more lines displayed on the screen when you use a style displaying 1 foot contours as opposed to 5 foot contours.

You may have even noticed a change in the performance of the program as we changed to the *Surface Style* displaying 1 foot contours in the previous exercises. This depends on the size and geometry of your surface as well as your computer hardware. To speed up your drawing you can simply change the *Surface Style* to one that displays a smaller number of objects, and you can always switch the style back if you need to see more detail or if you are ready to plot.

1. On the *Prospector* tab of the *Toolspace*, right-click on **Surface EG** and select ⇒ **Properties**.

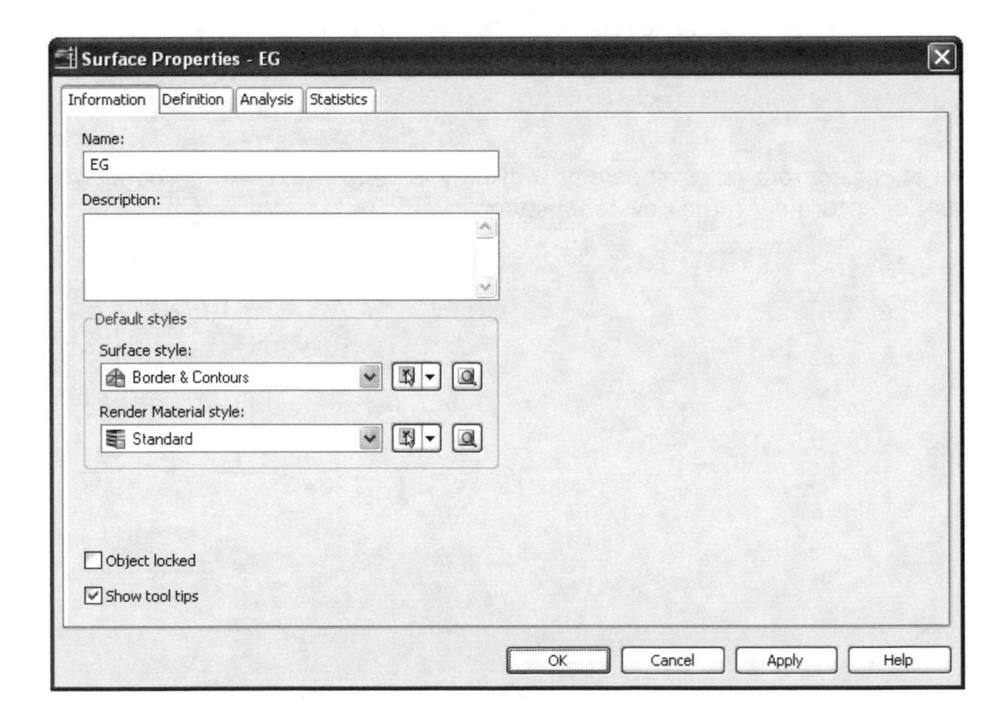

2. Select **Border & Contours** from the **Surface style** list on the *Information* tab of the *Surface Properties* dialog box.

3. Click **<<OK>>**.

The surface is now displayed showing 5 foot contours.

5.6 Chapter Summary

In this chapter you used points and breaklines from the survey data to create a survey quality existing ground surface. You learned various ways of editing and analyzing surfaces including the use of the preliminary surface to add extra data beyond the limits of the survey with the *Surface Paste* command. You also displayed and labeled contours using *Surface Styles* and *Contour Label Styles*. This *Surface* will be a critical part of your project as you continue with the design. The *Profiles, Sections, Corridor Models*, and *Grading* as well as *Volume Calculations* that you create in future chapters are all based on this *Surface*.

Chapter 6

Working with Proposed Horizontal Alignments

This chapter will focus on laying out and editing *Alignments* and *Parcels*. You will also explore how *Alignments* and *Parcels* interact with each other in a *Site*. The *Alignments* you create in this chapter will be used in future chapters to create the *Profile*, *Corridor Model*, and *Sections*.

- **Laying Out the Horizontal Alignment**
- **Editing Alignments**
- **Stationing and Offsets**
- **Laying Out Parcels**
- **Working with Parcel Styles and Labels**

Dataset:

To start this chapter you will continue working in the drawing named **Design.dwg**. You can continue with the drawing that you currently have from the end of the previous chapter or, if you are starting in the middle of the book, you can open the drawing **CH-06.dwg** located in the folder **C:\Cadapult Training Data\Civil 3D 2007\Level 1\Chapter Drawings.** Opening the drawing from the dataset provided will ensure that you have the drawing set up correctly for the exercises in the following chapter and overwrite any mistakes that you may have made in previous exercises.

6.1 Overview

This chapter will focus on laying out and editing *Alignments* and *Parcels*. You will also explore how *Alignments* and *Parcels* interact with each other in a *Site*. You will learn to use the *Parcel Layout Tools* to automatically subdivide and size parcels based on area and frontage criteria. The *Alignments* you create in this chapter will be used in future chapters to create the *Profile*, *Corridor Model*, and *Sections*.

6.1.1 Concepts

The dynamic nature of the alignment object in *Civil 3D* allows you to easily make changes to the geometry of the alignment. This allows you to approach the design process differently than you may have in the past with other programs. Now you can set a default curve radius for the alignment and lay out a rough draft in *Civil 3D*, them modify and revise it in real time.

Parcel layout is also interactive and dynamic. In this chapter you will learn to subdivide a large parcel automatically based on area and frontage criteria. You will also learn how parcels interact with each other when they are part of the same site. This makes editing easier and helps avoid mistakes because parcels are not allowed to overlap.

6.1.2 Styles

Styles are saved in the drawing and can be created and edited on the Settings tab of the *Toolspace* or on the fly during many object creation and editing commands.

Alignment Styles control the display of the alignment geometry. You have individual control over many components of the alignment including lines, curves, spirals, direction arrows, and PI points.

Alignment Label Sets control the group of styles used to label the alignment. This includes displaying stationing, ticks, geometry information, and much more.

Parcel Styles control the display of the parcel segments and the parcel area fill. *Parcel Styles* also control the *Parcel Name Template* which can be used to automatically name the parcels as they are created or renumbered.

Parcel Label Styles control the display of the parcel label text. This typically includes things like the parcel name and/or number, area in square feet and/or acres, and perimeter. *Parcel Labels* can be configured to display many other types of information as well, such as tax lot number, address, and user defined fields.

6.2 Laying Out the Horizontal Alignment

The dynamic nature of the alignment object in *Civil 3D* allows you to easily make changes to the geometry of the alignment. This can change the way that you think about the design process. Rather than sketching things out on paper and then entering them into the program you can now lay out a draft of the alignment in *Civil 3D* and easily modify it to try several alternatives on the fly. The point is that you don't have to get it perfect the first time.

6.2.1 Default Curve Settings

In this exercise you will set the default curve settings for the alignment layout commands. Once the alignment is laid out you can edit individual curves as needed.

1. Continue working in the drawing **Design.dwg**.
2. Thaw the layers **C-PROP** and **C-PROP-Line.**
3. Freeze the layers **C-TOPO-EG** and **Contour Label Line**.
4. **Regen** to display the parcel lines and labels.

The drawing should currently display the cyan parcel lines with green parcel labels.

5. Zoom to the large parcel at the center of the site.
6. Select **Alignments ⇒ Create By Layout**.

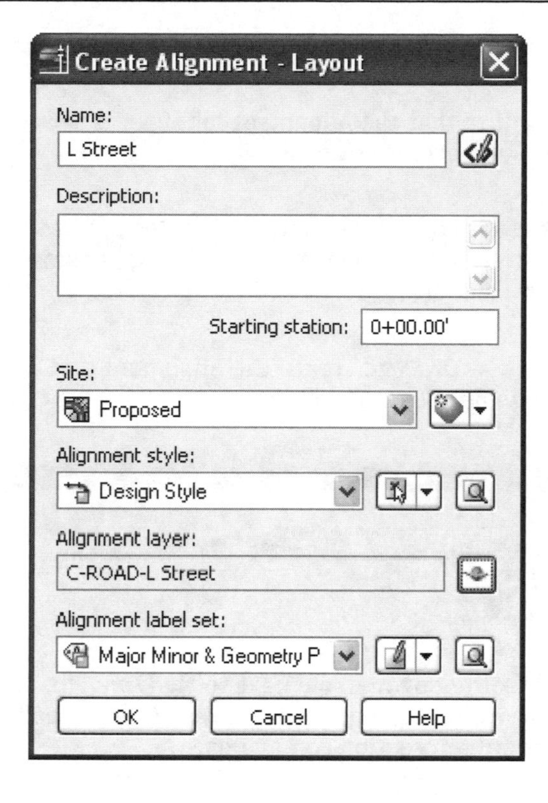

7. Enter **L Street** for the name of the **Alignment**.

8. Confirm that the **Site** is set to **Proposed**.

Alignments and parcels that are in the same *Site* interact with each other. When an alignment divides a parcel a new parcel is created. By placing the new alignment in the *Site* with the parcels you created in *Chapter 4* you will get to see this interaction. If you do not want an alignment to impact a set of parcels just create a new *Site* for the alignment.

9. Click the **Alignment Layer** button to open the *Object Layer* dialog box.

10. Set the **Modifier** to **Suffix**.

11. Enter **-*** as the **Modifier** *value*.

12. Click **<<OK>>** to close the **Object Layer** dialog box.

13. Confirm that the Alignment label set is set to **Major Minor & Geometry Point.**

The alignment label set controls the group of styles used to label the alignment. This includes displaying stationing, ticks, geometry information, and much more.

14. Click **<<OK>>** to create the alignment and open the *Alignment Layout toolbar*.

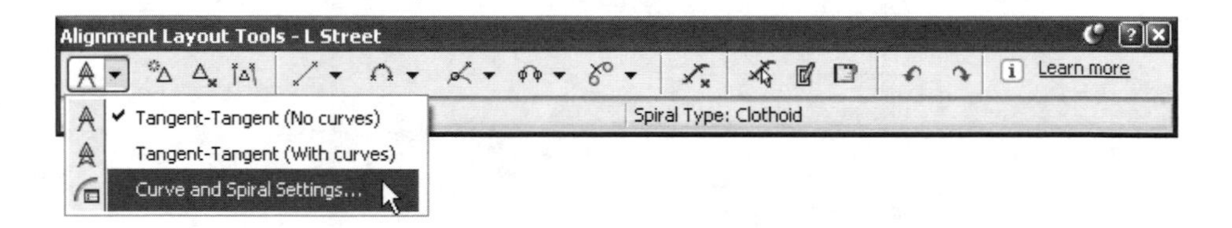

15. Click the **down arrow** next to the *Tangent-Tangent (No curves)* button on the *Alignment Layout* toolbar to display the other creation commands.

16. Click the **Curve and Spiral Settings** button.

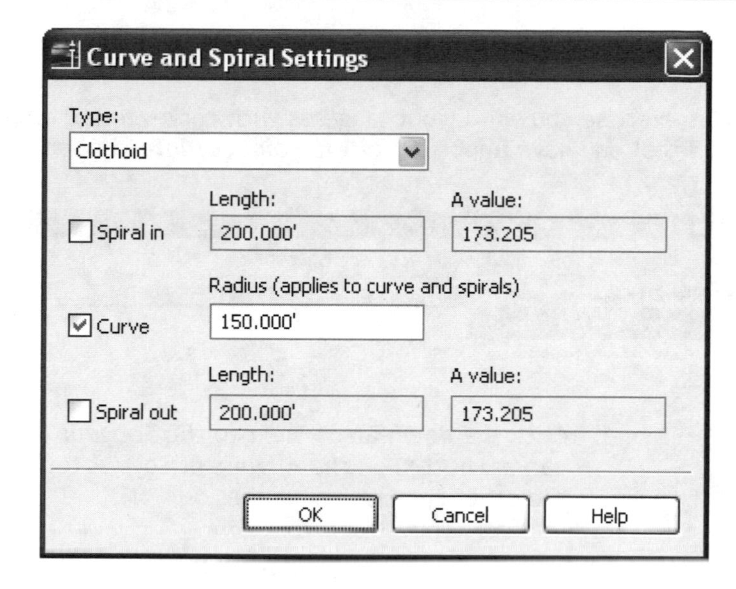

17. Confirm that the default **Radius** is set to **150'**.

You can also enable the use of spirals and set their defaults if you are laying out an alignment for a highway that requires them.

18. Click **<<OK>>**.

6.2.2 Creating Tangents with Curves

In this exercise you will layout tangents with curves based on locations of points that you have imported for the *Points of Intersection*.

1. Click the **down arrow** next to the *Tangent-Tangent (No curves)* button on *the Alignment Layout* toolbar to display the other creation commands.

2. Click the **Tangent-Tangent (With curves)** button.

3. At the command line enter **`'PN`**.

4. Enter the **Point Number Range: 1337-1340**.

The new alignment is drawn using the default curve settings that you configured in the previous exercise. The parcel that the alignment passes through is also divided and a new parcel is created. This is because the alignment and the parcel are in the same Site.

5. **<<Escape>>** to end the **`Point Number`** prompt.

6. **<<Enter>>** to end the alignment.

7. Save the drawing.

6.3 Editing Alignments

Alignments can be edited graphically with standard *AutoCAD* commands or with the *Alignment Layout* tools which include a tabular option. This allows you to graphically get the alignment close to the desired geometry and then fine tune it in a table.

6.3.1 Editing Alignments Graphically

Civil 3D alignments can be grip edited just like basic *AutoCAD* objects.

1. Pick the alignment in the drawing editor to display the grips.

2. Select the **grip at PI** number **2** at the northeast corner of the alignment.

3. At the command line, enter "@0,10" to stretch the PI 10 feet north.

You can also stretch any of the grips on the curves to change the curve radius and you can stretch the grip at the middle of a tangent to move that tangent while holding the bearing and length. This also updates the sizes of the two parcels divided by the alignment. The *AutoCAD Undo* command can be used to undo these changes.

6.3.2 Editing Alignments in Grid View

1. Select **Alignments ⇒ Edit Alignment Geometry**.

2. Select the **L Street alignment**.

3. Click the **Grid View** button .

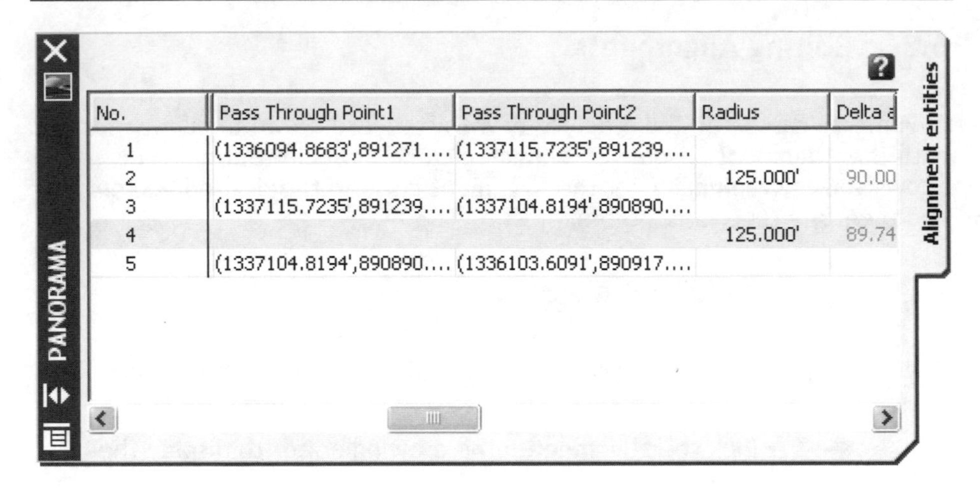

4. Scroll through the alignment information to find the curve radius.

5. Change the **Radius** of the first curve to **125**.

6. Change the **Radius** of the second curve to **125**.

The alignment will update graphically after each change. Also, you are not allowed to set a radius that does not fit the geometry of the alignment. So you can not create overlapping curves.

7. Close the **Grid View**.

8. Close the **Alignment Layout** toolbar.

9. Save the drawing.

6.4 Stationing and Offsets

Once an alignment is created stationing is displayed according to the alignment label set you selected when the alignment was first created. The alignment label set is a collection of alignment label styles that can be changed in the alignment properties.

6.4.1 Working with Alignment Labels

Alignment Labels are most commonly used to display stationing. However, they can also display *Northings, Eastings, Directions*, and *Design Speeds*.

1. On the *Prospector* tab of the *Toolspace*, expand **Sites, Proposed**, and **Alignments**.

2. Right-click on the **Alignment L Street** and select ⇒ **Properties.**

3. Select the **Labels** tab.

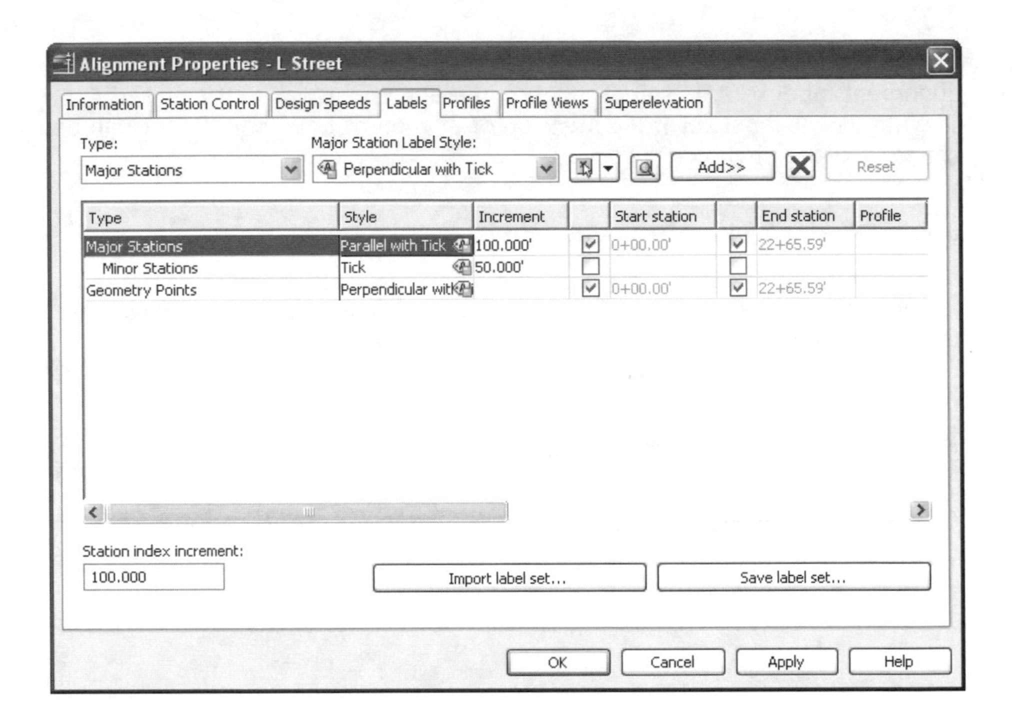

4. Change the **Major Station Style** to **Parallel with Tick** by clicking on the symbol in the *Style* field of the table.

You can also save and import Label Sets which are just saved collections of label styles so that you don't have to set each one individually.

5. Click **<<OK>>**.

6. Zoom in to review the changes to the stationing from the new alignment label style.

The alignment will be redisplayed with parallel stationing at the major stations. You can edit alignment label styles in an editor that is very similar to the one you used to create point label styles and contour label styles in the previous chapters.

6.4.2 Changing the Stationing of an Alignment

1. Confirm that the **Alignment** node under the *Site* **Porposed** is expanded on the *Prospector* tab of the *Toolspace*.

2. Right-click on the **Alignment L Street** and select ⇒ **Properties.**

3. Select the **Station Control** tab.

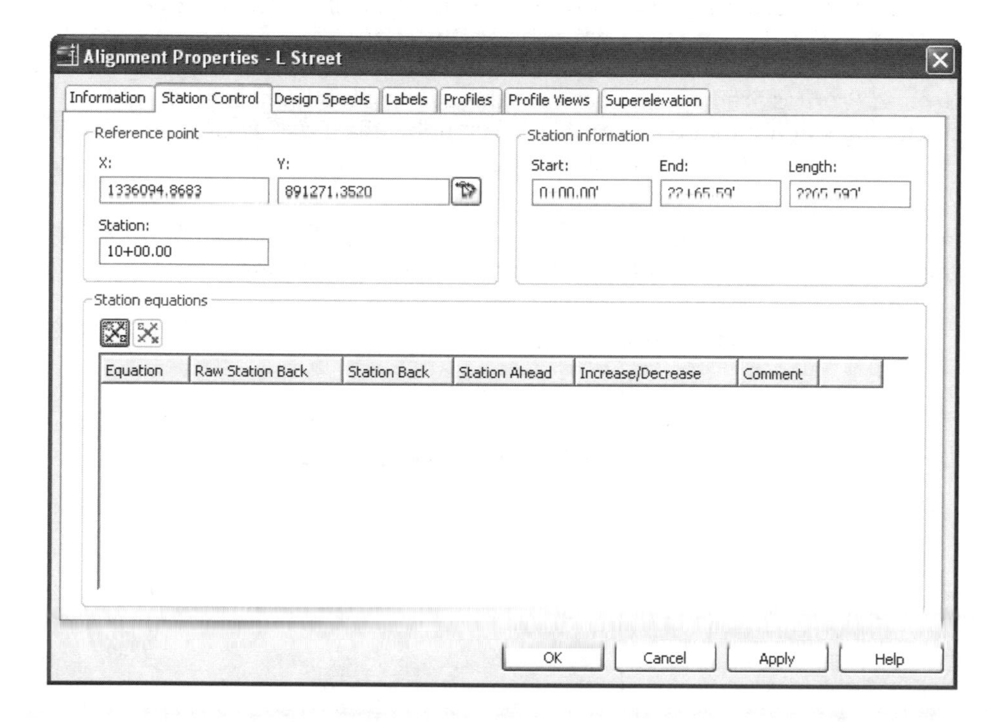

4. Change the **Station of the Reference Point** (by default the beginning of the alignment) to **1000.**

5. Click **<<OK>>.**

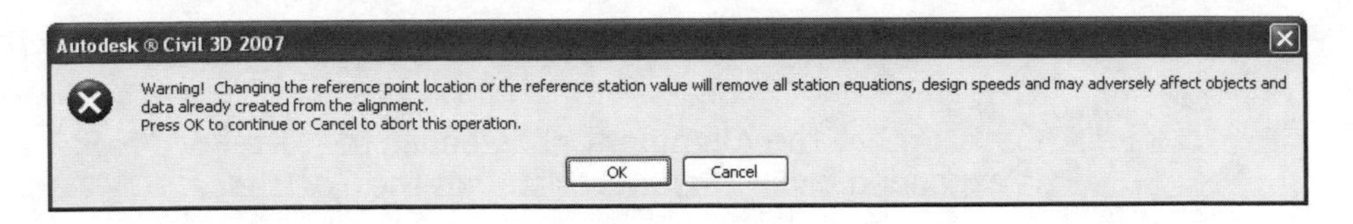

This warning notifies you that changing the stationing will remove any station equations and design speeds since they are based on stationing. You can reset any of these values to match the current stationing later.

Changing the stationing of an alignment will also impact any *Finished Grade Profiles* that you have defined for this alignment. These are also defined by station values so they would need to be updated. *Existing Ground Profiles* are based directly on the alignment so a change in stationing will not cause any problem for them, the *Existing Ground Profiles* will automatically be updated.

In this exercise you have not created any of these station based features at this point. So you will not need to worry about this warning.

6. Click <<**OK**>> to clear the warning.

7. Click <<**OK**>> to close the *Alignment Properties* dialog box and apply the new stationing.

8. Zoom in to review the updated station values.

6.4.3 Creating Alignment Offsets

Offsets can be created with the standard *AutoCAD OFFSET* command. This creates a Polyline on the current layer.

1. Create a new layer named **"C-ROAD-L Street-EP"**, set the color to **yellow** and set it **current**.

2. Enter `"O"` at the command line to start the *AutoCAD* **offset** command.

3. Enter `"12"` for the **offset distance**.

4. Select the **L Street alignment** and offset it to the right and left side of the road.

5. Enter to end the offset command.

6. Save the drawing.

These offsets are for drafting purposes only. They are not *Civil 3D* objects, displayed in the *Prospector*, or tied to our design in any way. They are simply *AutoCAD* polylines.

6.5 Laying Out Parcels

When the alignment was created it automatically split the parcel it is passing through and acts as another parcel line. As you edit the alignment the parcels are also edited.

6.5.1 Creating Right of Way Parcels

Right of Way parcels can be created from an alignment. However, the *Right of Way* lines are not dynamically linked to the alignment so if the alignment is changed the *Right of Way* parcel lines will need to be deleted and recreated.

1. Select **Parcels ⇒ Create ROW.**

2. Select the two parcels separated by the alignment. Be sure to pick the **parcel label** at the center of the parcel rather than the parcel lines.

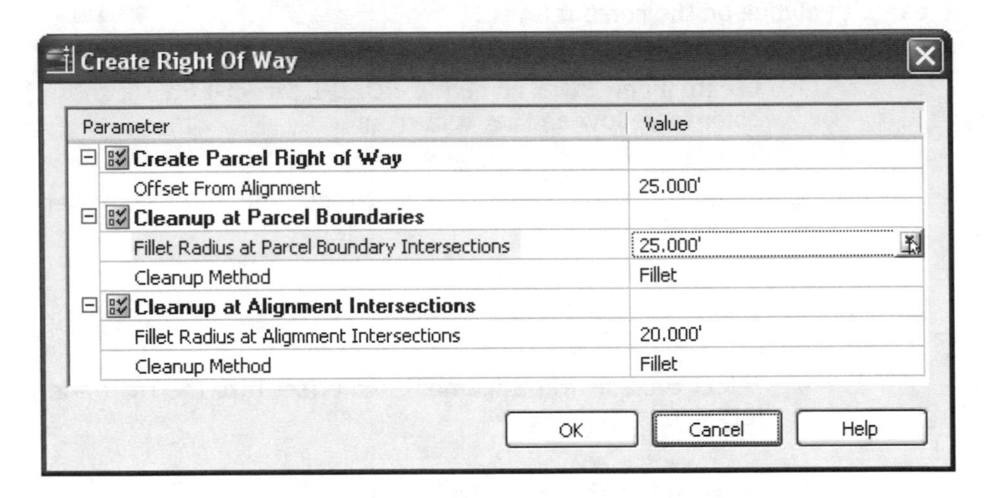

3. Enter **"25"** for the **Offset from alignment.**

4. Enter **"25"** for the **Fillet radius at parcel boundary intersections**.

By using the *Cleanup* method option you have the option to fillet or chamfer the boundary intersections as well as an option for no cleanup at all.

5. Click **<<OK>>** to create to *Right of Way* parcels.

The new cyan *ROW* parcel lines will be displayed along with a parcel label on each side of the centerline for the two new *ROW* parcels.

6.5.2 Merging Parcels

When using the *Right of Way* command at least two *Right of Way* parcels are created, one on each side of the alignment or alignment intersection. You may want to merge these parcels together into a single *Right of Way* parcel to make them easier to manage and display. You can do this by editing the parcels and using the Union command.

1. Select **Parcels ⇒ Edit Parcel Segments.**

2. Pick one of the parcel lines from one of the *ROW* parcels to start the edit command and display the *Parcel Layout Tools* toolbar.

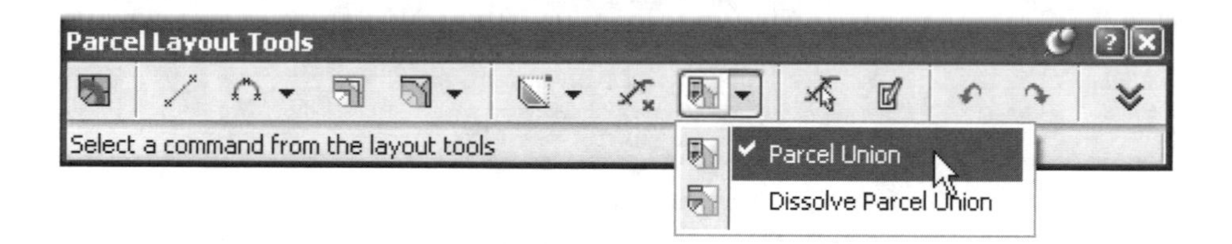

3. Click the **Parcel Union** button on the *Parcel Layout Tools* toolbar.

4. Pick the two *ROW* parcels from the drawing editor. It is easiest to select the parcels by picking the parcel label.

5. **<<Enter>> to merge the parcels.**

The two ROW parcels are now merged together and one of the parcel labels will disappear.

6. Enter "**X**" at the command line to exit the *Edit Parcel* command.

6.5.3 Creating Parcels Manually

Parcels can be laid out and divided by graphically picking the location of the new parcel line.

1. Select **Parcels ⇒ Create By Layout.**

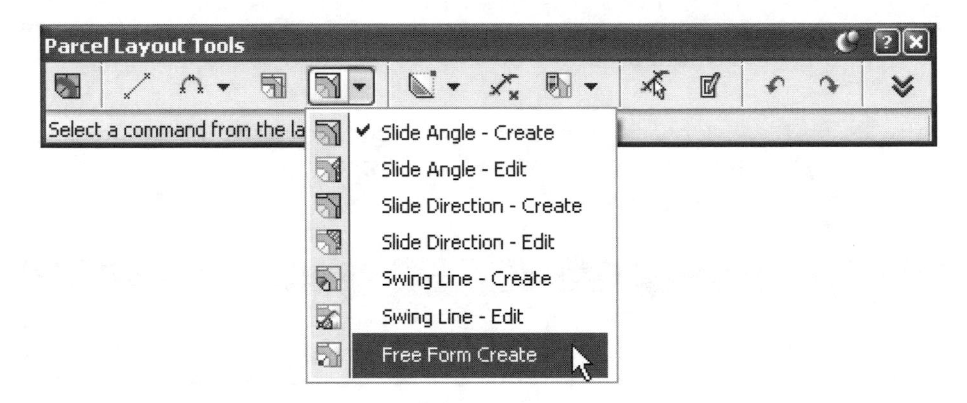

2. Click the **down arrow** button next to the **Slide Angle - Create** button on the *Parcel Layout Tools* toolbar to display the other creation commands.

3. Click the **Free Form Create** button.

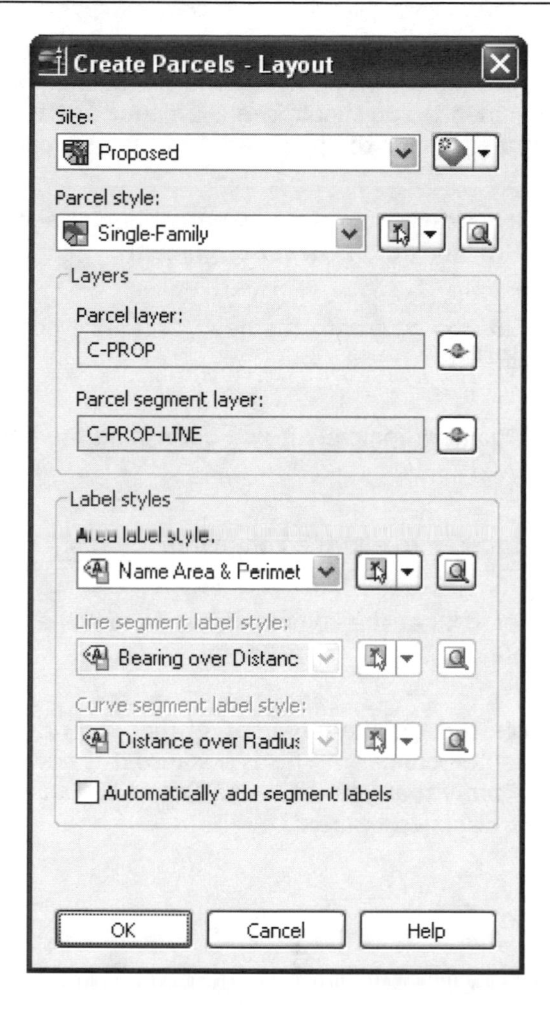

4. Confirm that the **Site** is set to **Proposed**.

5. Set the **Parcel** style to **Single Family**.

6. Set the **Area Label Style** to **Name Area and Perimeter**.

7. Click **<<OK>>**.

8. Move your cursor into the northwest corner of the main, U-shaped parcel wrapping around the outside of the alignment. You should see a dynamic preview line showing suggestions for the placement of the new parcel line.

9. Pick a location for the new parcel line at about **station 11+50** along the *L Street alignment*.

10. **<<Enter>>** to create the new parcel line perpendicular to the *ROW*.

If you pick a second point graphically it will determine the bearing of the new parcel line.

11. **<<Enter>>** to end the command.

12. Enter **"X"** at the command line to exit *the Parcel Layout Tool*.

The new parcel is created and the large parcel that you just divided now displays with a magenta boundary. This is because it is now using the parcel style *Single Family* that you set in the *Parcel Layout* dialog box. **Regen** if it does not display properly.

6.5.4 Creating Parcels with the Slide Bearing Tool

In this exercise you will create the next parcel at an exact size of 20,000 square feet and a minimum frontage of 100 feet. After entering these parameters *Civil 3D* will find the location of the new parcel line for you rather than you selecting it manually.

1. Select **Parcels ⇒ Create By Layout.**

2. Expand the **Parcel Layout Tools** toolbar if it is not already expanded by clicking the **expand** button on the far right of the toolbar.

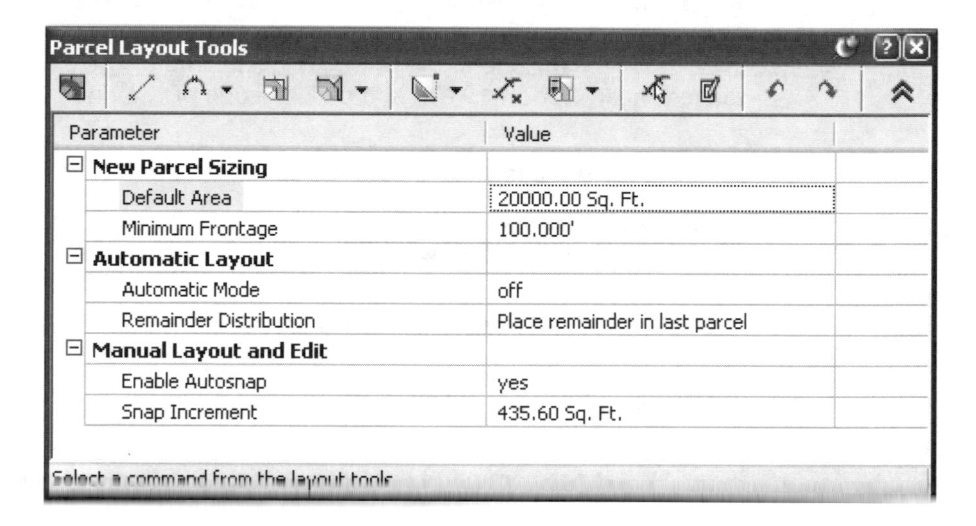

3. Set the default area to **20000.**

4. Confirm that the **Minimum Frontage** is set to **100.**

5. Click the **Slide Angle - Create** button on the *Parcel Layout Tools* toolbar.

6. Confirm that the **Site** is set to **Proposed.**

7. Set the **Parcel** style to **Single Family**.

8. Set the **Area Label Style** to **Name Area and Perimeter**.

9. Click <<**OK**>>.

10. Pick a point in the large parcel to be subdivided to the right of the parcel you created in the previous exercise.

11. When prompted to `Select start point on frontage` use an **End Point** snap to select the end of the parcel line that you created in the previous exercise where it intersects the ROW.

12. Drag the highlight line along the ROW to the end of the tangent near **station 19+00** and pick the end of the line.

13. Enter **90** for the **angle** at the frontage.

14. Accept the default area of **20000**.

The new parcel is created with an area of 20000 square feet.

15. <<**Escape**>> to end the command.

16. Enter "`X`" at the command line to exit the *Parcel Layout Tool*.

6.5.5 Creating Parcels with the Slide Bearing Tool Automatically

The Slide Bearing command can also be used to divide an entire parcel into as many parcels that fit into it according to the area and frontage parameters that you set.

1. Select **Parcels** ⇒ **Create By Layout**.

2. Expand the **Parcel Layout Tools** toolbar if it is not already expanded by clicking the **expand** button on the far right of the toolbar.

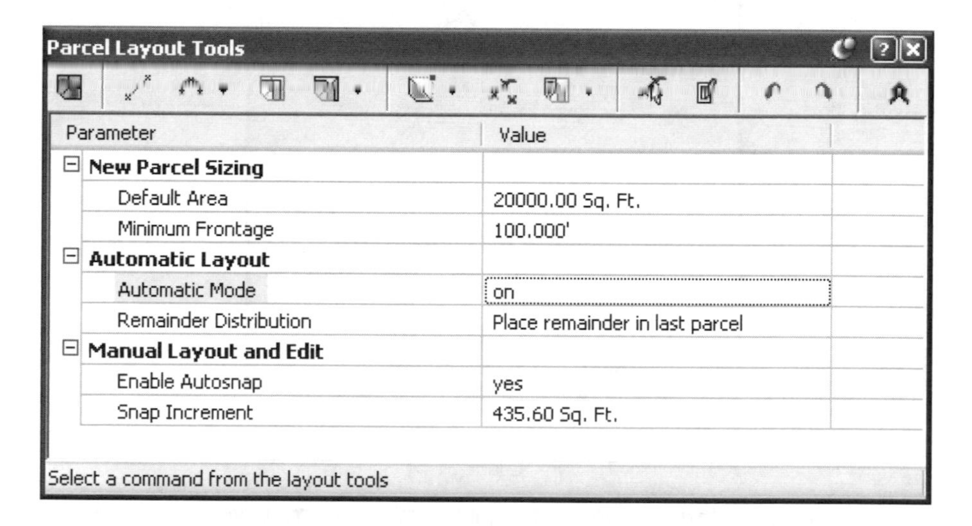

3. Confirm that the **Default Area** is set to **20000**.

4. Confirm that the **Minimum Frontage** is set to **100**.

5. Set the **Automatic Mode** to **On**.

6. Confirm that the **Remainder Distribution** will **Place the remainder in the last parcel.**

7. Click the **Slide Angle - Create** button.

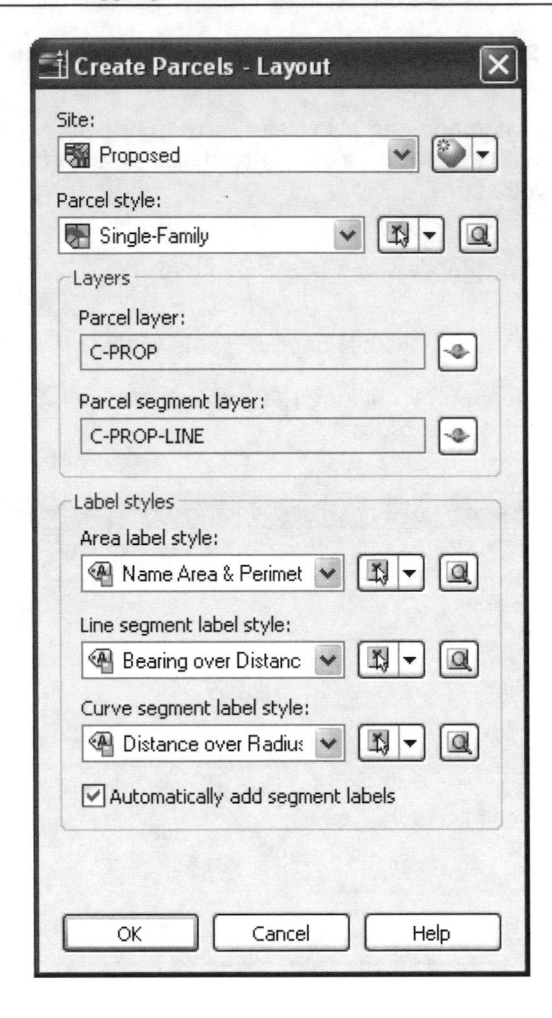

8. Confirm that the **Site** is set to **Proposed**.

9. Set the **Parcel** style to **Single Family**.

10. Set the **Area Label Style** to **Name Area and Perimeter**.

11. Enable the option to **Automatically add segment labels** and review the Line and Curve segment label options.

12. Click <<**OK**>>.

13. Pick a point in the large parcel to be subdivided to the right of the parcel you created in the previous exercise.

14. When prompted to `Select start point on frontage` use an **End Point** snap to select the end of the parcel line that you created in the previous exercise where it intersects the ROW.

15. Drag the highlight line along the *ROW* to the end of the tangent near **station 32+00** and pick the end of the line.

16. Enter **90** for the **angle** at the frontage.

17. Accept the default area of **20000**.

The large parcel is now divided into parcels with a minimum area of 20000 square feet and a minimum frontage of 100 feet with parcel lines perpendicular to the *ROW*. The last parcel is larger because you used the default option to place the remaining area in the last parcel.

18. **<<Escape>>** to end the command.

19. Enter "`X`" at the command line to exit the *Parcel Layout Tool*.

The newly created parcel segments are automatically labeled with the line and curve segment styles that you selected in the *Create Parcels - Layout* dialog box when you enabled the *Automatically add segment labels* option.

6.5.6 Editing Parcels

1. Select that last parcel line created by the automatic layout in the previous exercise. You will see that it only displays a single grip at the *ROW*.

2. Select the grip and drag it along the *ROW* to a position of your choice.

3. Notice that the parcel line stays perpendicular to the *ROW* and that the properties of both parcels are updated.

6.5.7 Deleting Parcels

Parcels are deleted by erasing a parcel segment or editing a parcel segment so that it is no longer closed. Do not try to delete a parcel by erasing the parcel label, even though it highlights the entire parcel when you select it, nothing will happen when you try to erase it.

1. Use the *AutoCAD* **ERASE** command to erase one of the parcel lines that you created during the parcel sizing exercises.

2. Notice that the parcels separated by that line are merged together and that the old parcel is removed from the *Prospector*.

6.5.8 Renumbering Parcels

Civil 3D makes renumbering parcels easy by allowing you to trace a path that will select parcels in the order that you would like them to be numbered. So it is best to not worry about parcels that may get numbered incorrectly as you go through the design process of adding, deleting, and editing parcels because you can easily renumber them at any time.

1. Select **Parcels ⇒ Renumber/Rename Parcels.**

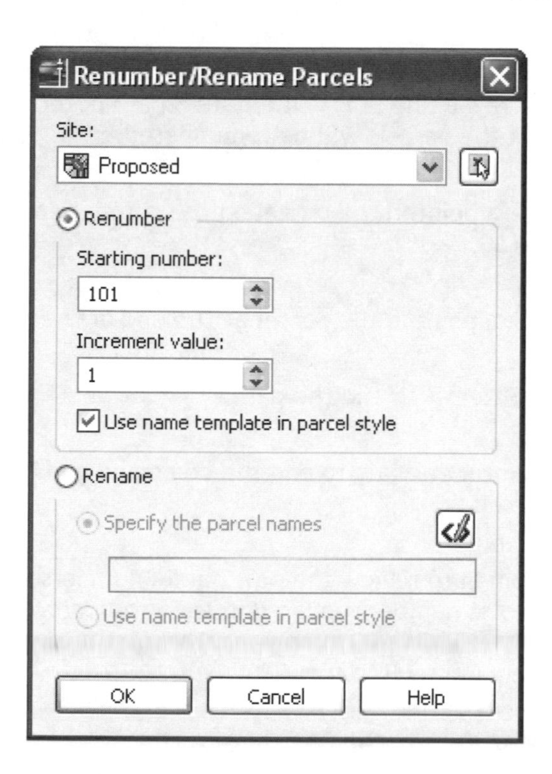

2. Confirm that the **Site** is set to **Proposed.**

3. Confirm that the **Renumber** option is selected.

4. Set the **Starting number** to **101**.

5. Confirm that the **Increment value** is set to **1**.

6. **Enable** the **Use name template in parcel style** option.

This will use the parcel name template that is defined in the parcel style. In this case it will name the parcels *SINGLE-FAMILY: 101* rather than just *1*.

7. Click **<<OK>>**.

8. Pick a point inside the first parcel that you created at the beginning of the alignment.

9. Pick a point in the parcel on the corner near **station 20+00** to draw a line that will create a path determining the order that the parcels will be renumbered.

10. Pick a point in the parcel on the corner near **station 23+00**.

11. Pick a point in the parcel at the end of the alignment.

12. **<<Enter>>** to end the path.

13. **<<Enter>>** again to end the command and renumber the parcels.

14. Zoom in to review the new parcel numbers. You will also see the numbers updated in the *Prospector*. If the parcels are not visible in the prospector right-click on the parcels node and select refresh.

15. Save the drawing.

6.6 Working With Parcel Styles and Labels

In this section you will create a new Parcel Style and Parcel Label Style to control the display of the parcel lines and the Labeling of the Parcel.

6.6.1 Controlling Parcel Display

1. On the *Prospector* tab of the *Toolspace*, expand the **Parcels** node under the *Site* **Proposed.**

2. Right-click on **PROPERTY: 101** under the **Parcels node** and select ⇒ **Properties.**

This is the first parcel that you subdivided and it is still using the original parcel style Property. You could also select the parcel label from the screen and right-click to select Parcel Properties.

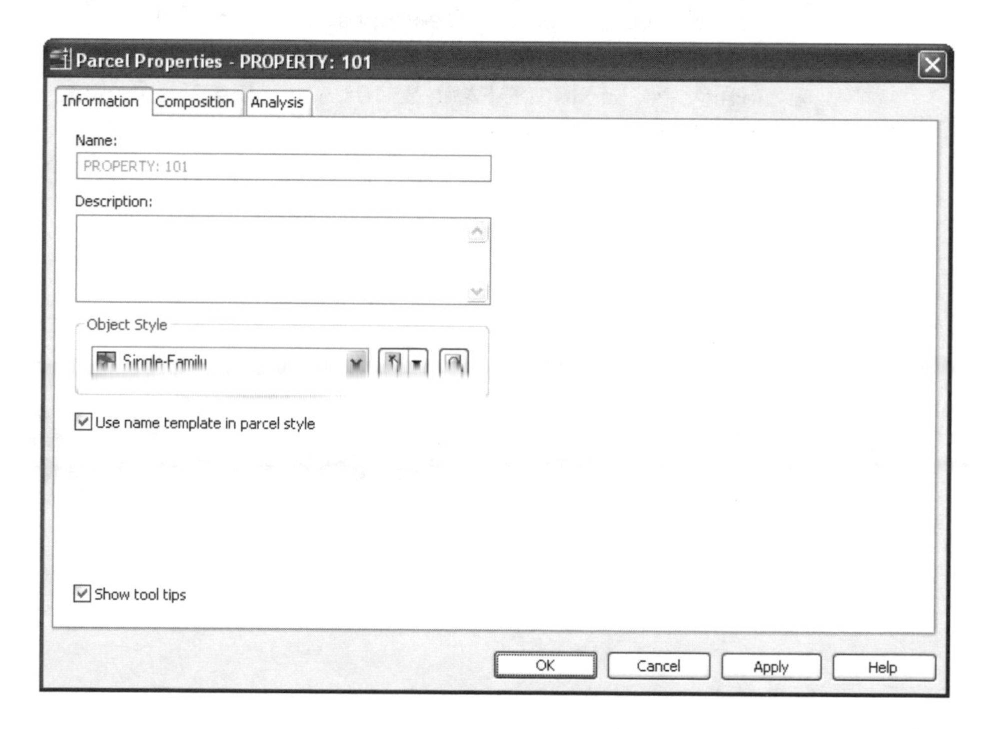

3. Change the **Object Style** to **Single Family.**

4. Click **<<OK>>** to display the parcel with the new style.

The parcel is now displayed with a magenta boundary line according to the properties of the parcel style. It is also renamed *SINGLE-FAMILY:101*. This is based on the name template in the parcel style.

6.6.2 Creating Parcel Styles

Parcel Styles control the display of the parcel segments and the parcel area fill. In this exercise you will create a new parcel style for Low Density Zoning that will display the parcels with yellow, phantom lines and a crosshatch buffer. You will create the parcel style on the fly through the *Parcel Properties* dialog box. However, you could also create the Parcel Style from the settings tab of the *Toolspace* as you did with the Point Styles in the *Chapter 4.*

1. On the *Prospector* tab of the *Toolspace*, confirm the **Parcels** node under the *Site* **Proposed** is expanded.

2. Right-click on **SINGLE-FAMILY: 101** under the *Parcels* node and select ⇒ **Properties**.

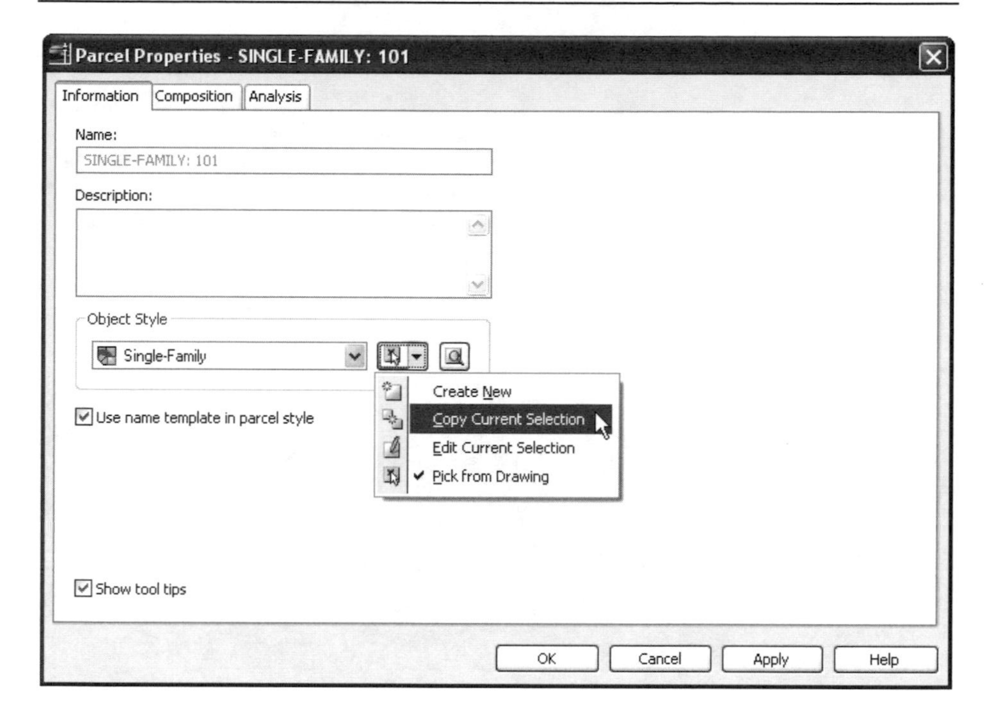

3. Click the **down arrow** button next to the **Object style** on the *Information* tab of the *Parcel Properties* dialog box to display the other creation commands.

4. Click the **Copy Current Selection** button.

This style can also be copied, edited, and created on the *Settings* tab of the *Toolspace* along with all other *Civil 3D Styles*.

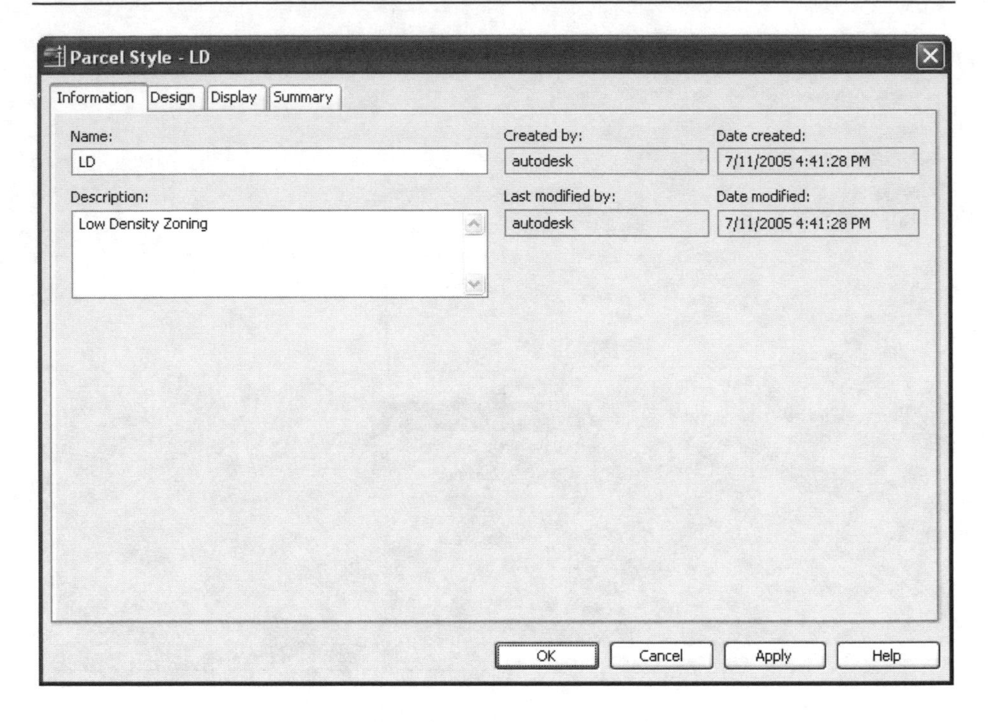

5. Enter **"LD"** for the **Name** of the new *Parcel Style*.

6. Enter **"Low Density Zoning"** for the **Description** of the new *Parcel Style*.

7. Select the **Design** Tab of the *Parcel Style* dialog box.

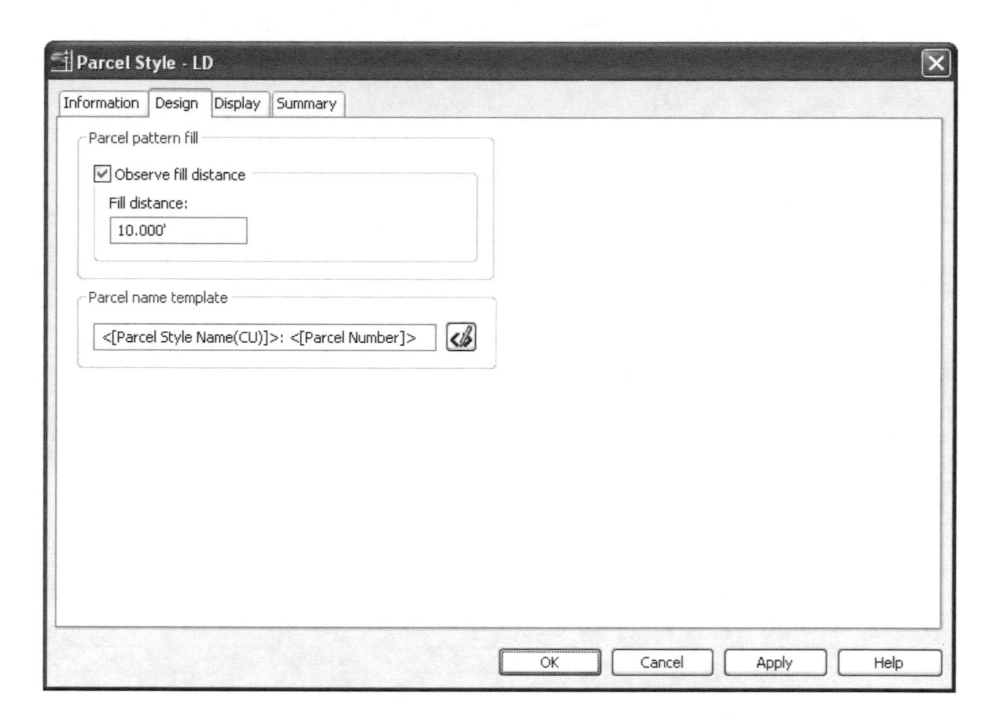

8. Enable the **Observe fill distance** option.

9. Set the **Fill distance** to 10.

This will limit the Area Fill component, or hatch, to a 10' buffer around the inside of the parcel boundary.

10. Select the **Display** Tab of the *Parcel Style* dialog box.

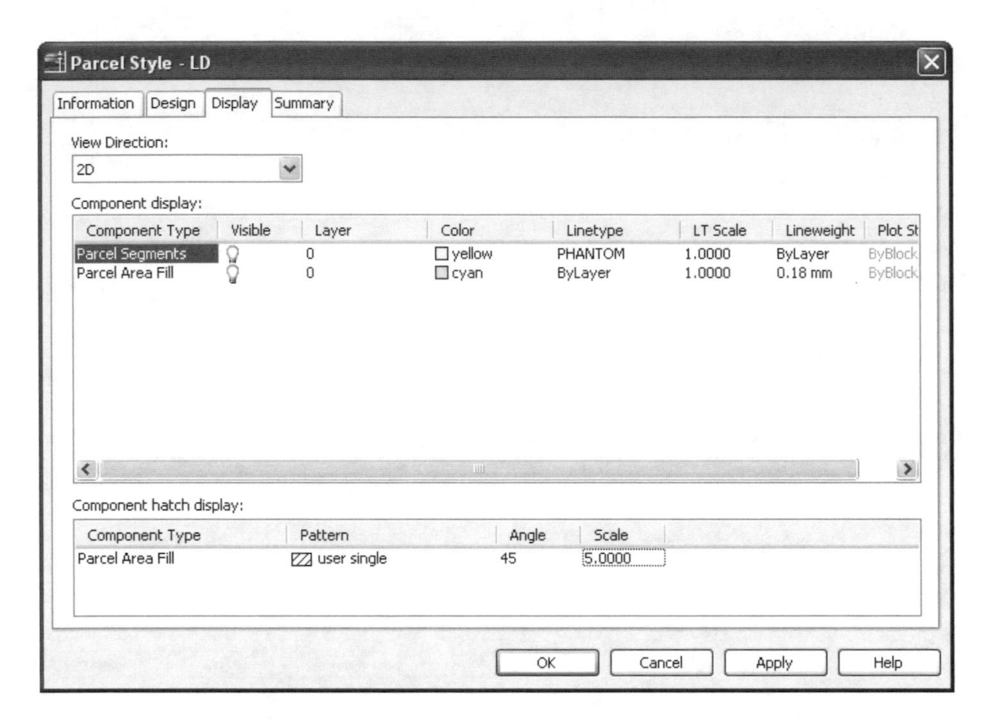

11. Confirm that the **View Direction** is set to **2D.**

12. Set the color of the **Parcel Segments** component to **Yellow.**

13. Set the *Linetype* of the **Parcel Segments** component to **Phantom.**

14. Turn on the display of the **Parcel Area Fill** component.

15. Set the color of the **Parcel Area Fill** component to **Cyan.**

16. In the *Component hatch display* section set the **Angle** to **45.**

17. In the *Component hatch display* section set the **Scale** to **5.**

This section allows you to set the hatch pattern, angle, and scale for the Parcel Area Fill component that you turned on in the previous steps.

18. Click **<<OK>>** to save the new **Parcel Style** and return to the *Parcel Properties* dialog box.

19. Click **<<OK>>** to close the *Parcel Properties* dialog box and apply the new *Parcel Style*.

20. At the command line enter "**LTSCALE**".

21. Set the **LTSCALE** to **20**.

The parcel is now displayed with a yellow, phantom boundary line according to the properties of the parcel style. It is also renamed *LD: 101*, which is based on the name template in the parcel style.

6.6.3 Creating Parcel Label Styles

Parcel Label Styles control the display of the parcel label text. This typically includes things like the parcel name and/or number, area in square feet and/or acres, and perimeter. *Parcel Labels* can be configured to display many other types of information as well, such as tax lot number, address, and user defined fields.

1. On the *Prospector* tab of the *Toolspace*, confirm the **Parcels** node under the *Site* **Proposed** is expanded.

2. Right-click on **LD: 101** under the **Parcels node** and select ⇒ **Properties.**

3. Select the **Composition** Tab of the *Parcel Properties* dialog box.

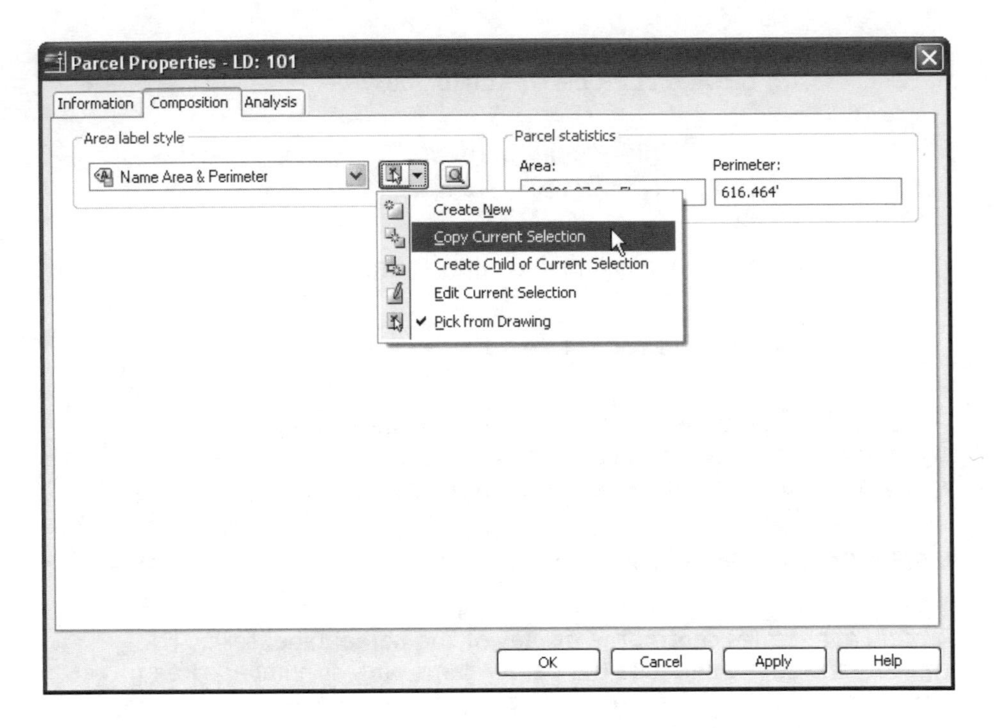

4. Click the **down arrow** button next to the **Area label style** to display the other creation commands.

5. Click the **Copy Current Selection** button.

This style can also be copied, edited, and created on the Settings tab of the Toolspace along with all other *Civil 3D Styles*.

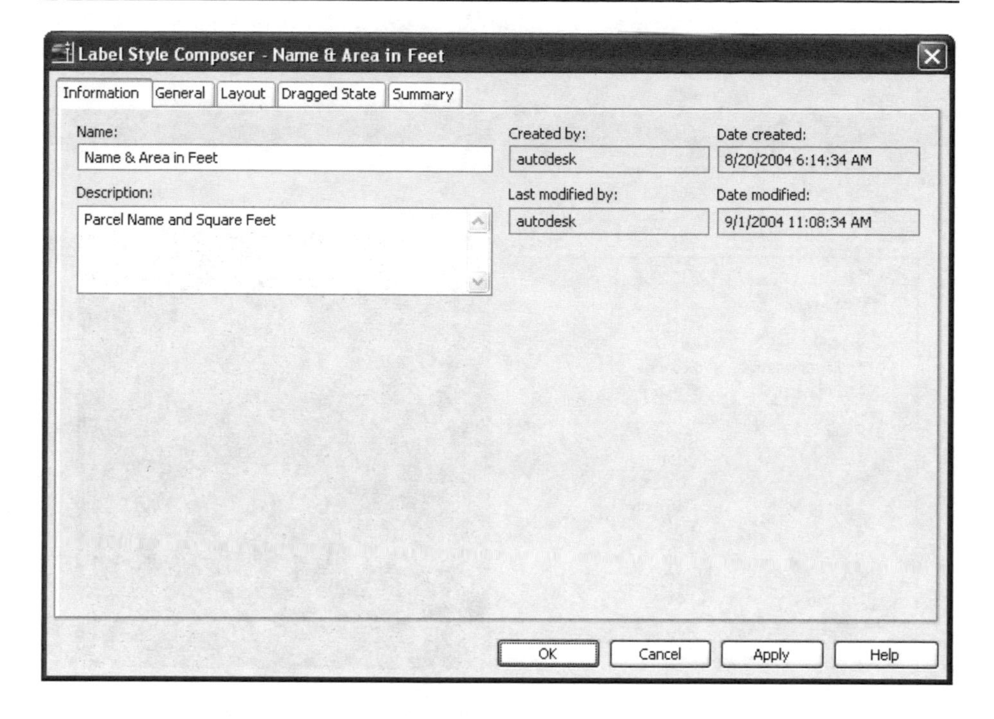

6. Enter "**Name & Area in Feet**" for the **Name** of the new *Parcel Label Style*.

7. Enter "**Parcel Name and Square Feet**" for the **Description** of the new *Parcel Label Style*.

8. Select the **Layout** tab.

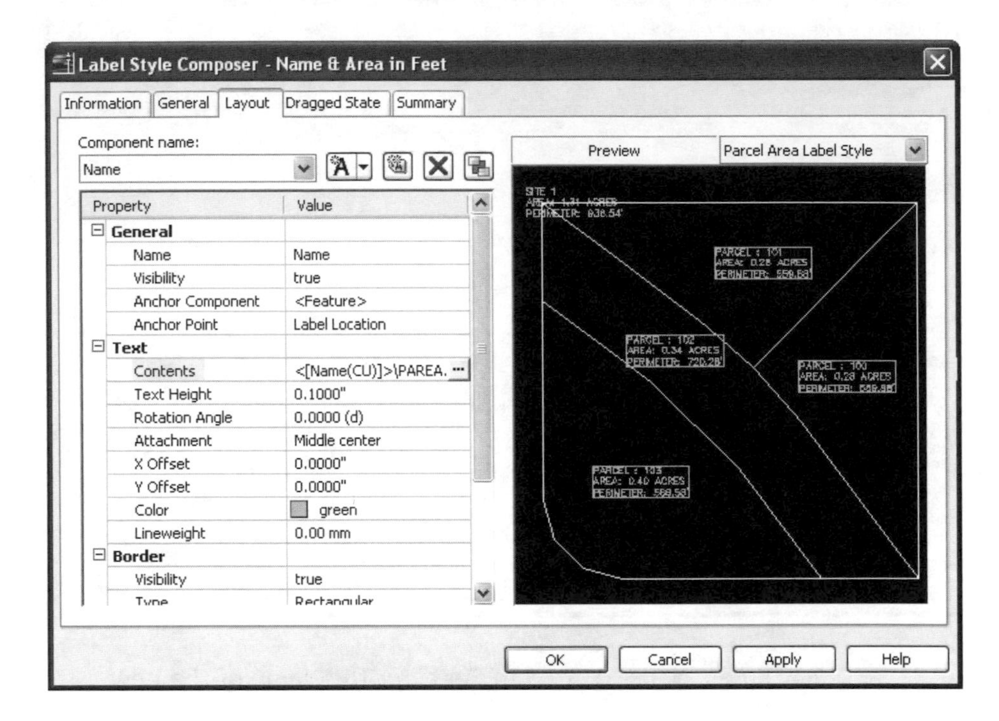

9. Click the **Text Contents** *Value* field to activate the ellipses <<...>> button.

10. Click the ellipses <<...>> button to open the *Text Component Editor*.

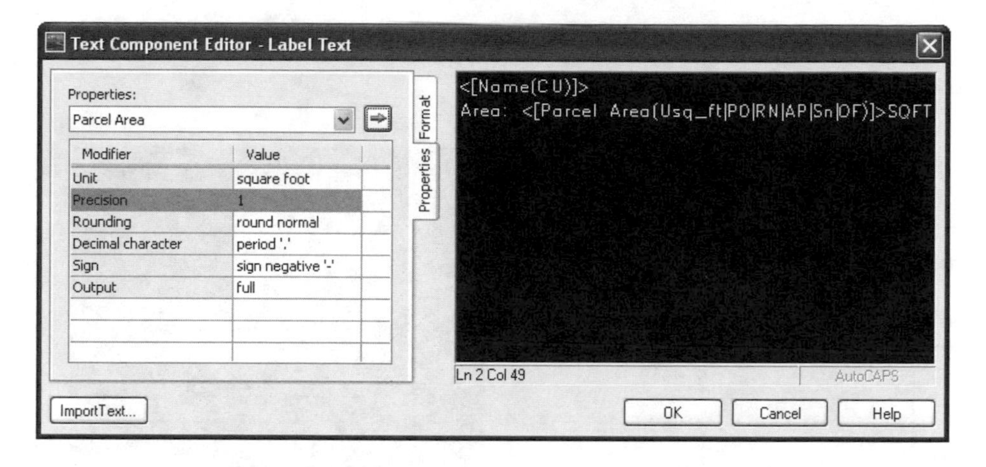

11. Delete the lines starting with **Area** and **Perimeter.**

12. Select **Parcel Area** from the **Properties** drop down list.

13. Confirm that the **Unit** value is set to **square foot**.

14. Set the **Precision** to **1** (no decimal places).

15. Click the **right arrow** [⇒] button to insert the code to display the area of the parcel in square feet.

16. Enter "**Area:** " in the editor to the left of the code to better describe the label.

17. Enter "**SQFT** " in the editor to the right of the code to describe the units of the label.

18. Click **<<OK>>** to save the contents of the new component.

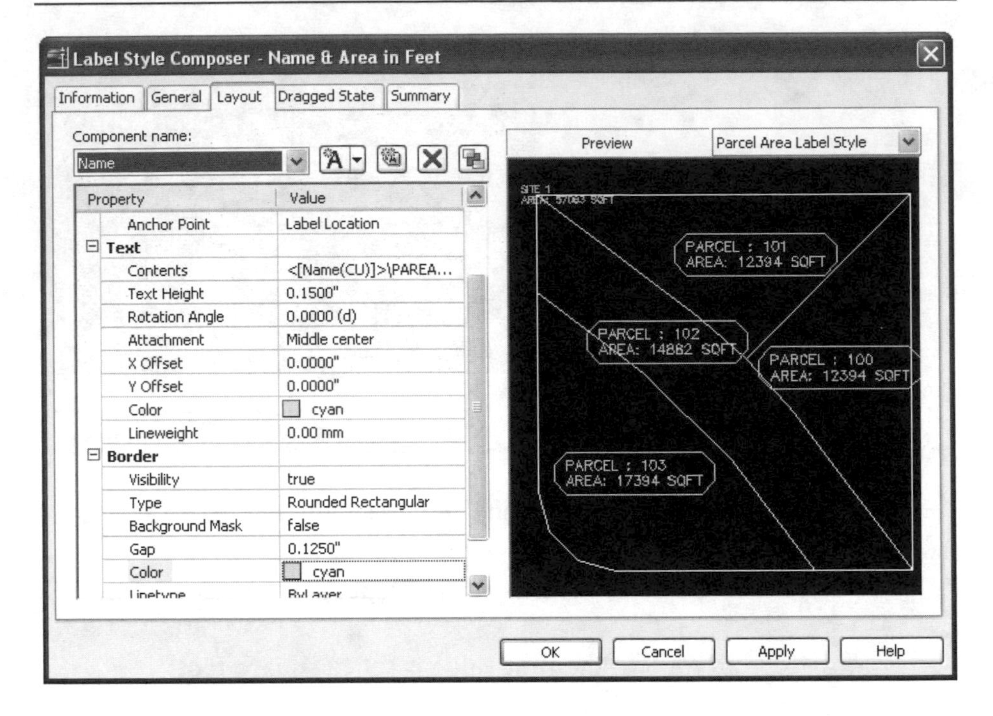

19. Back in the *Label Style Composer* set the **Text Height** to **0.15**.

20. Set the **Color** to **Cyan**.

21. Set the **Border Type** to **Rounded Rectangular**.

22. Set the **Border Gap** to **0.125**.

23. Set the **Border Color** to **Cyan**.

24. Click **<<OK>>** to save the new **Parcel Label Style** and return to the *Parcel Properties* dialog box.

25. Click **<<OK>>** to close the **Parcel Properties** dialog box and apply the new *Parcel Label Style*.

The parcel is now displayed with a cyan label showing the parcel name and area in square feet according to the properties of the *Parcel Label Style*.

6.6.4 Changing the Styles of Multiple Parcels

You can change the styles of a parcel in the *Parcel Properties* dialog box. However, this only allows you to select one parcel at a time. To change the styles of multiple parcels at the same time you must use the *Prospector*.

1. On the *Prospector* tab of the *Toolspace*, confirm the **Parcels** node under the *Site* **Proposed** is expanded.

2. Select the **Parcels** node in the *Prospector*.

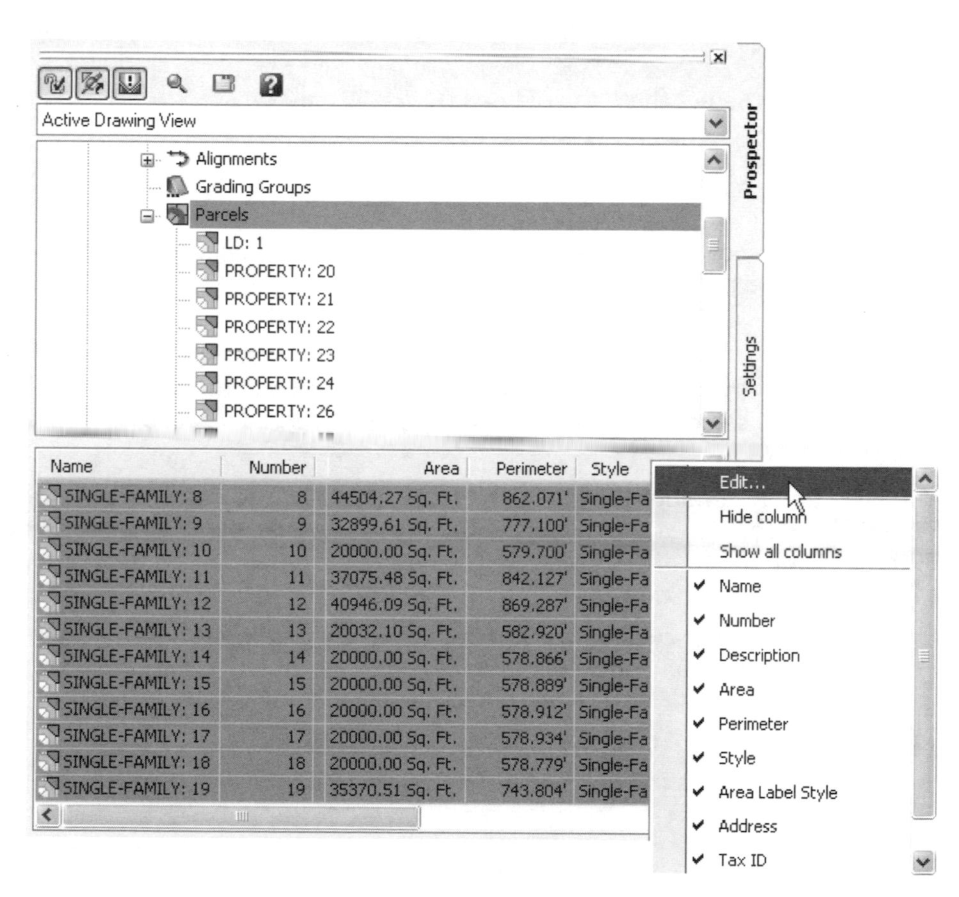

3. In the preview area of the *Prospector*, use the shift key to select all the parcels which use the *Single Family* style.

You may want to stretch out the *Toolspace* so that the preview area of the *Prospector* is wider to make it easier to see the table.

4. Right-click on header at the top of the **Style** column and select **Edit**.

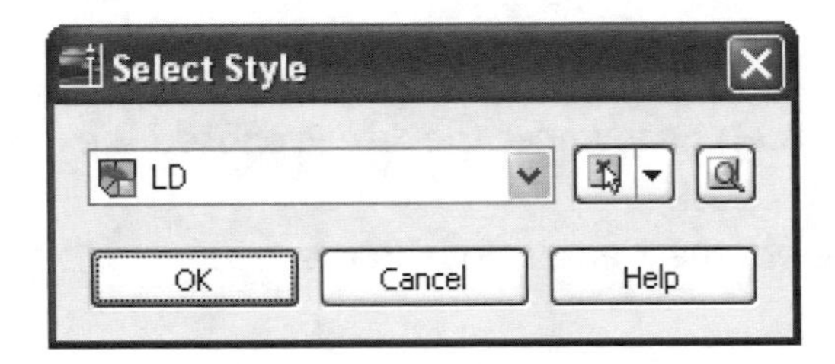

5. Select the *Parcel Style* **LD**.

6. Click **<<OK>>** to assign the style to the selected parcels.

The parcels are now displayed with a yellow, phantom boundary lines according to the properties of the parcel style. They are also renamed to start with *LD:*. This is based on the name template in the parcel style.

7. In the preview area of the *Prospector*, use the shift key to select all the parcels with a Name starting with *LD:*.

You will need to reselect the parcels because changing the *Parcel Style* renamed and reordered the parcels in the preview window changing the current selection set.

8. Right-click on the header at the top of the *Area Label Style* column and select **Edit**.

9. Select the *Parcel Label Style* **Name & Area in Feet.**

10. Click **<<OK>>** to assign the label style to the selected parcels.

The parcels are now displayed with a cyan label showing the parcel name and area in square feet according to the properties of the parcel label style.

11. Save the drawing.

6.7 Additional Exercises

Layout and size the parcels on the inside of the *L Street Alignment*. Since the alignment is U shaped you may not have good results if you define the entire *ROW* line as the parcel frontage. To avoid this potential problem first split the large parcel horizontally. Then subdivide the two parcels separately.

6.8 Chapter Summary

In this chapter you created and edited *Alignments* and *Parcels*. You also learned to use the *Parcel Layout Tools* to automatically subdivide and size parcels based on area and frontage criteria. You edited *Parcels* and explored how *Parcels* and *Alignments* in the same *Site* interact with each other. The *Alignments* you created in this chapter will be used in future chapters to create the *Profile*, *Corridor Model*, and *Sections*.

Chapter 7

Working with Profiles

Now that you have a horizontal alignment and a surface you are ready to create a profile of the existing ground along L Street. Creating a profile of an existing surface takes the horizontal information from the alignment and the vertical information from the surface and combines them to generate the profile.

Once a profile of the existing ground has been created you will layout and define a Finished Ground Profile. This process will be very similar to the way that you laid out and defined the horizontal alignment.

- **Creating an Existing Ground Profile**
- **Creating Profiles of Finished Ground**
- **Editing the Profile Graphically**
- **Editing the Profile in Grid View**

Dataset:

To start this chapter you will continue working in the drawing named **Design.dwg**. You can continue with the drawing that you currently have from the end of the previous chapter or, if you are starting in the middle of the book, you can open the drawing **CH-07.dwg** located in the folder **C:\Cadapult Training Data\Civil 3D 2007\Level 1\Chapter Drawings.** Opening the drawing from the dataset provided will ensure that you have the drawing set up correctly for the exercises in the following chapter and overwrite any mistakes that you may have made in previous exercises.

7.1 Overview

In this chapter you will use the surface and alignment data created in precious chapters to create an existing ground profile. You will create a *Profile View* to display the profile and explore how to control its display with styles. You will also layout and edit a finished ground profile that will be used in the next chapter to create the corridor model.

7.1.1 Concepts

The dynamic nature of the profile object in *Civil 3D* allows you to easily make changes to the geometry of the profile. This allows you to approach the design process differently than you may have in the past with other programs. Now you can set a default curve length or K value for the profile and lay out a rough draft in *Civil 3D*, then modify and revise it in real time either graphically or in a tabular format.

7.1.2 Styles

Styles are saved in the drawing and can be created and edited on the Settings tab of the *Toolspace* or on the fly during many object creation and editing commands.

Profile Style controls the display of the profile line.

Profile Label Styles control the profile label text. *Profile Label Styles* can be customized to display labels for stationing, grade breaks, curves, lines, and horizontal geometry points.

Profile Label Sets control the group of styles used to label the profile.

Profile View Style controls the display of *Profile View* which includes the grid, title, annotation, and vertical exaggeration.

Profile View Bands are bands of labels at the top or bottom of the profile. *Civil 3D* has the ability to create bands labeling station and elevation information as well as other information like the depth of cut or fill at the centerline, geometry information, and superelevation information.

7.2 Creating an Existing Ground Profile

The *Existing Ground Profile* is based on the *Surface* and *Alignment* that you created in the previous chapters. The *Profile* is an object and is dynamically linked to the *Alignment* and *Surface* so it will automatically update if there are any changes to either of them.

7.2.1 Sampling and Drawing the Profile

1. Continue working in the drawing **Design.dwg.**

2. Select **Profiles ⇒ Create From Surface.**

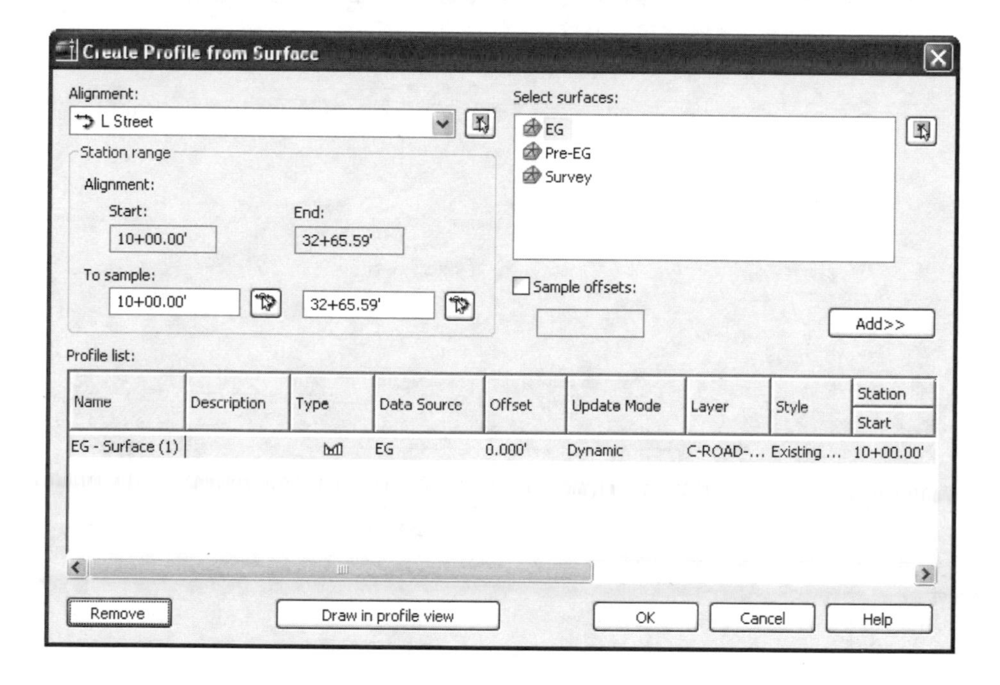

3. Select the **L Street Alignment.**

4. Select the *Surface* **EG.**

5. Click **<<Add>>.**

You can use the *Sample offsets* option to create offset profiles at the offset distance you enter in the field below. Multiple offsets can be entered in a comma delimited format. You can also set the layer that the profile will be drawn on and the style used to display the profile line in the table at the bottom of the dialog box.

6. Click **<<Draw in profile view>>**.

To display the *Profile* in the drawing you must draw it in a *Profile View*. If you just click <<OK>> the *Profile* will be sampled but not displayed in the drawing.

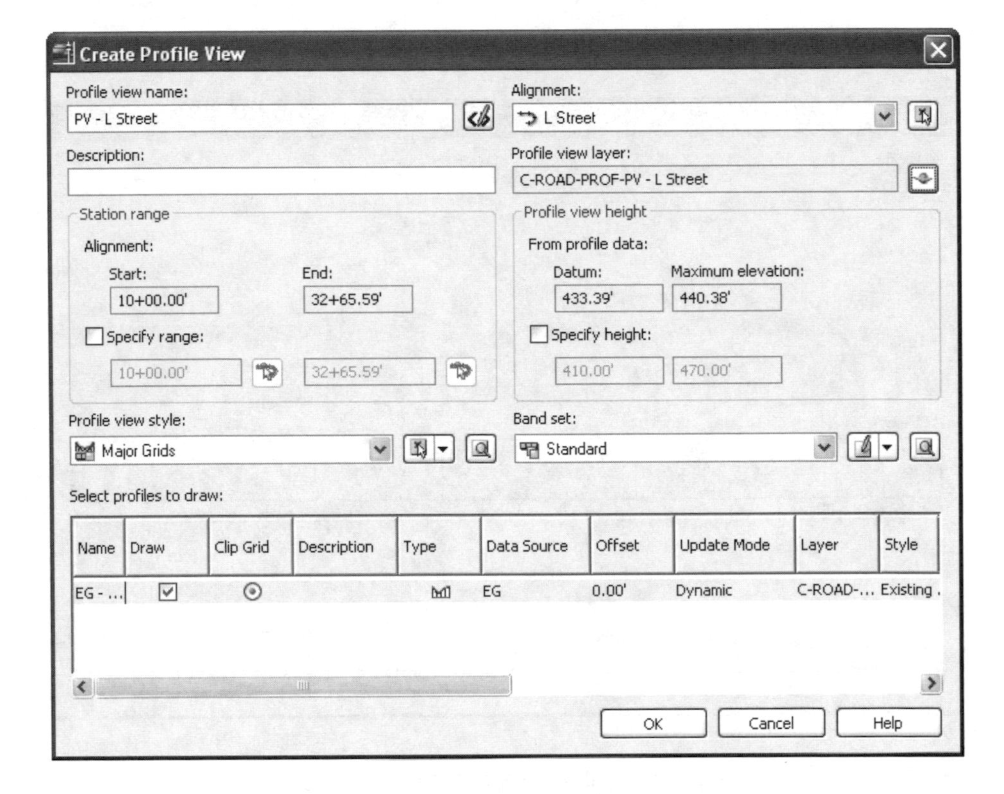

7. Enter a Profile view name of **PV-L Street.**

8. Click the **Profile view layer** button to open the *Object Layer* dialog box.

9. Set the **Modifier** to **Suffix**.

10. Enter **-*** as the **Modifier** *value*.

11. Click **<<OK>>** to close the *Object Layer* dialog box.

12. Back in the *Create Profile View* dialog box confirm that the **Profile view style** is set to **Major Grids**.

The *Profile View Style* controls the display of *Profile View* which includes the grid, title, annotation, and vertical exaggeration. You may want to create a *Profile View Style* for each vertical exaggeration that you typically use.

13. Confirm that the **Band set** is set to **Standard**.

Profile View Bands are bands of labels at the top or bottom of the profile. In Autodesk Land Desktop the standard profile grid was created with bands of labels for stationing and elevations at the bottom of the profile. *Civil 3D* has the ability to create bands labeling station and elevation information as well as other information like the depth of cut or fill at the centerline, geometry information, and superelevation information.

14. Click **<<OK>>**.

15. At the command line you will be asked for a *Starting Point* for the profile. You can either pick a point on the screen or type in an X,Y coordinate. For this exercise you will type in a coordinate. The starting point will define the lower left corner of the profile. At the command line enter **`1339100,890600`**.

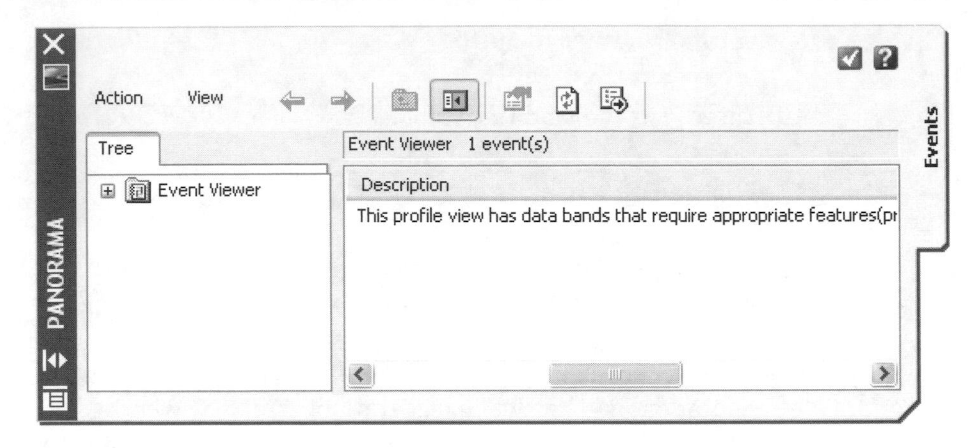

You will see an information message in the Event Viewer that notifies you that one of the Profile Bands requires information from a feature that has not yet been assigned to the band. In this case it is referring to the finished ground profile. After you create the finished ground profile you will need to assign it to the appropriate band to display the finished ground elevations.

16. Close the *Event Viewer* by clicking the **green check button.**

17. **Zoom Extents** to find and review the profile.

The vertical blue lines display the location of the beginning and end of the horizontal curves. The profile view is a single object and can be moved and erased with *AutoCAD* commands without damaging the profile data. The *Existing Ground Profile* is linked to the alignment. So if you edit the alignment the profile will automatically be updated. This allows you a tremendous amount of flexibility during the design process to make changes and try many alternatives.

7.2.2 Changing the Profile View Style

1. Select the **Profile View** by selecting the grid not the profile line.

2. Right-click and select ⇒ **Profile View Properties.**

You can also access the *Profile View Properties* through the *Prospector*.

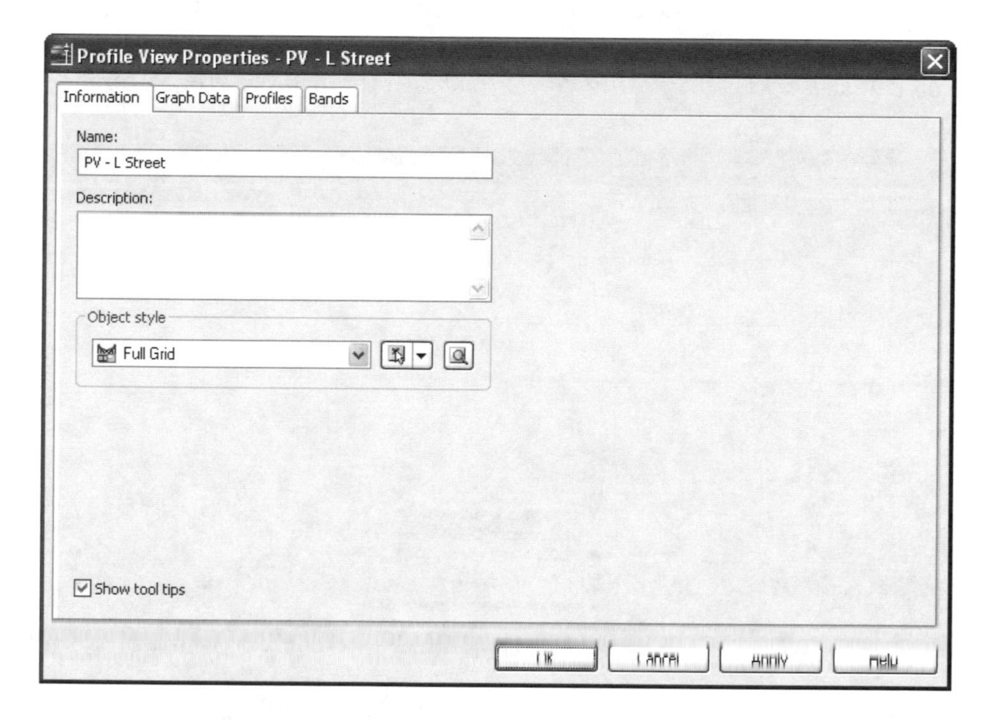

3. Set the **Object Style** to **Full Grid**.

4. Click <<**OK**>> to apply the new style.

The profile view grid spacing changes and the minor grid lines are added.

7.2.3 Creating a Profile View Style

In this exercise you will create a new *Profile View Style* that uses a vertical exaggeration of 5, along with major and minor grid lines at intervals of 50' and 10' horizontally and 10' and 2' vertically.

1. Select the **Profile View.**

2. Right-click and select ⇒ **Profile View Properties.**

You can also access the *Profile View Properties* through the *Prospector*.

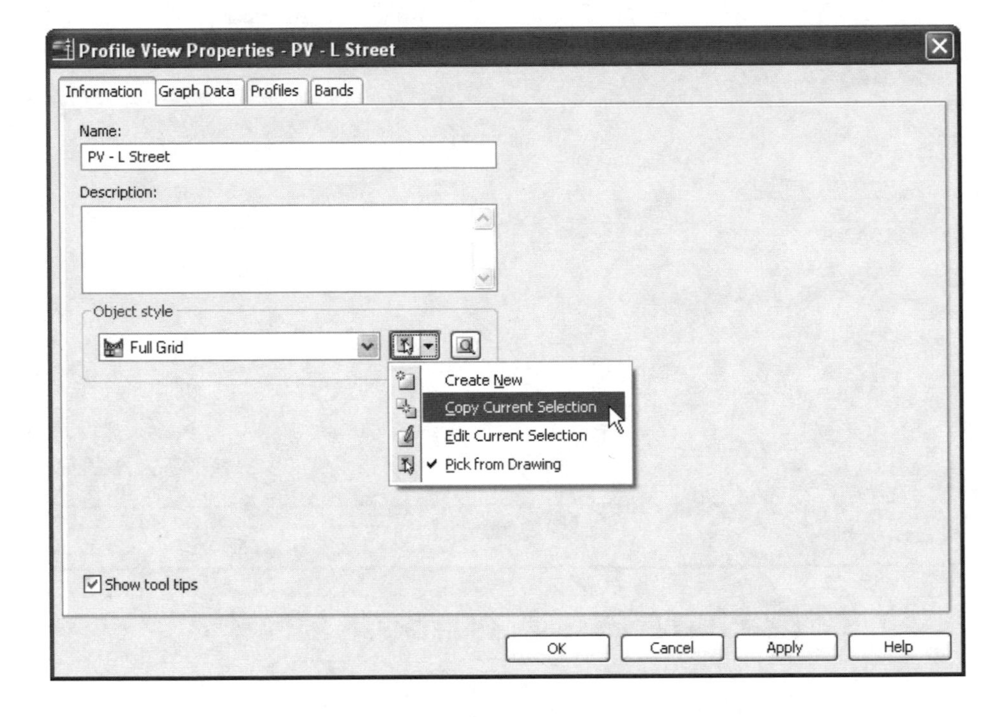

3. Click the **down arrow** button next to the **Object style** on the *Information* tab of the *Profile View Properties* dialog box to display the other creation commands.

4. Click the **Copy Current Selection** button.

This style can also be copied, edited, and created on the *Settings* tab of the *Toolspace* along with all other *Civil 3D Styles*.

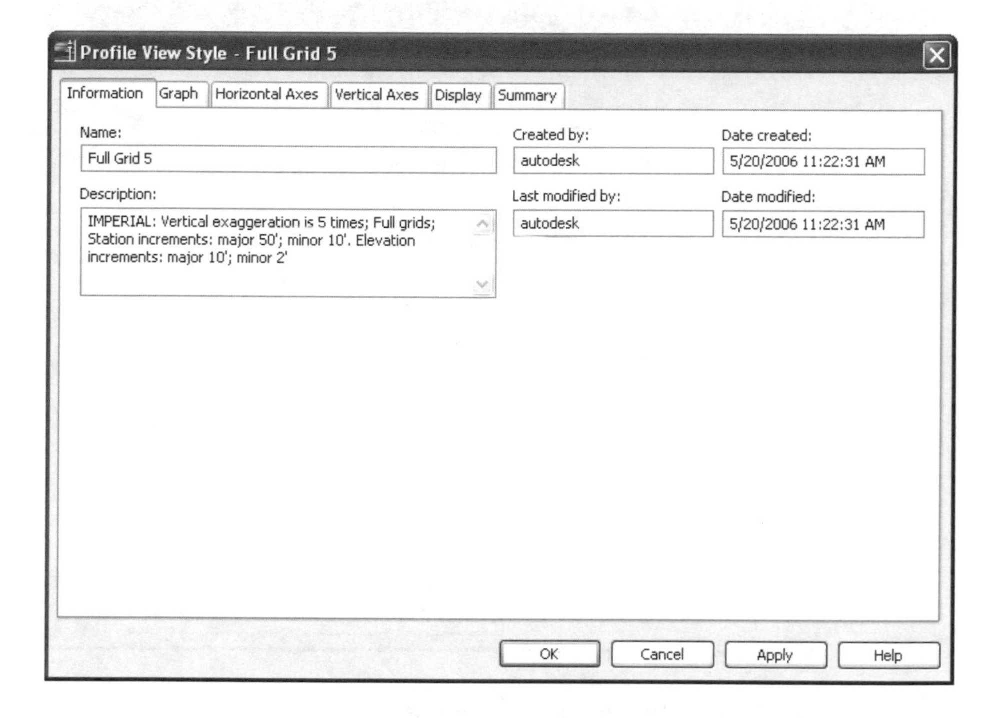

5. Name the new style "**Full Grid 5**".

6. Enter a description of "**IMPERIAL: Vertical exaggeration is 5 times; Full grids; Station increments: major 50'; minor 10'. Elevation increments: major 10'; minor 2'**".

The description is optional so you can be as detailed as you like, or skip it completely.

7. Select the **Graph** tab of the *Profile View Style* dialog box.

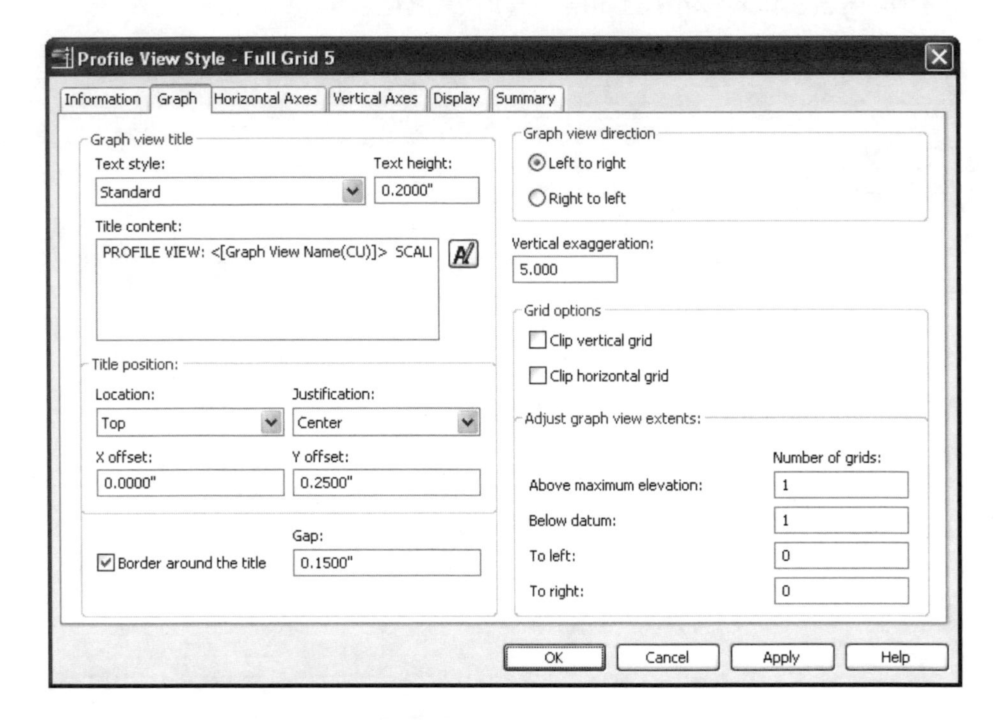

8. Change the **Vertical exaggeration** to **5**.

9. Select the **Horizontal Axes** tab of the *Profile View Style* dialog box.

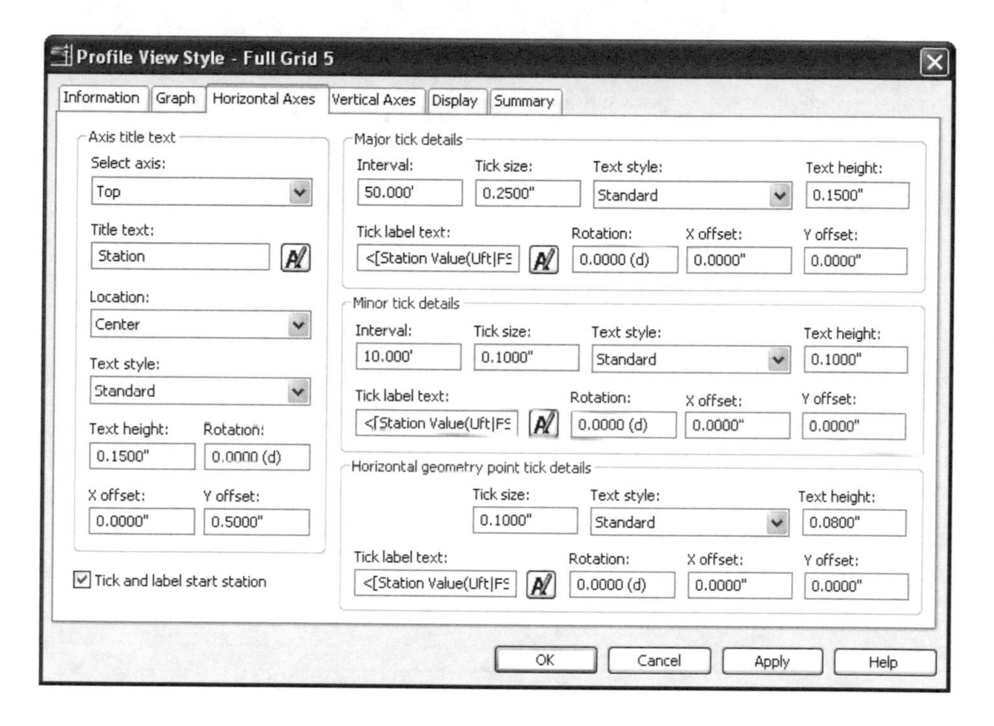

10. Confirm that the **Select Axis** option is set to **Top**.

11. Set the **Major tick details Interval** to **50**.

12. Set the **Minor tick details Interval** to **10**.

13. Set the **Select Axis** option to **Bottom**.

14. Set the **Major tick details Interval** to **50**.

15. Set the **Minor tick details Interval** to **10**.

The tick interval controls the grid spacing as well as the spacing of the labels.

16. Select the **Vertical Axes** tab of the *Profile View Style* dialog box.

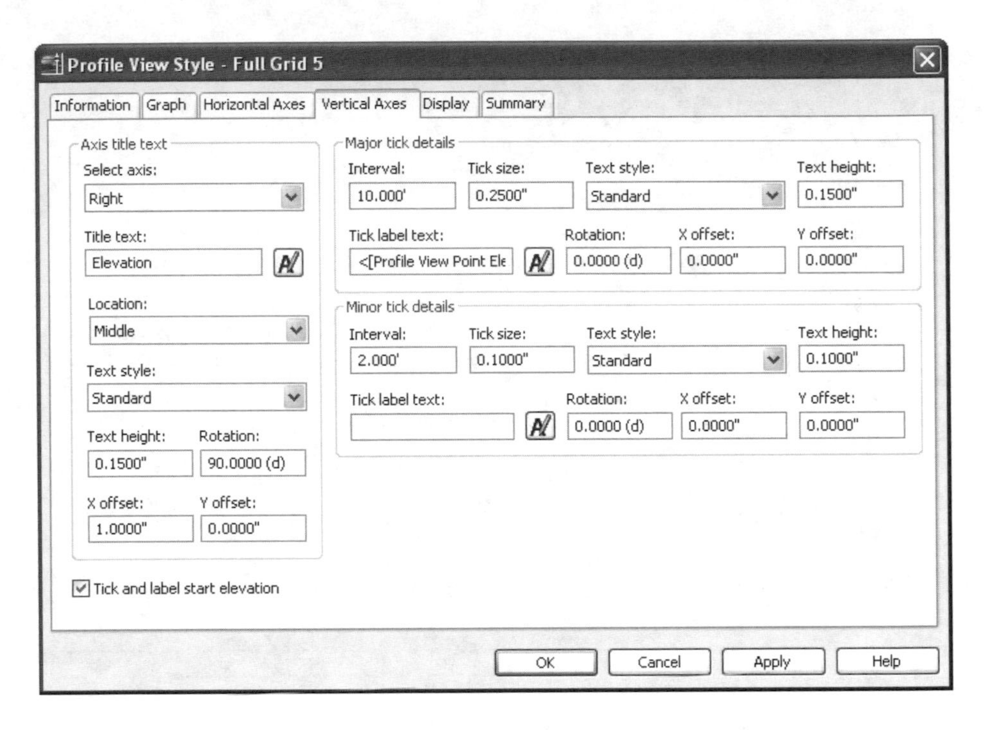

17. Confirm that the **Select Axis** option is set to **Right.**

18. Confirm that the **Major tick details Interval** is set to **10**.

19. Set the **Minor tick details Interval** to **2.**

20. Set the **Select Axis** option to **Left.**

21. Confirm that the **Major tick details Interval** is set to **10**.

22. Set the **Minor tick details Interval** to **2.**

The tick interval controls the grid spacing as well as the spacing of the labels.

23. Select the **Display** tab of the *Profile View Style* dialog box.

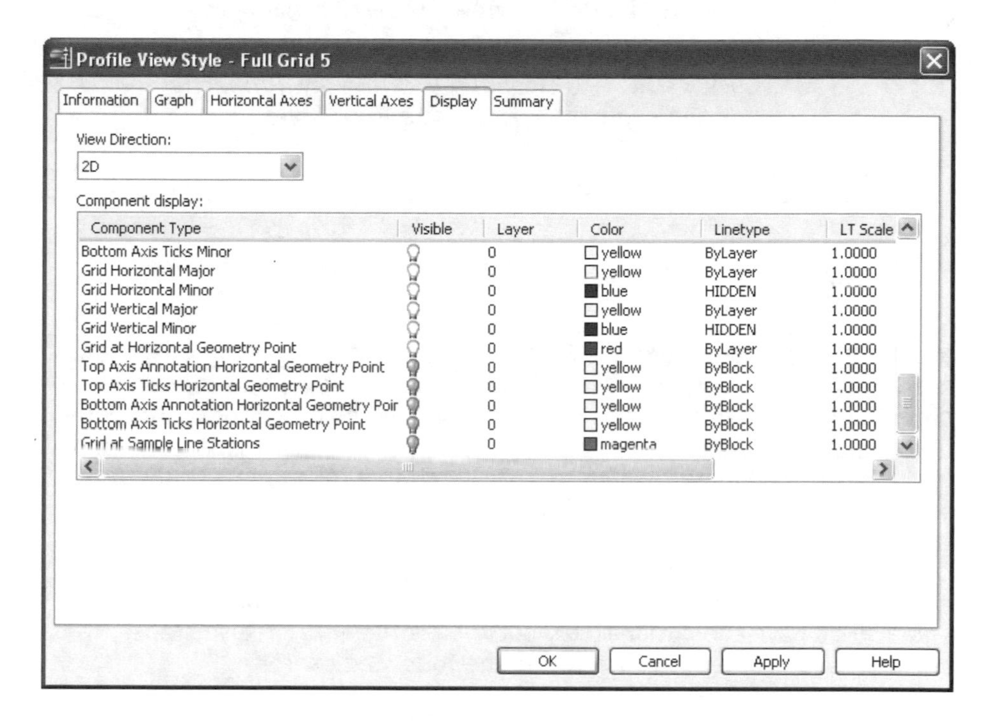

24. Confirm that the **View Direction** is set to **2D**.

25. Set the color of the **Grid Horizontal Major** component to **Yellow**.

26. Set the color of the **Grid Horizontal Minor** component to **blue** and the linetype to **HIDDEN**.

27. Set the color of the **Grid Vertical Major** component to **Yellow**.

28. Set the color of the **Grid Vertical Minor** component to **blue** and the linetype to **HIDDEN**.

29. Turn on the **Grid Horizontal Geometry Point** component and set its color to **red**.

30. Click **<<OK>>** to save the new **Profile View Style** and return to the *Profile View Properties* dialog box.

31. Click **<<OK>>** to close the *Profile View Properties* dialog box and apply the new *Profile View Style*.

32. Save the drawing.

7.3 Creating Profiles of Finished Ground

Many of the *Profile Layout* and *Editing* commands have a similar look and feel to the *Alignment Layout* and *Editing* tools. This type of interactive object allows you to approach profile design and modifications differently than you may be used to with other programs like *Land Desktop*. You may find it faster and more flexible to just sketch in a rough draft of the profile with the profile layout tools and then refine it either graphically or in a tabular format. You don't have to lay out the profile perfectly the first time, this method also allows you to look at many different alternatives quickly and even undo if you want to go back to the previous iteration. You also do have the option to lay out the profile with precise grades, elevations, and stations as we will in the following exercises.

7.3.1 Constructing the Finished Ground Centerline

In this exercise you will lay out the *Finished Grade Profile* using the transparent commands to enter information by *Grade and Stations* as well as by *Station and Elevation*. If you have difficulty with the transparent commands or if you would rather just sketch in the profile, you can do that as well and then use the *Profile Editing* commands to match the geometry of the exercise. *Civil 3D* is very flexible this way and you should take advantage of that to work the way that you are most comfortable.

1. Confirm that **Dynamic Input** is turned off. If it is on Click the DYN button on the *Status Bar* to disable it.

SNAP GRID ORTHO POLAR OSNAP OTRACK DUCS DYN LWT MODEL

Using *Dynamic Input* with the *Transparent Commands* can be confusing and cause unexpected results.

2. Select **Profiles ⇒ Create by Layout.**

3. Select the **L Street Profile View.**

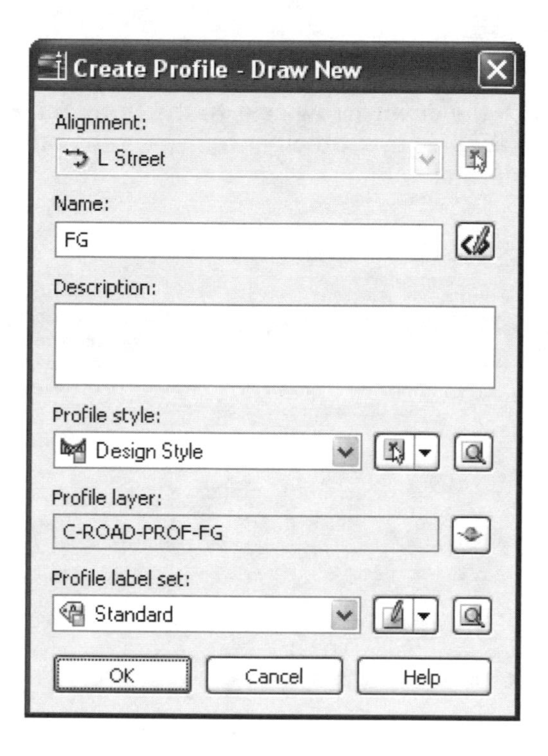

4. Name the Profile **FG.**

5. Click the **Profile layer** button to open the *Object Layer* dialog box.

6. Set the **Modifier** to **Suffix.**

7. Enter **-*** as the **Modifier** *value.*

8. Click **<<OK>> to close the** *Object Layer* dialog box.

9. Click **<<OK>>** to save the Profile and open the *Profile Layout Tools* toolbar.

10. Click the **down arrow** next to the *Draw Tangents button* on the *Profile Layout* toolbar to display the other creation commands.

11. Click the **Curve Settings** button.

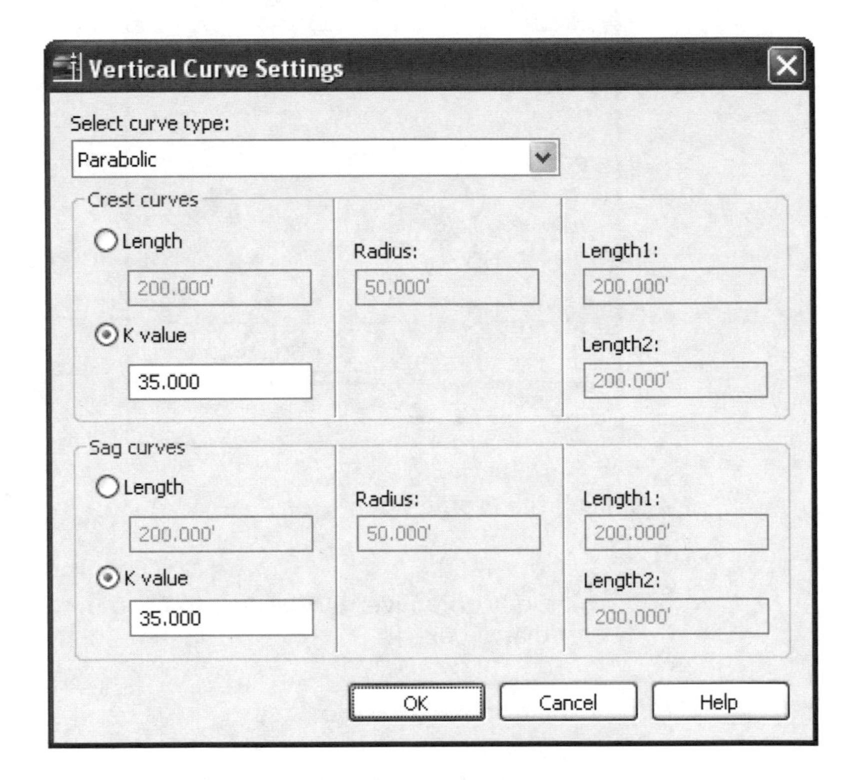

12. Set the **Curve Type** to **Parabolic**.

13. Set the **K value** to **35** for both the *Crest* and *Sag* curves.

14. Click **<<OK>>**.

15. Click the **Draw Tangents With Curves** button.

16. Use an **Endpoint** OSNAP to snap to the beginning of the existing ground profile.

If you cancel the command without creating the first segment of the profile you will need to delete the *FG* profile from the *Prospector* and start the exercise again.

17. At the command line enter `'PGS` to change the prompt to specify a **`Grade and Station`** on the Profile.

The transparent profile commands can also be issued from the *Transparent Commands* toolbar.

18. Select the Profile View when prompted.

19. Enter a **Grade** of **-3**.

20. Enter a **Station** of **1090**.

21. **<<Escape>>** to end the *Grade and Station* prompt.

22. Enter `'PSE` to change the prompt to specify a **`Station and Elevation`** on the Profile.

23. Enter a **Station** of **2040**.

24. Enter an **Elevation** of **436**.

25. Enter a **Station** of **2300**.

26. Enter an **Elevation** of **438**.

27. Enter a **Station** of **2900**.

28. Enter an **Elevation** of **434**.

29. Enter a **Station** of **3140**.

30. Enter an **Elevation** of **435**.

31. **<<Escape>>** to end the `Station and Elevation` prompt.

32. Use an **Endpoint** Osnap and snap to the end of the existing ground centerline.

33. **<<Enter>>** to end the command.

34. Save the drawing.

7.3.2 Editing the Profile Graphically

Civil 3D Profiles can be grip edited just like basic *AutoCAD* objects.

1. Select the **Finished Ground** profile to display the grips.

2. Zoom in to **PVI number two** at the location of the first vertical curve.

You will see several different types of grips. The circular grip allows you to drag the curve changing its length and K value. The triangular grip pointing up allows you to drag the *PVI* to a new location while maintaining the curve properties. The triangular grips extending from the incoming and outgoing tangents allow you to slide the PVI along the existing grade. This is a very powerful and useful option. All of the graphical edits will prevent you from editing the profile in a way that would create overlapping vertical curves.

3. Select the **triangular grip** extending from the first tangent in the profile.

4. Drag it along the grade to a new location of your choice.

5. When you are finished grip editing the profile, press **<<Escape>>** to clear the grips.

7.3.3 Editing the Profile in Grid View

1. Select **Profiles ⇒ Edit Profile Geometry**.

2. Select the **Finished Ground** profile for the L Street alignment to open the *Profile Layout Tools* toolbar.

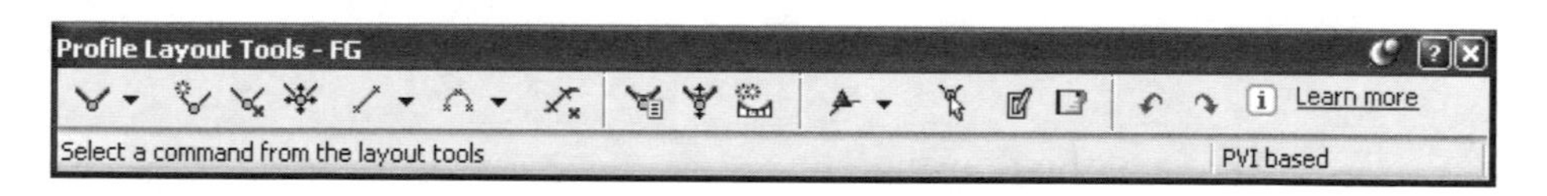

3. Click the **Grid View** button.

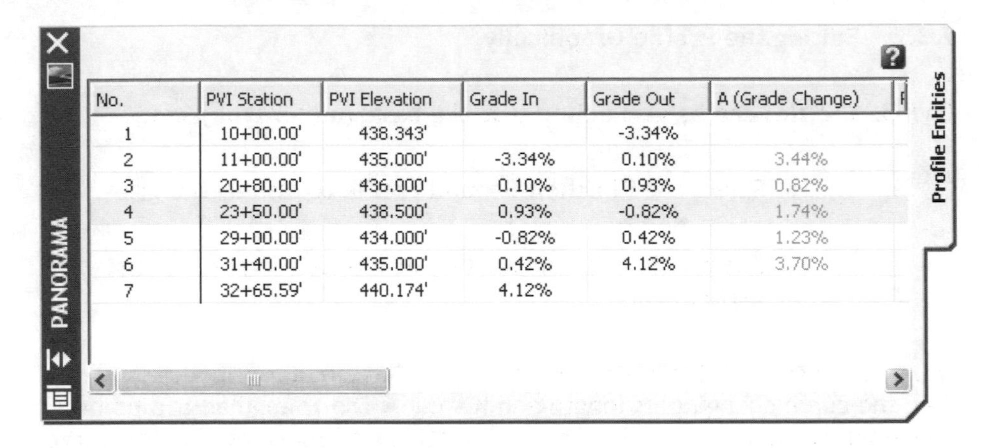

No.	PVI Station	PVI Elevation	Grade In	Grade Out	A (Grade Change)
1	10+00.00'	438.343'		-3.34%	
2	11+00.00'	435.000'	-3.34%	0.10%	3.44%
3	20+80.00'	436.000'	0.10%	0.93%	0.82%
4	23+50.00'	438.500'	0.93%	-0.82%	1.74%
5	29+00.00'	434.000'	-0.82%	0.42%	1.23%
6	31+40.00'	435.000'	0.42%	4.12%	3.70%
7	32+65.59'	440.174'	4.12%		

4. Change the **Elevation PVI #2** to **435**.

The profile will update graphically after each change. Also, you are not allowed to make a change that does not fit the geometry of the profile. So you can not create overlapping curves.

5. Change the **Station PVI #2** to **1100**.

6. Change **the Station PVI #3** to **2080**.

7. Change the **Station PVI #4** to **2350**.

8. Change the **Elevation PVI #4** to **438.50**.

9. Confirm the remaining PVI Stations and Elevations match the table above.

10. Close the **Grid View**.

11. Close the **Profile Layout** toolbar.

12. Save the drawing.

7.3.4 Working with Profile Labels

Profile labels are controlled by the profile label styles. These styles were originally selected when you selected the profile label set in the *Profile - Create By Layout* command. In this exercise you will edit the Profile Curve Label style to add a background mask.

1. Select the **Finished Ground Profile Line**.

2. Right-click and select ⇒ **Edit Labels**

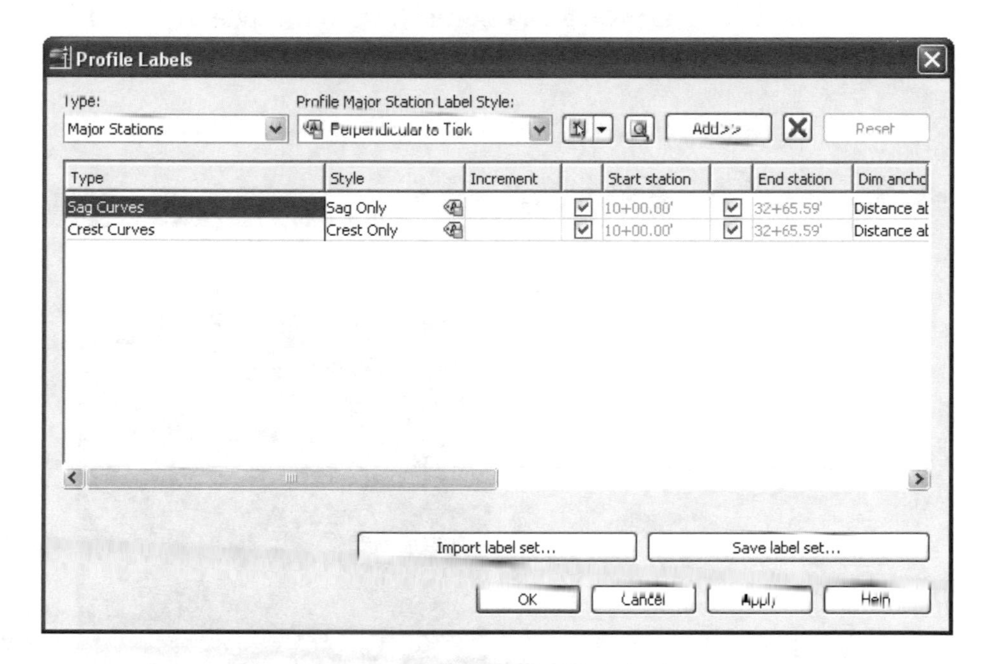

3. Click the symbol in the *Style* field for the **Sag Curves** to display the *Pick Label Style* dialog box.

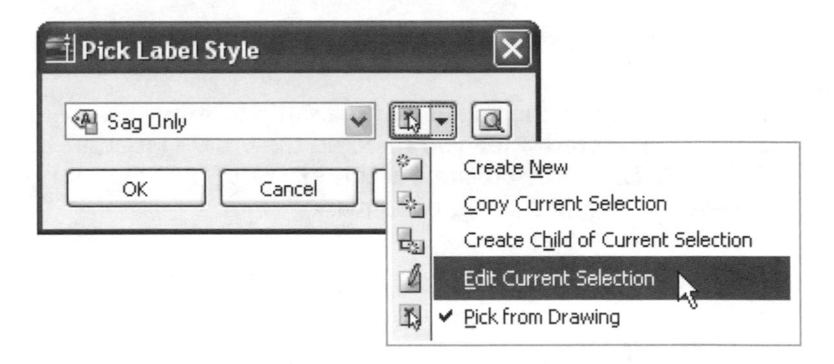

4. Click the **down arrow** button next to the label style and select the **Edit Current Selection** button.

5. Select the **Layout** tab of the *Label Style Composer*.

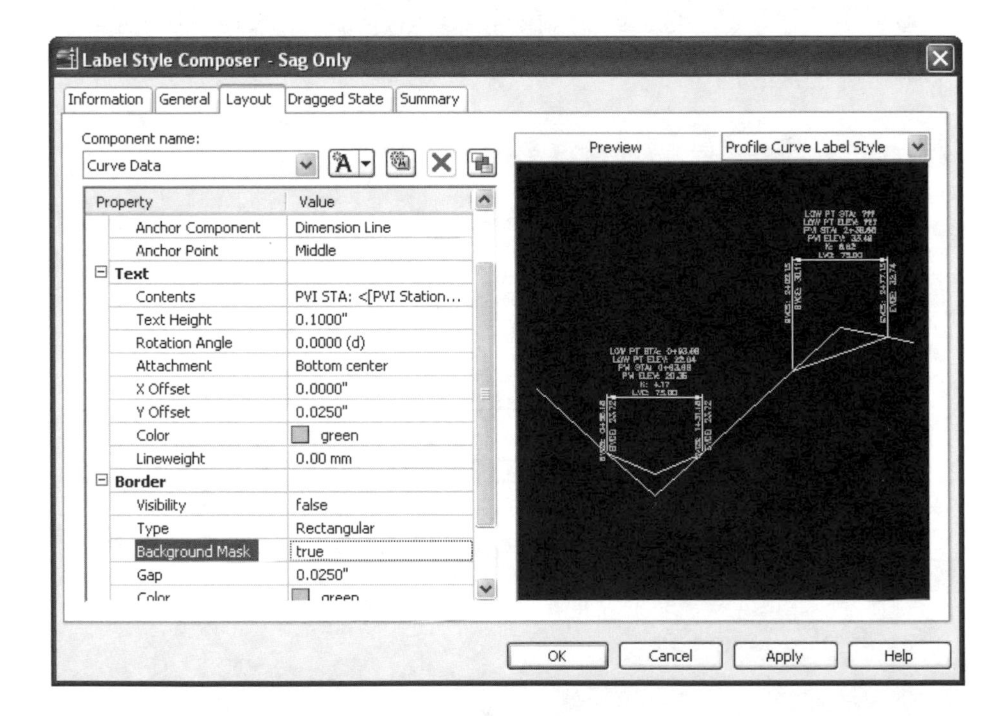

6. Confirm that the **Component name** is set to **Curve Data**.

7. Set the **Background Mask** to **true**.

8. Set the **Component name** to **Sag Data**.

9. Set the **Background Mask** to **true**.

10. Click <<**OK**>> to save the changes to the label style.

11. Click <<**OK**>> to return to the *Profile Labels* dialog box.

12. Repeat the process to edit the *Crest Curves* label style and enable the **Background Mask** for the **Curve Data** and **Crest Data** components.

The profile curve labels will now display with a background mask making them easier to read.

7.3.5 Working with Profile View Bands

1. Zoom in to the bottom of the profile and examine the band displaying the two elevation values. These are for the EG and FG profiles. However, currently they are both displaying the same numbers.

2. Select the **Profile View**.

3. Right-click and select ⇒ **Profile View Properties**.

You can also access the *Profile View Properties* through the *Prospector*.

4. Select the **Bands** tab of the *Profile View Properties* dialog box.

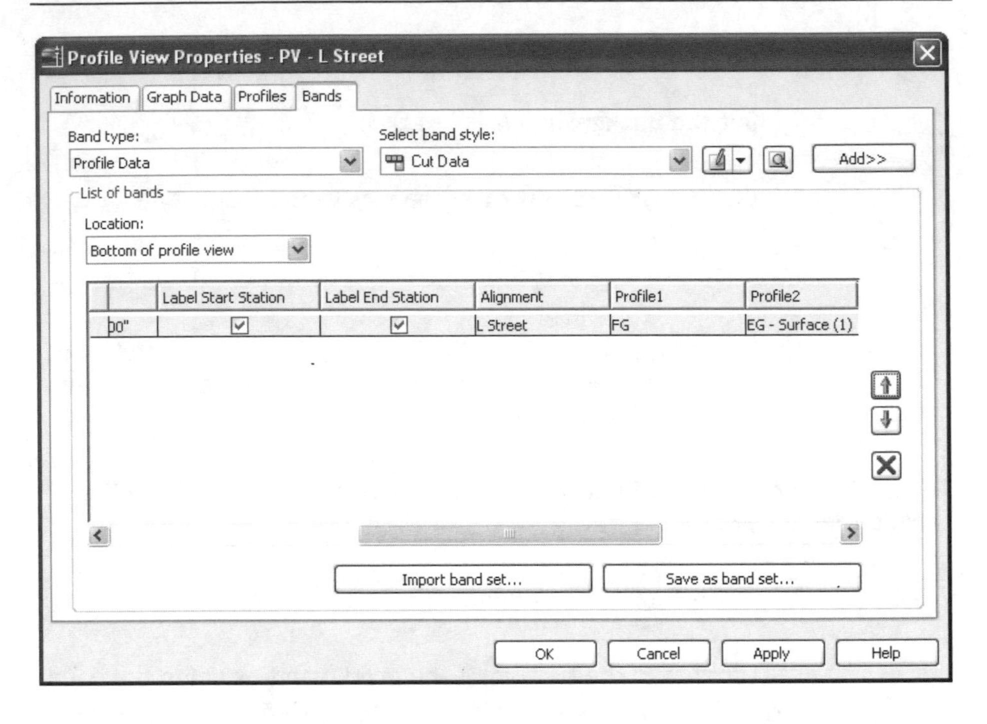

5. Scroll to the right to find the **Label Start Station** setting for the default band.

6. Enable the **Label Start Station** option.

7. Enable the **Label End Station** option.

8. Set **Profile1** to **FG**.

This band displays the elevations of the EG and FG profiles. By default both profile settings for the band are set to EG. This is the setting that you were warned about in the *Event Viewer* when you created the *Profile View*. If you do not change this setting the band will display the elevations from the EG profile in both labels.

9. Click **<<OK>>** to update the band.

10. Zoom in to the bottom of the profile and examine the band to see the changes.

7.3.6 Adding Profile View Bands

Profile View Bands can also display other information like the depth of cut or fill at the centerline, geometry information, and superelevation information.

1. Select the **Profile View**.

2. Right-click and select ⇒ **Profile View Properties.**

3. Select the **Bands** tab of the *Profile View Properties* dialog box.

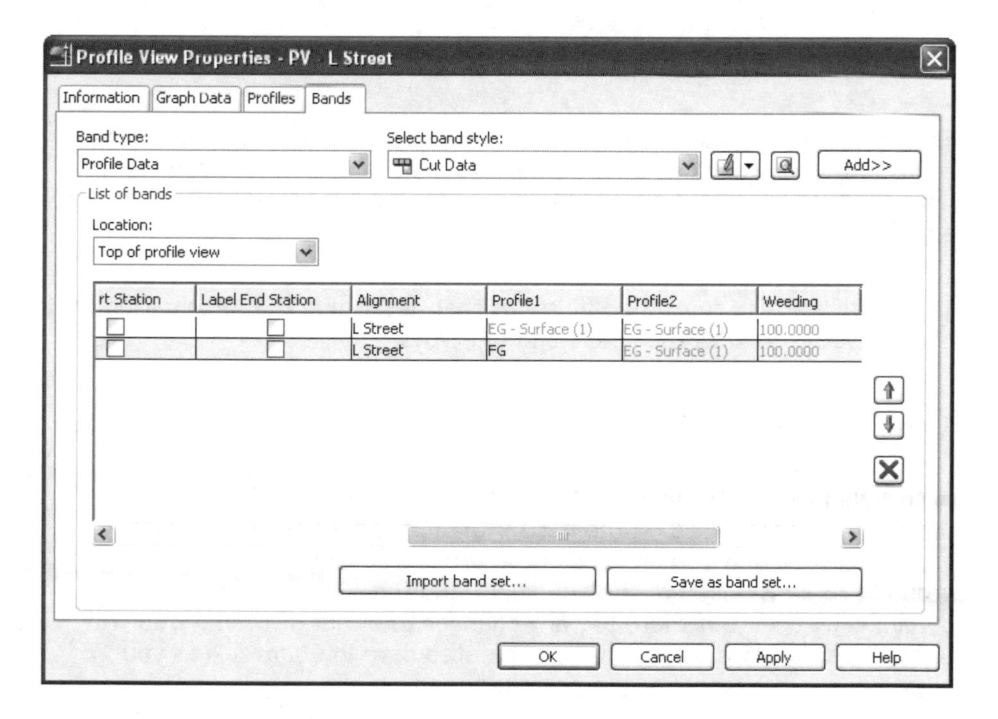

4. Set the **Location** to **Top of profile view.**

5. Set the **Band type** to **Horizontal Geometry**.

6. Set the **Select band style** setting to **Geometry.**

7. Click <<**Add**>>.

8. Set the **Band type** to **Vertical Geometry**.

9. Set the **Select band style** setting to **Geometry**.

10. Click <<**Add**>>.

11. Scroll to the right to find the **Profile1** setting for the *Vertical Geometry* band.

12. Set **Profile1** to **FG**.

13. Click <<**OK**>>.

The profile view is now displayed with the original band now showing the elevations of the existing and finished ground profiles. There are also two new bands at the top of the profile view showing the horizontal geometry and vertical geometry information.

The Bands are also controlled by styles that determine the information that is displayed as well as the colors and linetypes.

7.4 Chapter Summary

In this chapter you have created both an existing and finished ground profile. They are based on the horizontal alignment and the existing ground surface that you created in the previous chapters. The existing ground profile is linked to the alignment and the surface so it will dynamically update if there are any changes to either of them. You will use the profile information that you created here in *Chapter 8* as you create a corridor model of the proposed road.

Chapter 8

Corridor Modeling

In this chapter you will create an assembly to define the components and geometry of a road. Then you will combine the assembly with, an alignment, profile, and surface that you created in previous chapters to create a corridor model. The corridor model is used to create surfaces and sections that you will plot as well as use for quantity takeoffs.

- **Working with Assemblies**
- **Working with Corridors**
- **Working with Sections**
- **Creating a Surface Showing Final Site Conditions**

Dataset:

To start this chapter you will continue working in the drawing named **Design.dwg**. You can continue with the drawing that you currently have from the end of the previous chapter or, if you are starting in the middle of the book, you can open the drawing **CH-08.dwg** located in the folder **C:\Cadapult Training Data\Civil 3D 2007\Level 1\Chapter Drawings.** Opening the drawing from the dataset provided will ensure that you have the drawing set up correctly for the exercises in the following chapter and overwrite any mistakes that you may have made in previous exercises.

8.1 Overview

In this chapter you will create an assembly to define the components and geometry of a road. Then you will combine the assembly with, an alignment, profile, and surface that you created in previous chapters to create a corridor model. The corridor model is used to create surfaces and sections that you will plot as well as use for quantity takeoffs. You will conclude the chapter by exporting a corridor surface and merging it with a copy of the existing ground surface to create a surface that represents the final site conditions after the road is built.

8.1.1 Concepts

The *Corridor* model combines and dynamically links the assembly, alignment, profile, and surface. With all the components of the road design dynamically linked together in the *Corridor* object you have the flexibility to experiment with many design alternatives, review the way that they impact the model, and even undo the changes to try something else.

8.1.2 Styles

Styles are saved in the drawing and can be created and edited on the Settings tab of the *Toolspace* or on the fly during many object creation and editing commands.

Section Style controls the display of the section line.

Section Label Styles control the profile label text. *Section Label Styles* can be customized to display labels for offsets, segments, and grade breaks.

Section Label Sets control the group of styles used to label the section.

Section View Style controls the display of *Section View* which includes the grid, title, annotation, and vertical exaggeration.

Section View Bands are bands of labels at the top or bottom of the section.

8.2 Working with Assemblies

An *Assembly* is a collection of *Subassemblies* that make up the components of a road or any other design based on an *Alignment*. *Assemblies* are combined with *Alignments* and *Profiles* to create a *Corridor* model of your design.

8.2.1 Subassembly Catalogs

Civil 3D is packaged with several catalogs of *Subassemblies* that you can configure to create *Assemblies* for your design. In this exercise you will browse through some of those catalogs.

1. Continue working in the drawing **Design.dwg**.

2. Select **General** ⇒ **Catalog** to open your catalog window with a list of available catalogs in your catalog library.

3. Select **Corridor Modeling Catalogs (Imperial)** to display the list of catalogs.

C3D Imperial Channel and Retaining Wall Subassembly Catalog
This catalog contains

C3D Imperial Generic Subassembly Catalog
This catalog contains Generic subassemblies for corridor

C3D Imperial Getting Started Subassembly Catalog
This catalog contains the

C3D Imperial Rehab Subassembly Catalog
This catalog contains subassemblies for higway

C3D Imperial Subdivision Roads Subassembly Catalog
This catalog contains all

C3D Imperial Transportation Design Subassembly Catalog
This catalog contains all

4. Select any of these catalogs and browse through the lists of available subassemblies.

You can create *Tool Palettes* from the subassemblies in the catalog.

5. Close the *Content Browser* when finished.

8.2.2 Creating an Assembly

In this exercise you will create a basic *Assembly* using the *Subassemblies* from the *Imperial -Basic* tool palette. You can create much more complex *Assemblies* using the other *Subassembly* catalogs after you get comfortable with the process.

1. Select **Corridors ⇒ Create Assembly.**

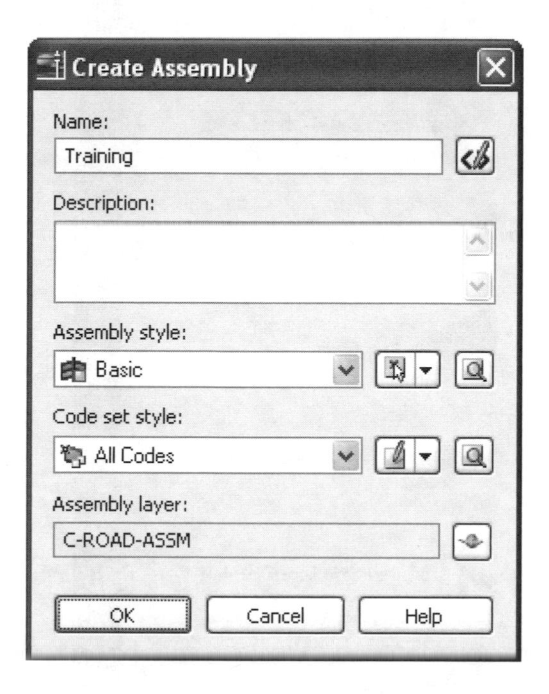

2. Name the *Assembly* "**Training**".

3. Click **<<OK>>** to create the *Assembly* using the default styles and layer.

4. Select a location for the assembly in a blank area of the drawing.

5. Zoom in to the *Assembly*.

If you have trouble finding the *Assembly* you can select it in the *Prospector* and use the `Zoom to` command. You may need to refresh the *Prospector* if it does not display the *Assembly*.

6. Select **General ⇒ Tool Palettes Window.**

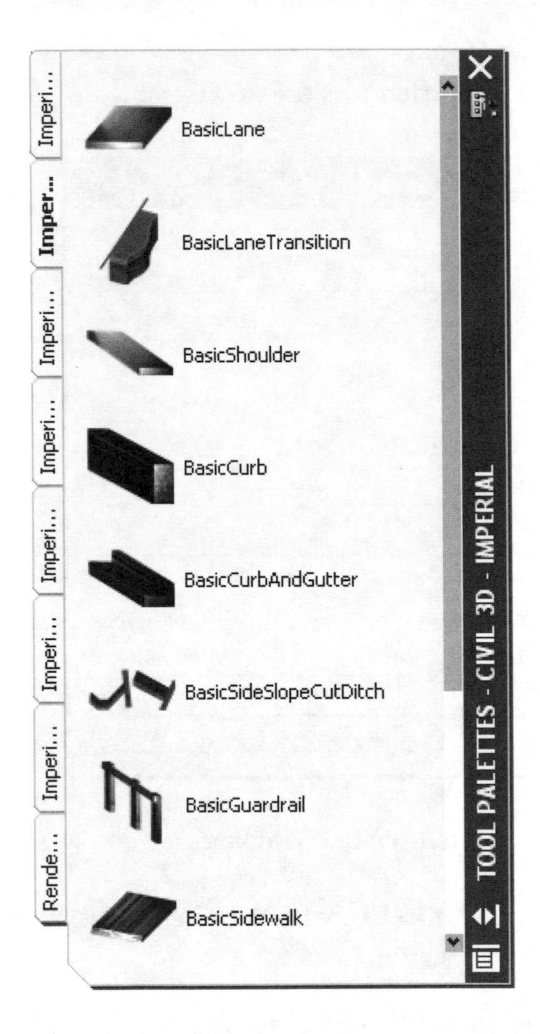

7. Select the **Imperial - Basic** tab.

8. Click the **Basic Lane** tool to open the *Properties* window displaying the *Parameters* that you can configure for the *Basic Lane*.

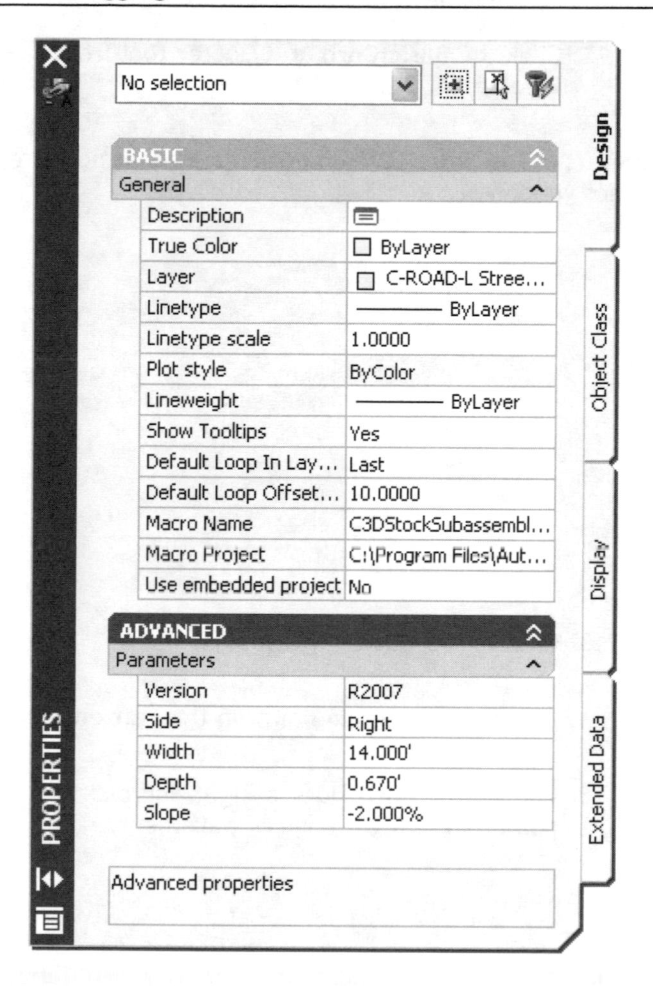

9. Set the **Lane Width** to **14.**

10. Pick the vertical line on the *Assembly* to add the *Subassembly* to the *Assembly*.

11. Change the **Lane Side** to **Left** in the *Properties* window and pick the vertical line on the assembly again.

12. Click the **Basic Curb and Gutter** tool from the **Tool Palette**.

The parameters of the *Basic Curb and Gutter* subassembly are displayed in the Properties window.

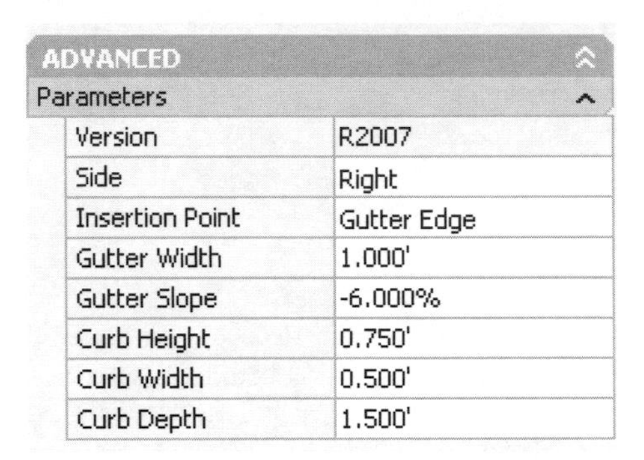

ADVANCED

Parameters

Version	R2007
Side	Right
Insertion Point	Gutter Edge
Gutter Width	1.000'
Gutter Slope	-6.000%
Curb Height	0.750'
Curb Width	0.500'
Curb Depth	1.500'

13. Change the **Gutter width** to **1**.

14. Pick the connection point on the Left edge of pavement.

15. Change the **Curb Side** to **Right** and pick the connection point on the Right edge of pavement.

16. Click the **Basic Sidewalk** tool.

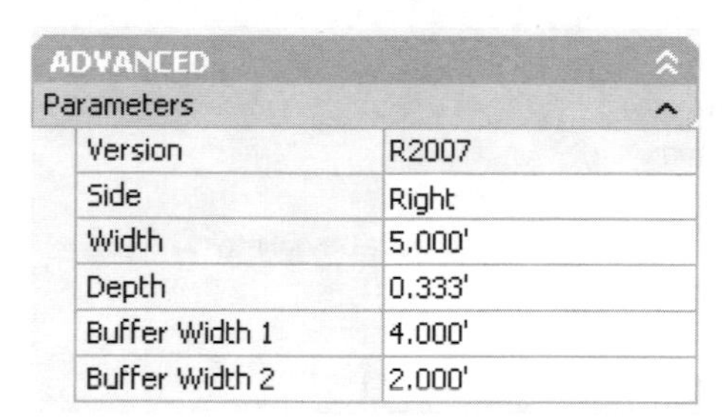

17. Set the sidewalk **Width** to **5.**

18. Set the **Buffer Width 1** to **4**.

19. Set the **Buffer Width 2** to **2**.

20. Pick the connection point on the Right back of curb.

21. Change the **Sidewalk Side** to **Left** and pick the connection point on the Left back of curb.

22. Click the **Basic Side Slope Cut Ditch** tool.

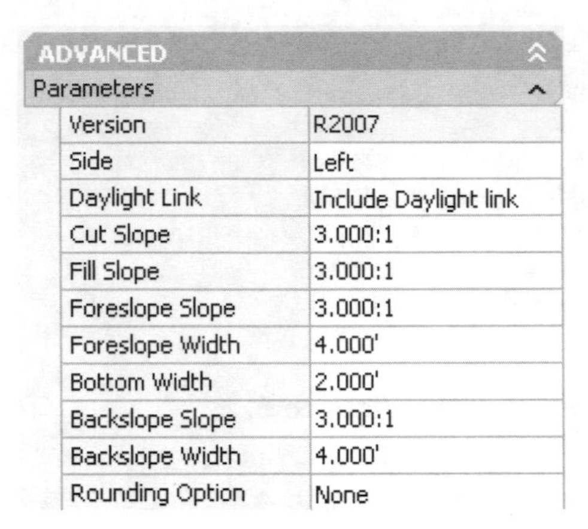

ADVANCED

Parameters

Version	R2007
Side	Left
Daylight Link	Include Daylight link
Cut Slope	3.000:1
Fill Slope	3.000:1
Foreslope Slope	3.000:1
Foreslope Width	4.000'
Bottom Width	2.000'
Backslope Slope	3.000:1
Backslope Width	4.000'
Rounding Option	None

23. Set the slopes to **3:1** by entering **3** in the fields.

24. Pick the connection point on the Left back of sidewalk.

25. Change the **Daylight Side** to **Right** and pick the connection point on the right back of sidewalk.

26. **<<Enter>>** twice to end the command.

The *Subassemblies* have been added to your *Assembly* and configured according to the parameters you entered in the *Properties* dialog box. You can modify any of those parameters by right-clicking on the *Assembly* in the *Prospector* and selecting ⇒ *Properties*.

27. Close the Tool Palette.

Be sure not to erase the Assembly or you will lose it. You may want to freeze the layer so that it is not accidentally selected and deleted.

8.3 Working with Corridors

The Corridor dynamically links the Alignment, Profile, Surface, and Assembly together to model your design.

8.3.1 Creating a Corridor

1. Select **Corridors ⇒ Create Corridor**.

2. When asked to `Select a baseline alignment` at the command line, **<<Enter>>** to select it from a list.

You also have the option to pick it graphically from the screen.

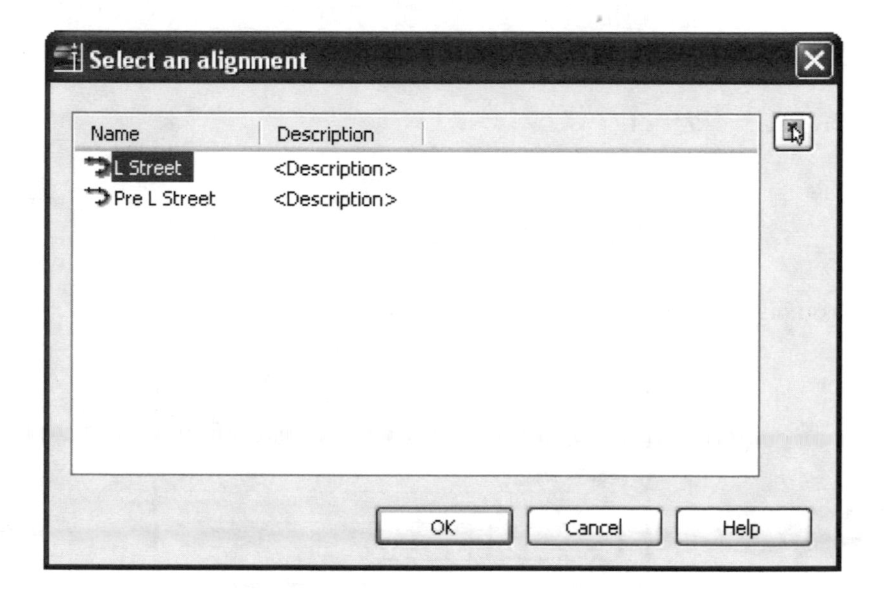

3. Select the *Alignment* **L Street.**

4. Click **<<OK>>**.

5. When asked to `Select a profile` at the command line, **<<Enter>>** to select it from a list.

You also have the option to pick it graphically from the screen.

6. Select the *Profile* **FG**.

7. Click **<<OK>>**.

8. When asked to `Select an assembly` at the command line, **<<Enter>>** to select it from a list.

You also have the option to pick it graphically from the screen.

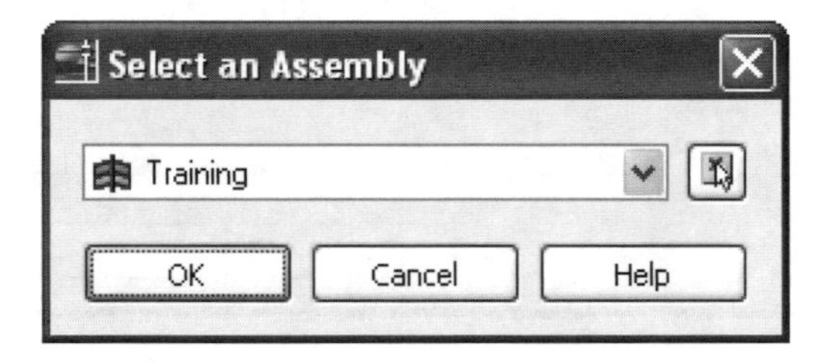

9. Select the *Assembly* **Training**.

10. Click **<<OK>>**.

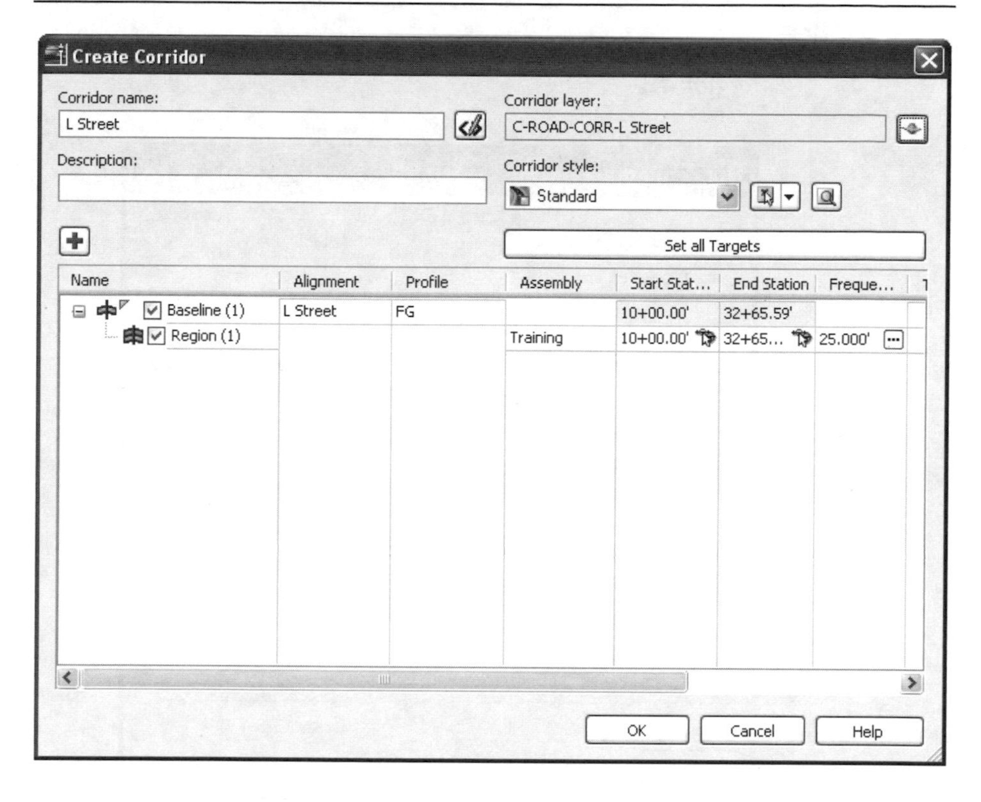

11. Name the corridor "**L Street**".

12. Click the **Profile view layer** button to open the *Object Layer* dialog box.

13. Set the **Modifier** to **Suffix**.

14. Enter **-*** as the **Modifier** *value*.

15. Click **<<OK>>** to close the *Object Layer* dialog box.

The *Corridor* model is based on *Sections* that it samples and updates as you make changes to the design. You can control the frequency of these *Sections* when you create the *Corridor*.

16. Click the ellipses <<...>> button in the **Frequency** field of *Region 1*.

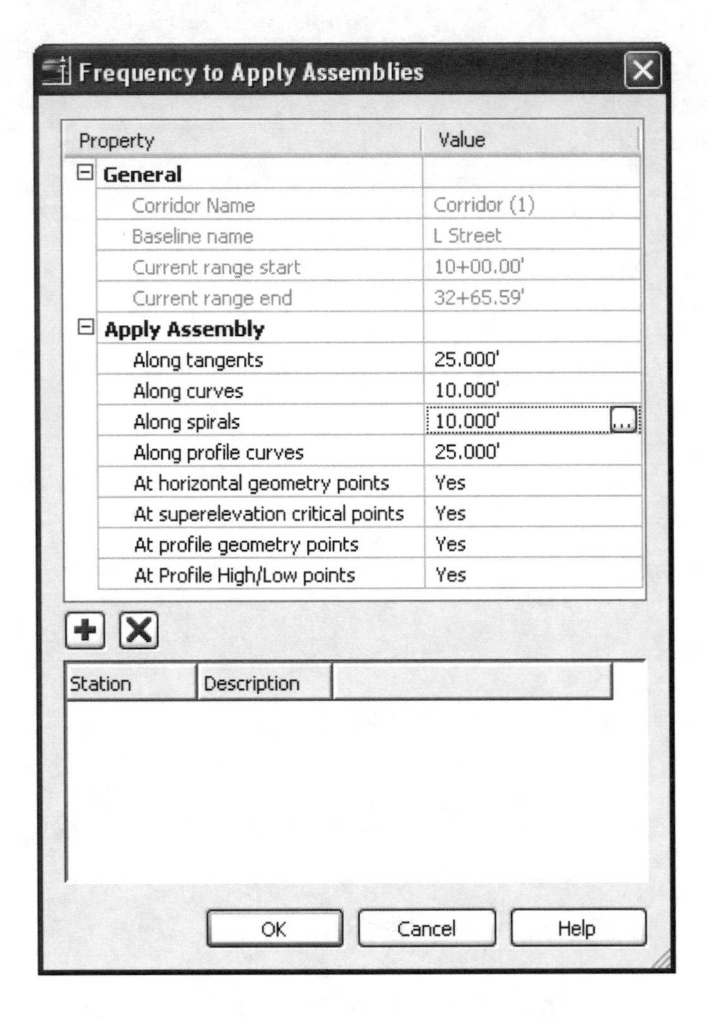

17. Confirm that the **Frequency Along tangents** is set to **25**.

18. Set the **Frequency Along curves** to **10**.

19. Set the **Frequency Along profile curves** to **10**.

There are not any spirals in your alignment so this setting does not matter for this exercise. However, you can control the frequency that the spirals are modeled in the corridor if needed. You can also specify additional stations by picking the blue plus button.

20. Click **<<OK>>** to return to the *Create Corridor* dialog box.

21. Click **<<Set all Targets>>.**

The Target Mapping dialog box is where you attach Subassemblies to Surfaces, Alignments, and Profiles. If a Subassembly was designed to attach to one of these objects it will display in the Target list. In this exercise the only Subassembly that you used that can be linked to an object is the *Basic Side Slope Cut Ditch* which will daylight to a surface.

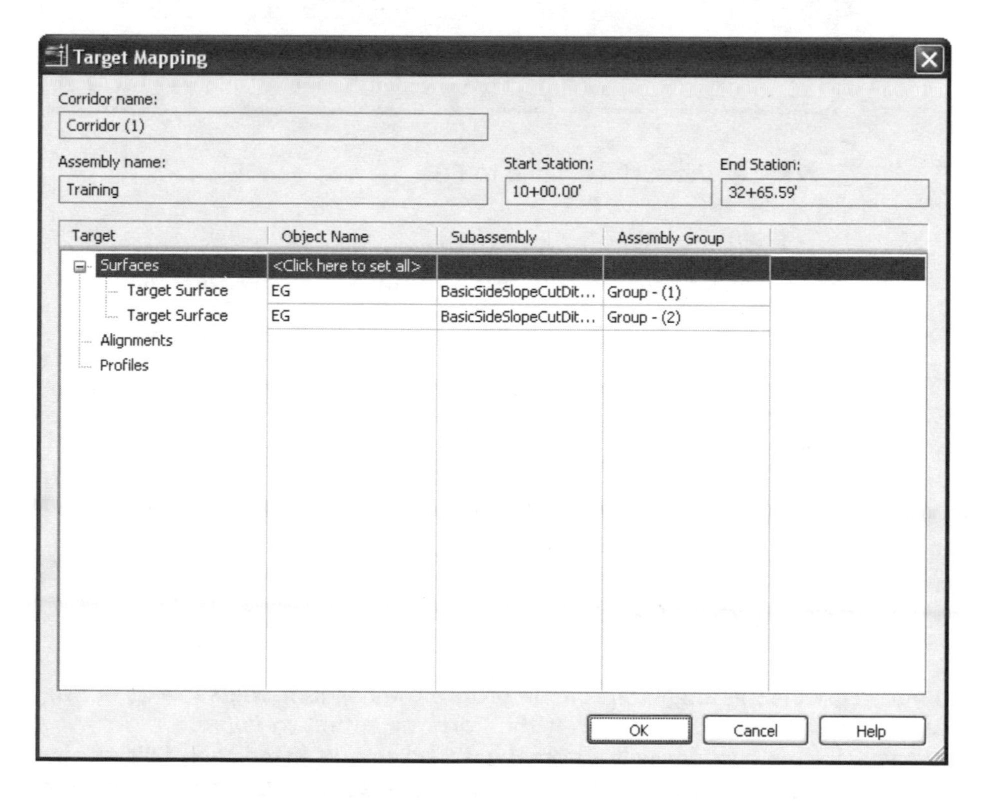

22. Click the **Surfaces** field at the top of the Object Name column to set all surfaces.

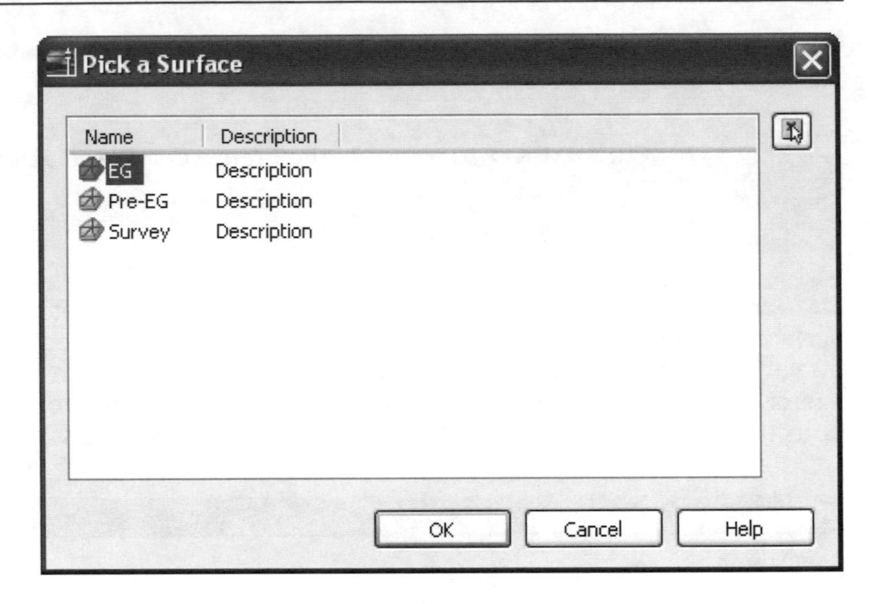

23. Set the surface name to **EG**.

24. Click <<**OK**>> to set the surface that the *Corridor* will use for daylighting and return to the *Target Mapping* dialog box.

25. Click <<**OK**>> to return to the *Create Corridor* dialog box.

26. Click <<**OK**>> to create the corridor.

27. Save the drawing.

8.3.2 Editing a Corridor

If you grip edit the alignment or the profile the corridor will show as out of date in the *Prospector*. If you set the Corridor option to *Rebuild Automatic*, the corridor will automatically be rebuilt after each edit. This is a nice feature but may be inconvenient because you will have to wait for the corridor to rebuild and redisplay after each edit. If this option is disabled you can make several changes and rebuild when you are ready to review them.

With all the components of the road design dynamically linked together in the Corridor object you have the flexibility to experiment with many design alternatives, review the way that they impact the model, and even undo the changes to try something else.

In this exercise you will make a change to the profile, rebuild the corridor and review the changes, then Undo back to the original state.

1. Zoom to the profile and select the **FG** profile line to display the grips.

2. Select one of the grips and stretch a PVI to a new location.

3. Notice how the corridor is displayed in the *Prospector*. It will have the orange out of date symbol next to it.

4. On the *Prospector* tab of the *Toolspace*, right-click on the *Corridor* **L Street** and select ⇒ **Rebuild**.

5. Review the changes to the corridor.

6. **Undo** twice to return the corridor and the profile to their original states.

You may need to undo more than 2 times if you have zoomed or issued any other commands during the process.

Editing the alignment and the assembly will also make the corridor out of date.

8.3.3 Creating Corridor Surfaces

Surfaces can be created and linked to the *Corridor* model. These surfaces display along with your other surfaces in the *Prospector*. Since they are linked to the *Corridor* model they are automatically updated and rebuilt any time that the *Corridor* is rebuilt.

These *Corridor* surfaces are created on the current layer so your first step will be to create a layer for the *Corridor* surfaces.

1. Create a new layer named **C-TOPO-L Street Corridor** and set it **current**.

2. On the *Prospector* tab of the *Toolspace*, right-click on the *Corridor* **L Street** and select ⇒ **Properties**.

3. Select the **Surfaces** tab.

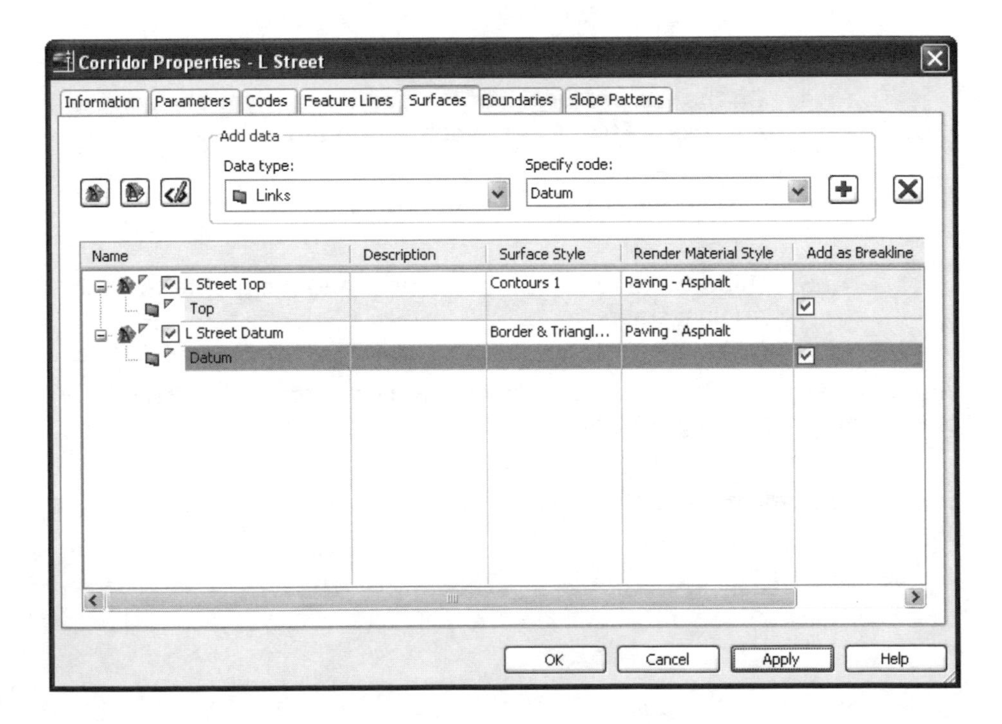

4. Click the **Create Corridor Surface** button.
5. Confirm that the **Data type** is set to **Links**.
6. Set the **Specify Code** option to **Top**.
7. Click the **Add Surface Item** button.
8. Rename the surface to **L Street Top**.
9. Set the **Surface Style** to **Contours 1**.
10. Enable the **Add as Breakline** option.
11. Click the **Create Corridor Surface** button.
12. Set the **Specify Code** option to **Datum**.
13. Click the **Add Surface Item** button.
14. Rename the surface to **L Street Datum**.
15. Set the **Surface Style** to **Border & Triangles & Points**.
16. Enable the **Add as Breakline** option.

17. Click the **Boundaries** tab.

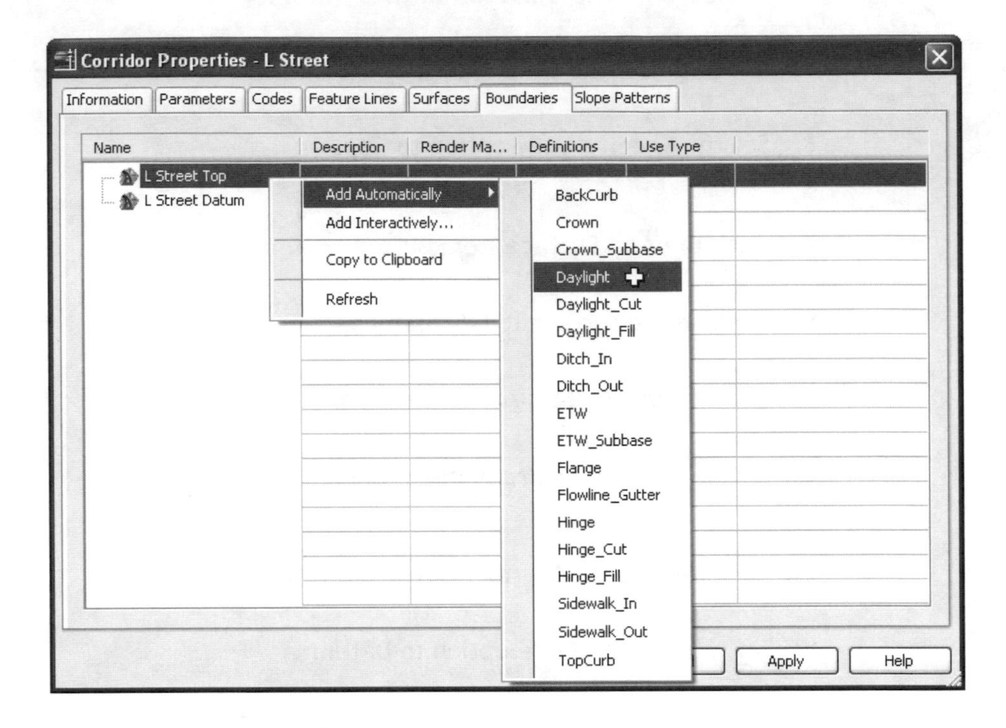

18. Right-click on the surface **L Street Top** and select **Add Automatically ⇒ Daylight.**

This will use the daylight line as the boundary for the top surface. Without a boundary the surface would triangulate outside of the extents of the corridor.

19. Right-click on the surface **L Street Datum** and select **Add Automatically ⇒ Daylight.**

20. Click **<<OK>>** to add the surfaces to the corridor.

The new surfaces linked to the corridor now display contours for the top surface and triangles for the datum surface. These surfaces are linked to the corridor model so they are automatically updated any time the corridor is changed. The surfaces are also displayed in the *Prospector* under the *Surface* node.

21. Save the drawing.

8.3.4 Viewing and Editing Corridor Sections

In this exercise you will view the *Corridor* one section at a time. You also have the option to edit the individual *Corridor Sections* through this command.

1. Select **Corridors ⇒ View/Edit Corridor Sections.**

2. When asked to `Select a Corridor` at the command line, **<<Enter>>** to select it from a list.

You also have the option to pick it graphically from the screen.

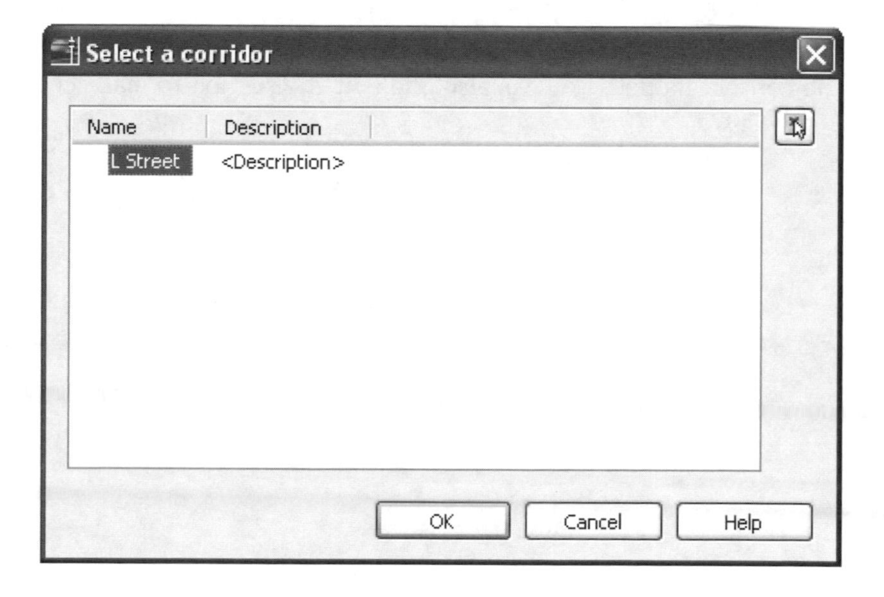

3. Select the **L Street** corridor.

4. Click **<<OK>>** to display the first corridor section and open the *View/Edit Corridor Section Tools* dialog box.

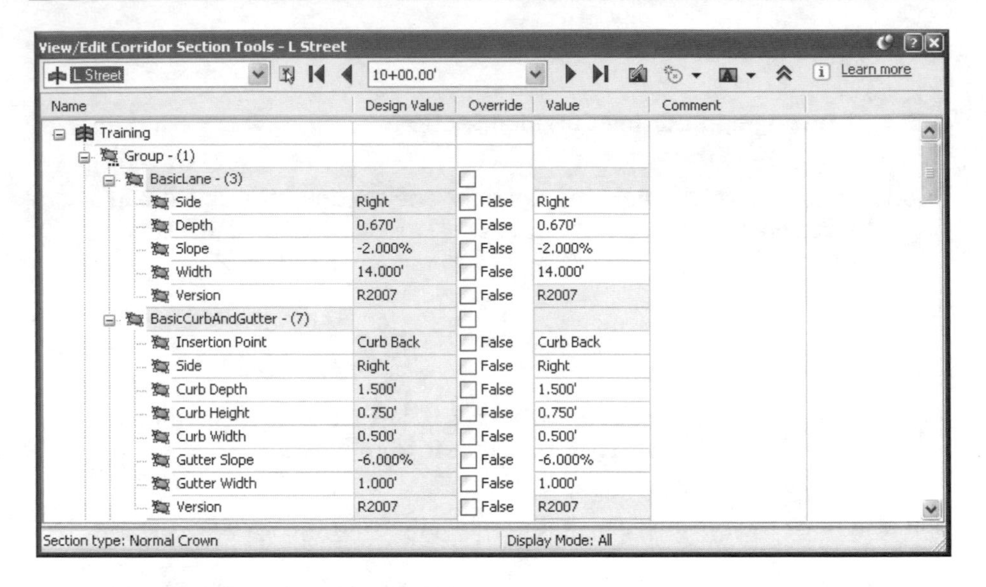

You can edit the parameters of each Assembly, section by section, to refine the corridor model. You can also grip edit the section to make changes graphically.

5. Scroll through several of the sections and review them on the screen.

6. Close the *View/Edit Corridor Section Tool* dialog box when finished.

8.3.5 Exporting Corridor Points

1. Select **Corridors ⇒ Utilities ⇒ COGO Points from Corridor.**

2. When asked to `Select a Corridor` at the command line, **<<Enter>>** to select it from a list.

You also have the option to pick it graphically from the screen.

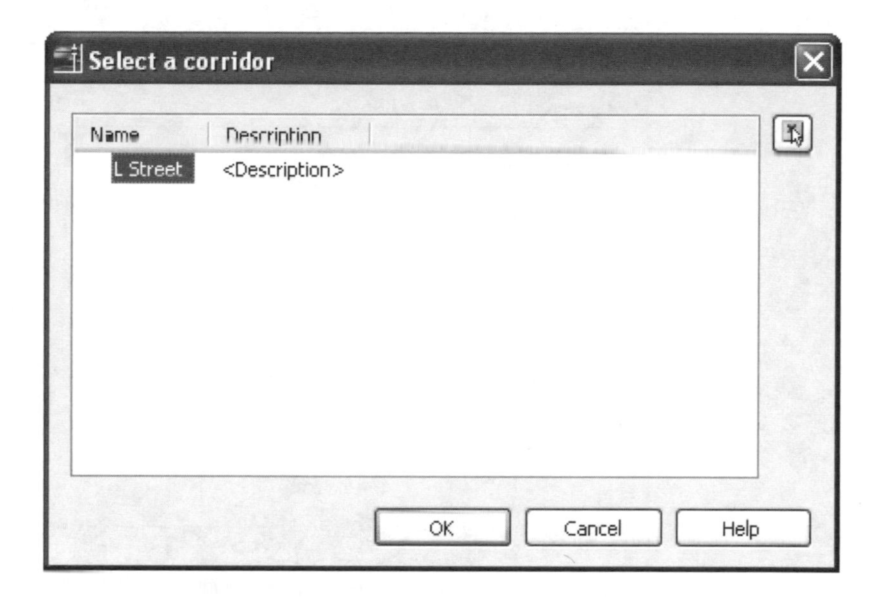

3. Select the **L Street** corridor.

4. Click **<<OK>>** to display the *Export COGO Points* dialog box.

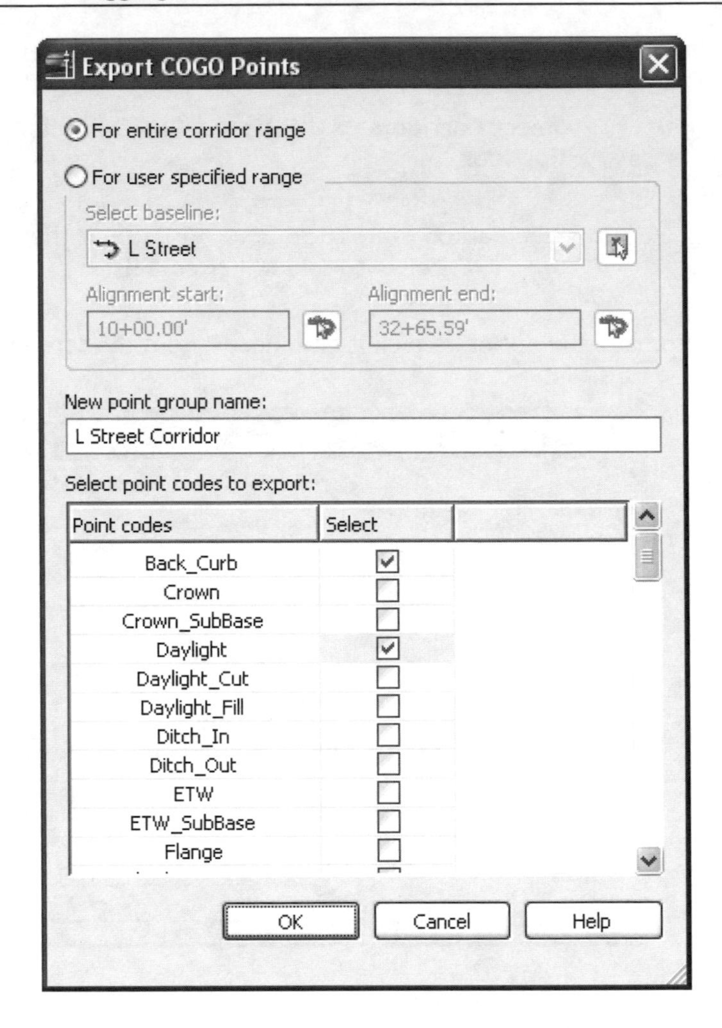

5. Confirm that the option is set to create points **For the entire corridor range**.

If needed, you can run this command several times as you specify specific ranges and make changes to the point group name and codes used.

6. Enter the **New point group** name **"L Street Corridor"**.

7. Select all the *Point Codes* using your shift key and disable them.

8. Select the **Point Codes Back_Curb** and **Daylight** to enable these two point codes.

The list of *Point Codes* displayed in the *Export COGO Points* dialog box is dependant on the subassemblies used to create the Assembly in Corridor model. This is only for example purposes; you may want more information than this if you were creating points to stake the project. *Civil 3D* gives you the flexibility to select exactly the information you want without forcing you to create more points than you need.

9. Click **<<OK>>** to create the points.

10. On the *Prospector* tab of the *Toolspace*, click on the *Point Group* **L Street Corridor** and review the list of points in the preview window.

11. Save the drawing.

8.4 Working with Sections

Sections can be samples and drawn for regular *Surfaces* as well as *Corridor Surfaces*. These *Sections* are used for output and can be plotted in *Section Views* as well as used for *Quantity Takeoffs*. These *Sections* are not used for design; the design of the model is handled by the *Corridor*.

8.4.1 Creating Sample Lines

In this exercise you will create *Sample Lines* that sample *Sections* for selected *Surfaces* at a specified interval.

1. Select **Sections ⇒ Create Sample Lines.**

2. When asked to `Select an alignment` at the command line, **<<Enter>>** to select it from a list.

You also have the option to pick it graphically from the screen.

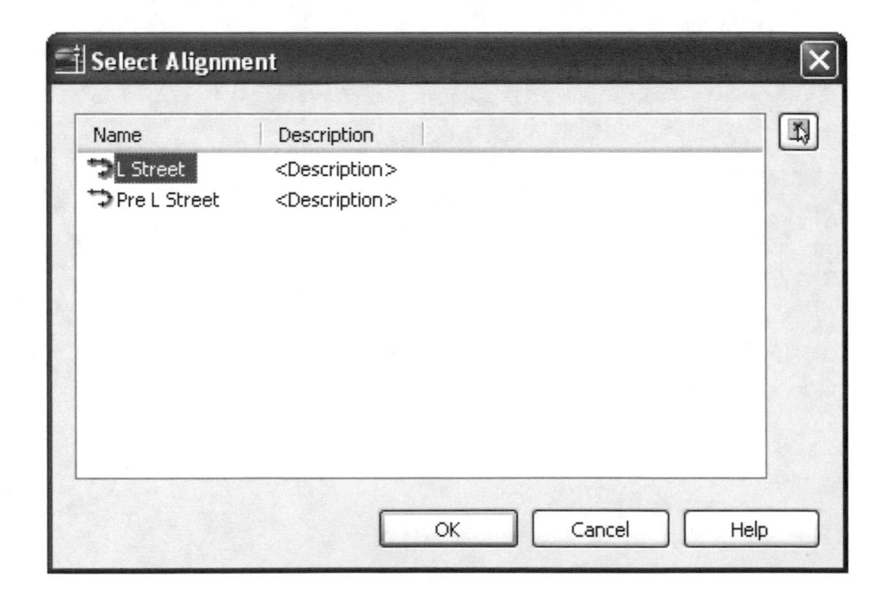

3. Select the Alignment **L Street.**

4. Click **<<OK>>** to define the Sample Line Group.

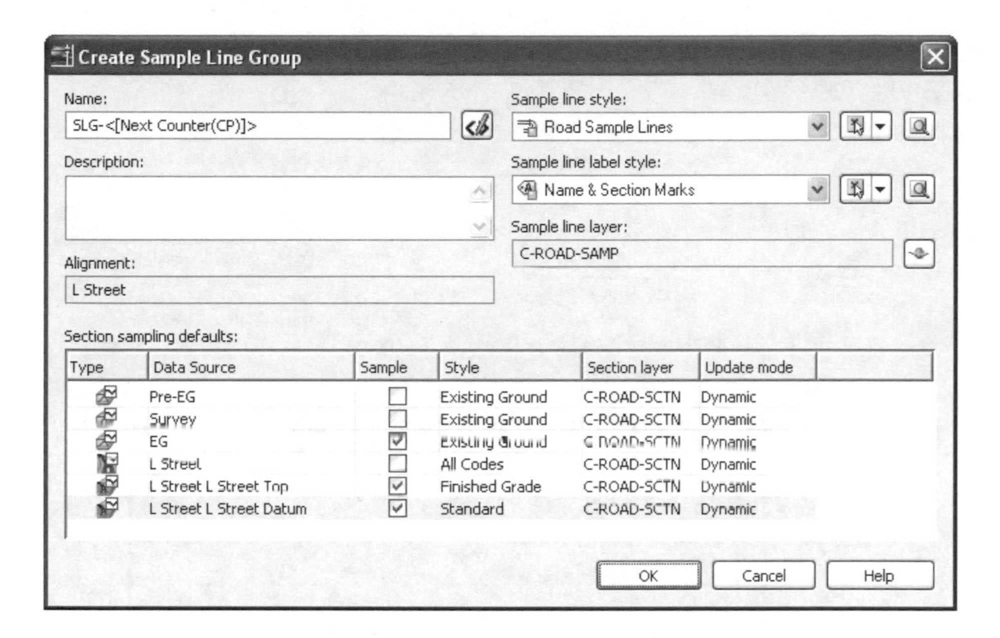

5. Leave the **Name** of the *Sample Line Group* unchanged. This will automatically name it according to the counter.

The Sample Line Group will be displayed under the Alignment L Street in the *Prospector* so a unique name does not help for descriptive purposes.

6. **Disable** the options to Sample Surfaces **Pre-EG**, **Survey**, and the codes from the **L Street** corridor.

You will sample the *Surface EG* and the *Top* and *Datum* corridor surfaces.

7. Confirm the **Style** for **EG** is set to **Existing Ground**.

8. Set the **Style** for **L Street Top** to **Finished Grade**.

9. Set the **Style** for **L Street Datum** to **Standard**.

This style controls the display of the section segments and points.

10. Click **<<OK>>** to open the *Sample Line Tools* toolbar.

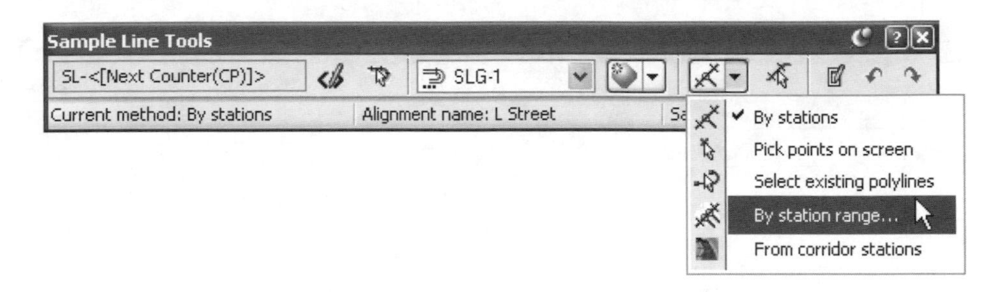

11. Click the **By Station Range** button.

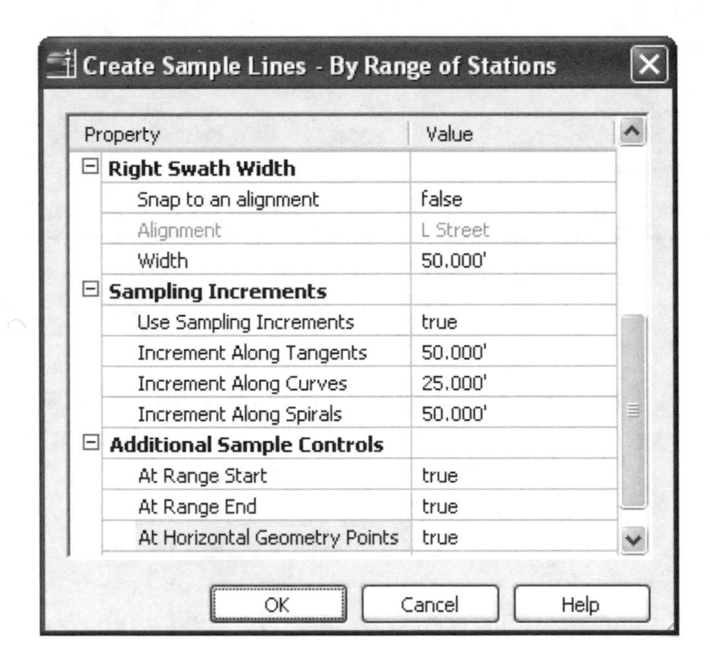

12. Set the **Left** and **Right** *Swath Widths* to **50**.

13. Confirm that the **Increment Along Tangents** is set to **50**.

14. Set the **Increment Along Curves** to **25**.

There are not any spirals in your alignment so this setting does not matter for this exercise. However, you can control the increment along the spirals that sections are sampled if needed.

15. **Enable** the option to sample a section at the **Range Start** and **Range End.**

16. **Enable** the option to sample **Horizontal Geometry Points**.

17. Click **<<OK>>** to sample the sections and return to the *Sample Line Tools* toolbar.

You can add critical sections at any point along the alignment by selecting it graphically or typing the station at the command line.

18. Close the *Sample Line Tools* toolbar.

8.4.2 Creating Section Views

To display the *Sections* in the drawing you must create *Section Views*. *Section View* are similar to *Profile Views*, they use styles to control the vertical exaggeration, grid, and labeling, as well as bands to display information along the top or bottom of the *Section View*.

1. Select **Sections ⇒ Create Multiple Views.**

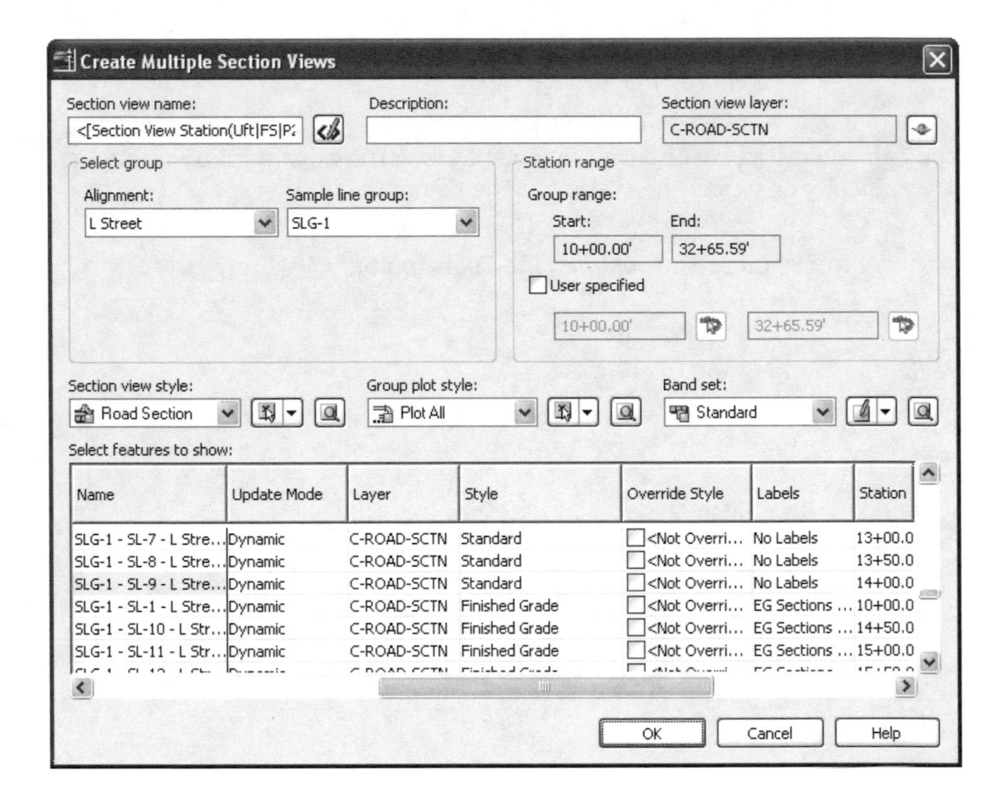

2. Leave the **Name** of the Section Views unchanged. This will automatically name each section view according to the counter.

Each *Section View* will be displayed under the corresponding *Sample Line* in the *Prospector* so a unique name does not help a lot for descriptive purposes.

3. In the table at the bottom of the dialog box scroll to find the **Data Source Column.**

4. Click on the header of the **Data Source Column** to sort the section lines according to the data that was sampled for that section line.

Each section line uses a Style to display the line and a Label Style to add text to the section. These can each be created and configured in style editors similar to ones you have used in previous exercises. By sorting the list according to the data source you can select all the EG section lines at once as well as all the Datum lines and Top Surface lines to assign the desired Label Styles.

5. Use the shift key to select all the section lines using the data source **EG**.

6. Click on one of the selected **Labels** fields and change the *Label Style* to **No Labels**. You may need to scroll to the right to find the *Labels* column.

7. Use the shift key to select all the section lines using the data source **L Street Datum**.

8. Click on one of the selected **Labels** fields and change the *Label Style* to **No Labels**.

9. Click **<<OK>>** to create the *Section Views*.

10. Select a blank area in the drawing to create the sections.

The Section Views will be created up and to the right of the point you pick.

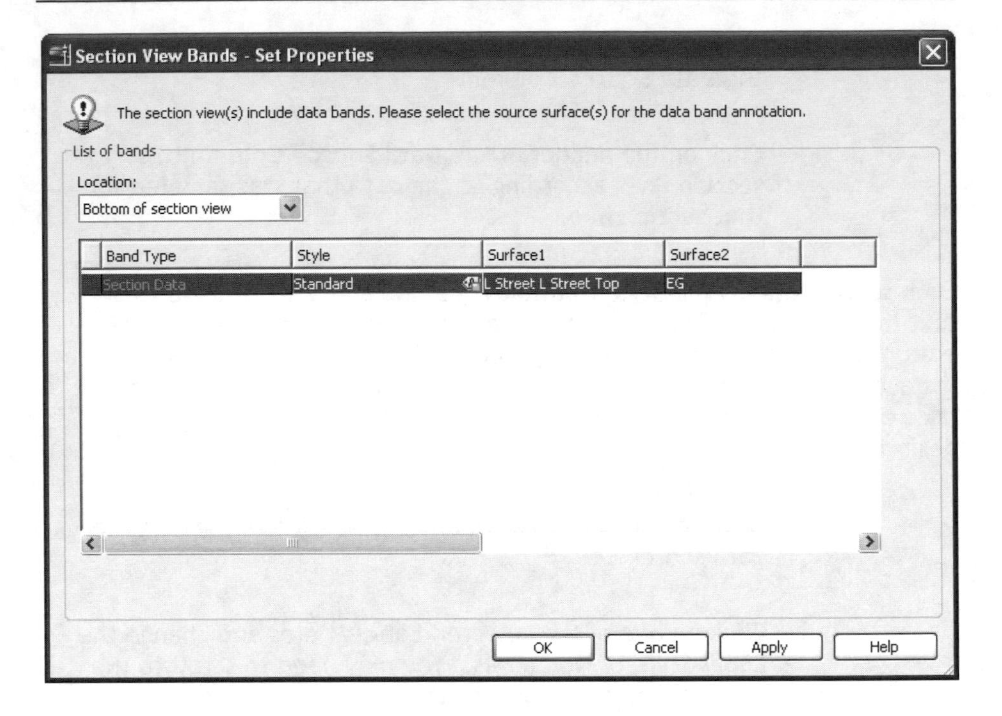

11. When the *Section View Bands* dialog box is displayed set **Surface 1** to **L Street Top.**

12. Zoom in and examine the sections.

13. Save the drawing.

8.4.3 Volume Calculations

In this exercise you will calculate earthwork volumes based on the sections you sampled in the previous exercises. The sections do not need to be displayed in *Section Views* to calculate volumes, they only need to be sampled.

1. Select **Sections** ⇒ **Define Materials.**

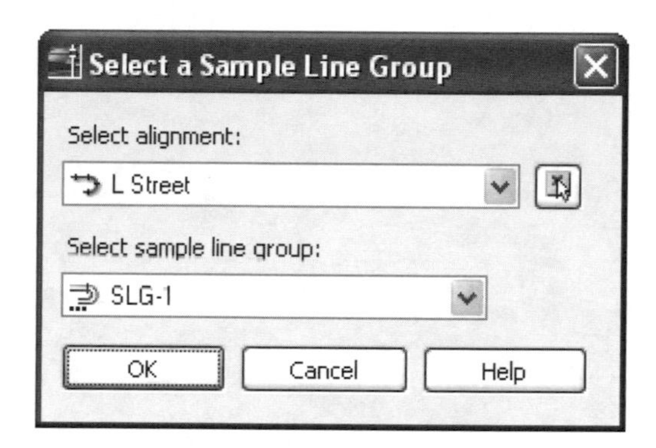

2. Confirm that the **alignment** is set to **L Street**.

3. Confirm that the **sample line group** is set to **SLG-1**.

4. Click **<<OK>>**.

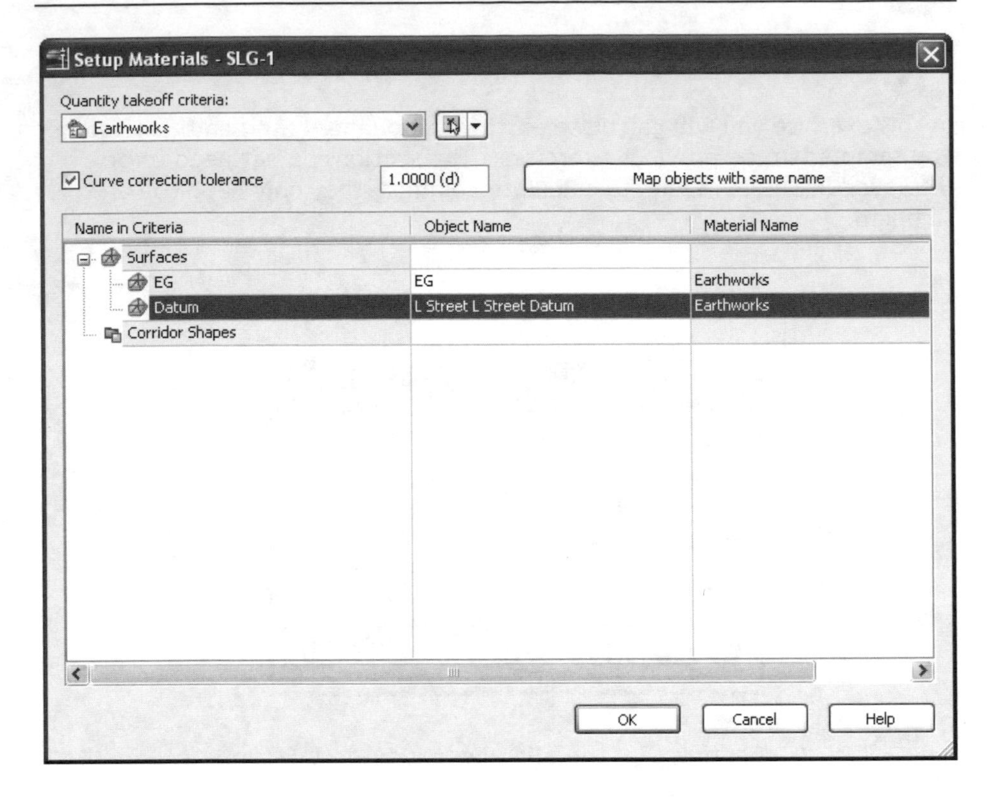

5. Set the **Quantity Takeoff Criteria** to **Earthworks.**

6. Set the **EG Surface** option to **EG**.

7. Set the **Datum Surface** option to **L Street Datum**.

8. Click **<<OK>>** to save the material setting.

9. Select **Sections ⇒ Add Tables ⇒ Total Volume.**

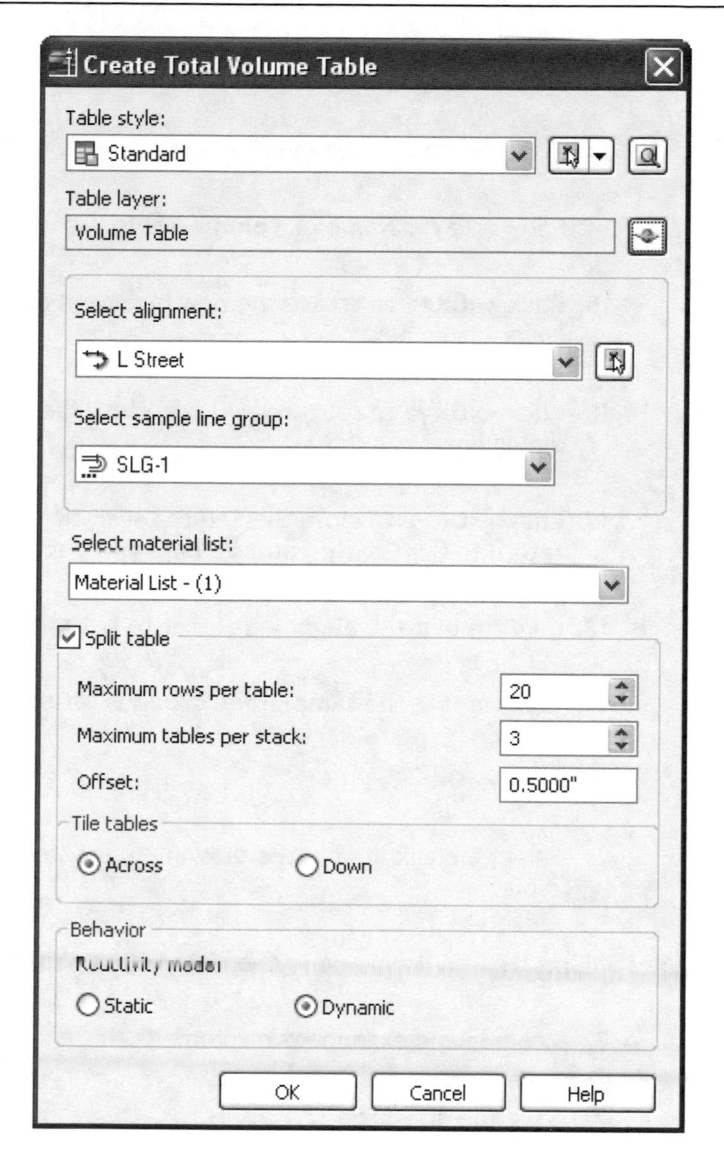

10. Confirm the **Table style** is set to **Standard**.

11. Click the **Table layer** button to open the *Object Layer* dialog box.

12. Click the **Layer button** in the *Select Layer* dialog box.

13. Click the **New button** in the *Layer Selection* dialog box.

14. Enter a **Layer Name** of **Volume Table**.

15. Click **<<OK>>** to create the new layer and close the *Create Layer* dialog box.

16. Click **<<OK>>** to set the layer and close the *Layer Selection* dialog box.

17. Click **<<OK>>** to close the *Select Layer* dialog box and return to the *Create Total Volume Table* dialog box.

18. Confirm that the **alignment** is set to **L Street**.

19. Confirm that the **sample line group** is set to **SLG-1**.

20. Click **<<OK>>**.

21. Select a blank area in the drawing to create the volume table.

The tables will be created down and to the right of the point you pick.

22. Zoom in and examine the volume table.

23. Save the drawing.

This volume table is dynamic and will update when there are changes to the corridor model.

8.5 Creating a Surface Showing Final Site Conditions

In this section you will *Paste* a *Corridor Surface* into a copy of the existing ground surface. This creates a surface showing the final site conditions after the road is built. You can use this surface to display contours of the finished project and also as the target surface for grading in the next chapter.

8.5.1 Pasting Surfaces

In this exercise you will make a copy of the surface *EG* and rename it to *Final*. Then you will paste the surface *L Street Top* into the new surface *Final* to create a surface showing the final site conditions after the road is built.

1. Set layer **0** current.

2. Freeze the layers **C-ROAD-CORR-L Street, C-ROAD-L Street, C-ROAD-SAMP,** and **C-TOPO-L Street Corridor.**

3. Thaw the layer **C-TOPO-EG.**

4. At the command line enter `"CP"` to start the AutoCAD `Copy` command.

5. Pick the surface **EG** and press **<<Enter>>**.

6. Type `"D"` for **displacement.**

7. **<<Enter>>** to accept the default displacement of `0,0,0`.

This will create a copy of the *Surface EG* at the same coordinates. You will also see the new surface displayed in the Prospector with the name *EG (1)*. The original surface and the copy are on the same layer and use the same surface style, so they will look exactly the same in the drawing editor. To separate them on to different layers you can first rename the new surface and change its style so that it is displayed differently.

8. On the *Prospector* tab of the *Toolspace*, right-click on the *Surface* **EG (1)** and select ⇒ **Properties**.

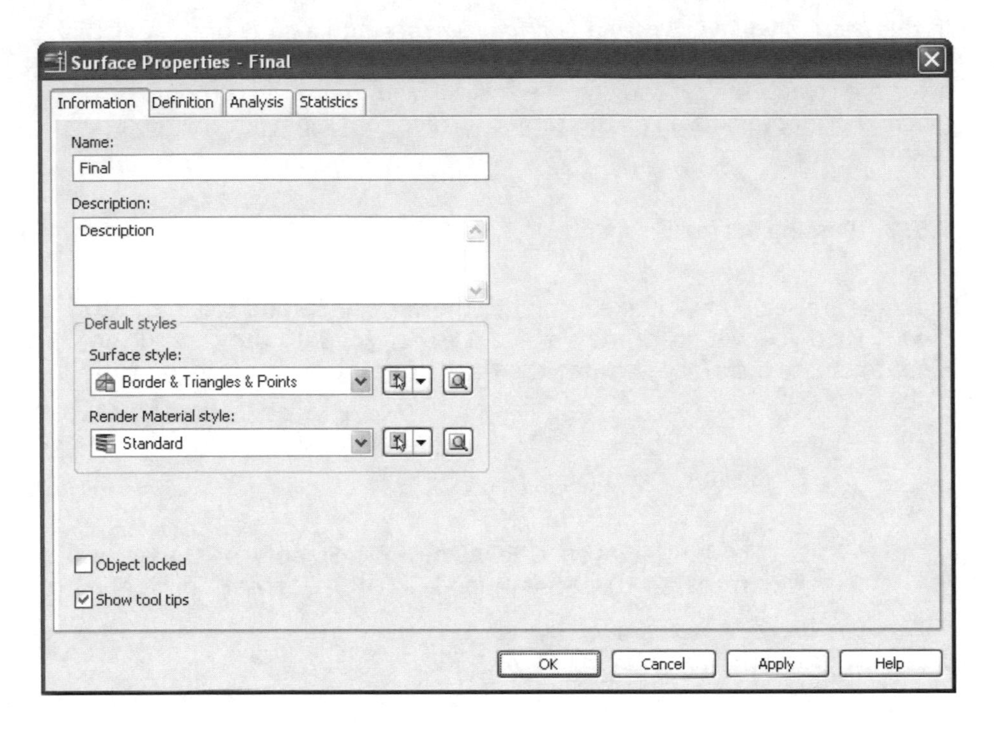

9. On the *Information* tab of the *Surface Properties* dialog box, enter a new **Name** of **Final**.

10. Select **Border & Triangles & Points** from the **Object style** list.

11. Click <<**OK**>> to rename the surface and display it as triangles.

12. Create a new layer named "**C-TOPO-Final**".

13. Use the AutoCAD **Properties** command to move the surface **Final** to the layer **C-TOPO-Final**.

14. Freeze the layers **C-TOPO-EG**.

15. **Regen** if needed to clean up the display.

16. Expand the *Surface* **Final** on the *Prospector* tab of the *Toolspace*.

17. Expand the **Definition** node under *Surface* **Final**.

Be sure that you are under the surface *Final* in the *Prospector*. As you continue to add surfaces to your drawing it can become easy to have the wrong surface expanded and edit the wrong surface.

18. Right-click on **Edits** under the *Definition* **node,** and select ⇒ **Paste Surface.**

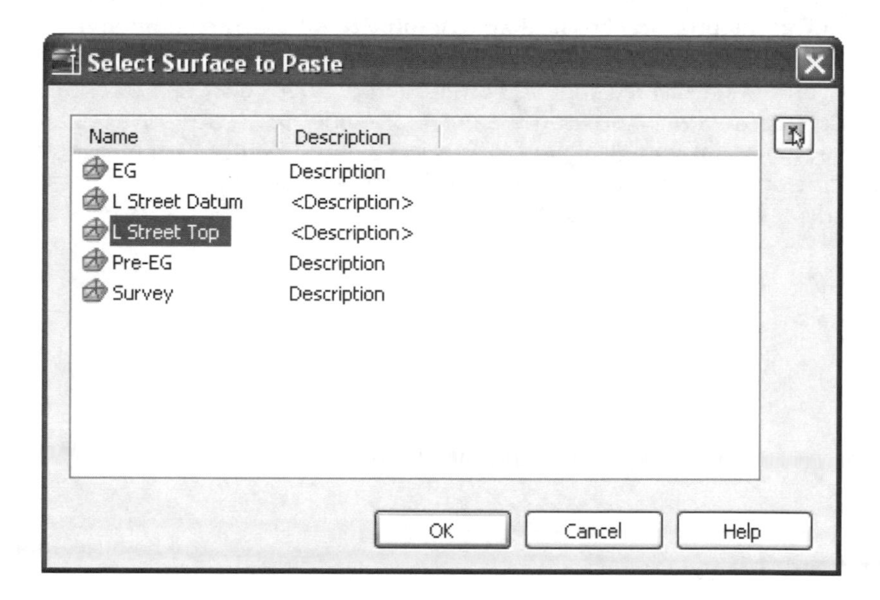

19. Select *Surface* **L Street Top** to paste into **Final.**

20. On the *Prospector* tab of the *Toolspace*, right-click on the *Surface* **Final** and select ⇒ **Properties.**

21. On the *Information* tab of the *Surface Properties* dialog box, select **Border & Contours** from the **Object style** list.

22. Click <<**OK**>> to display the *Surface* **Final** as contours.

23. Save the drawing.

8.6 Additional Exercises

Create a second Assembly and Corridor using some of the more complex Subassemblies found in the other Imperial Tool Palettes. You can create multiple corridor models using the same alignment and profile to explore different design alternatives without impacting the original corridor model.

8.7 Chapter Summary

In this chapter you created an assembly to define the components and geometry of a road. Then you combined the assembly with, an alignment, profile, and surface that you created in previous chapters to create a corridor model. You used the corridor model to create surfaces and sections that you displayed in *Section Views*. Finally, you merged a corridor surface with a copy of the existing ground surface to create a surface that represents the final site conditions after the road is built.

Chapter 9

Grading

In this chapter you will create a grading group to grade a pond. You will learn how the grading group can be dynamically linked to a surface. Finally, the pond surface will be compared with the existing surface to create a volume surface and calculate cut and fill quantities.

- **Working with Grading Groups**
- **Volume Calculations**

Dataset:

To start this chapter you will continue working in the drawing named **Design.dwg.** You can continue with the drawing that you currently have from the end of the previous chapter or, if you are starting in the middle of the book, you can open the drawing **CH-09.dwg** located in the folder **C:\Cadapult Training Data\Civil 3D 2007\Level 1\Chapter Drawings.** Opening the drawing from the dataset provided will ensure that you have the drawing set up correctly for the exercises in the following chapter and overwrite any mistakes that you may have made in previous exercises.

9.1 Overview

In this chapter you will create a grading group to grade a pond. The grading group will be dynamically linked to a surface so that the surface is updated automatically each time that you add a grading object to the grading group or edit a grading object within the group. Finally, you will create *Grid* and *TIN Volume* surfaces to calculate the volume of the pond and explore ways to use surface styles to display the volume surfaces showing depths of cut and fill.

9.1.1 Concepts

Grading Groups are a collection of *Grading* objects that interact with each other. *Grading Groups* are components of a *Site* and interact with other *Grading Groups*, *Alignments*, and *Parcels* contained in the same *Site*. In this chapter you will learn how to create grading objects, based on grading criteria. You will explore how these objects come together in the grading group and create a surface based on the group.

9.1.2 Styles

Styles are saved in the drawing and can be created and edited on the Settings tab of the *Toolspace* or on the fly during many object creation and editing commands.

Surface Styles control the display of surfaces. You have individual control over many components of the surface including contours, triangles, points, borders, slope analysis, elevation bands, watersheds, and flow arrows.

Grading Styles control the display of grading objects. You have individual control over many components of the grading object including the center marker, slope patterns, target lines, and projection lines.

9.2 Working with Grading Groups

Grading Groups are a collection of *Grading* objects that interact with each other. *Grading Groups* are components of a *Site* and interact with other *Grading Groups*, *Alignments*, and *Parcels* contained in the same *Site*.

9.2.1 Creating a Grading Group

In this exercise you will create a new *Site* and a *Grading Group*.

1. Continue working in the drawing **Design.dwg**.

2. On the *Prospector* tab of the *Toolspace*, right-click on **Sites** and select → **New**.

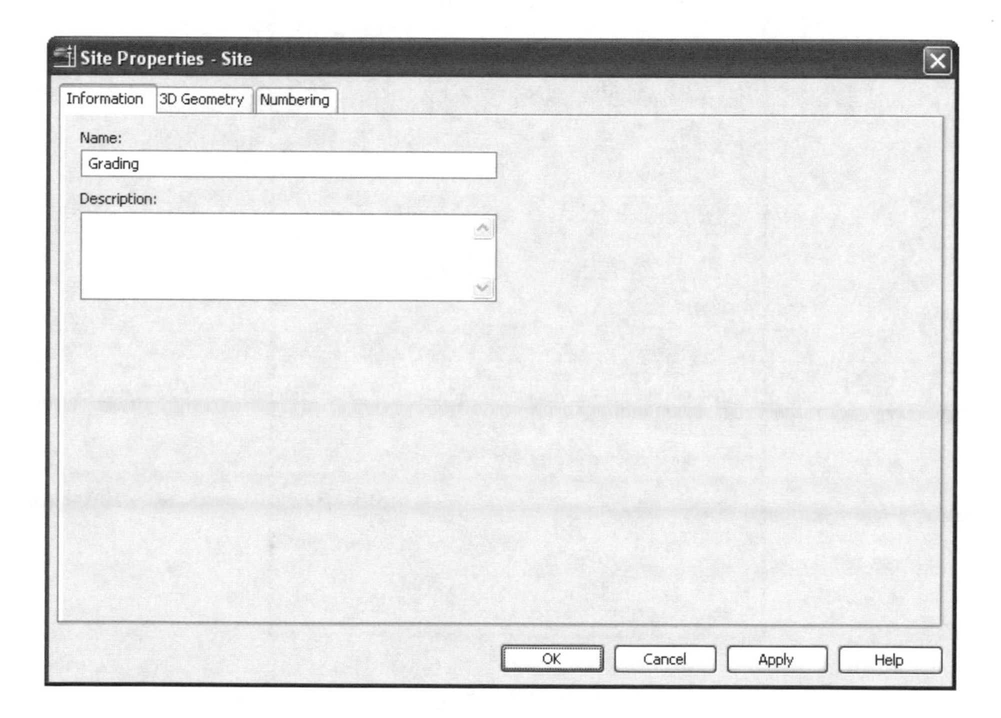

3. Enter **"Grading"** for the **Name**.

4. Click **<<OK>>** to create the *Site*.

5. On the *Prospector* tab of the *Toolspace*, expand the **Sites** node.

6. Expand the *Site* **Grading.**

7. Right-click on **Grading Groups** node under the Site *Grading* and select ⇒ **New.**

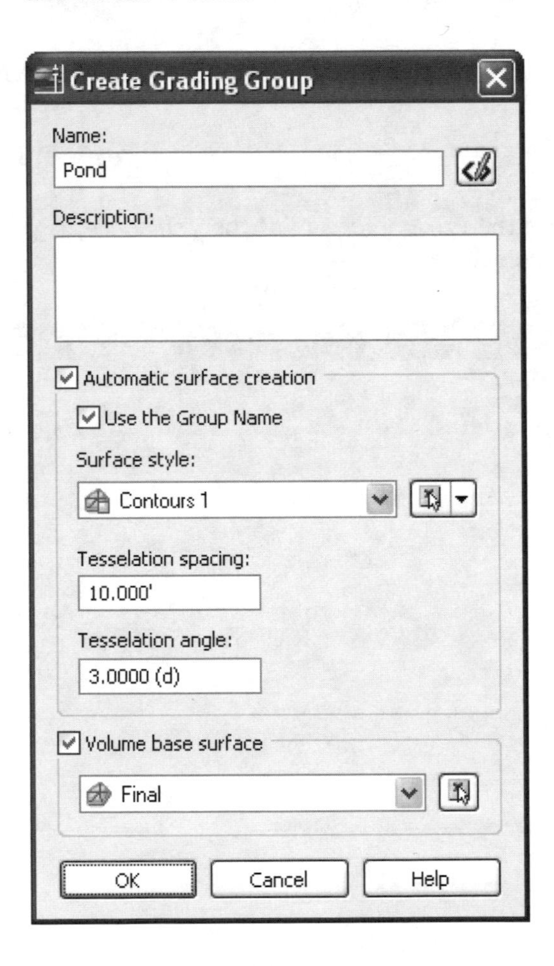

8. Enter **Pond** for the **Name** of the *Grading Group.*

9. **Enable** the **Automatic Surface Creation** option.

10. Set the **Surface style** to **Contours 1**.

11. Enable the **Volume base surface** option.

12. Set the **Volume base surface** to surface **Final**.

13. Click **<<OK>>** to create the *Grading Group* and open the *Create Surface* dialog box.

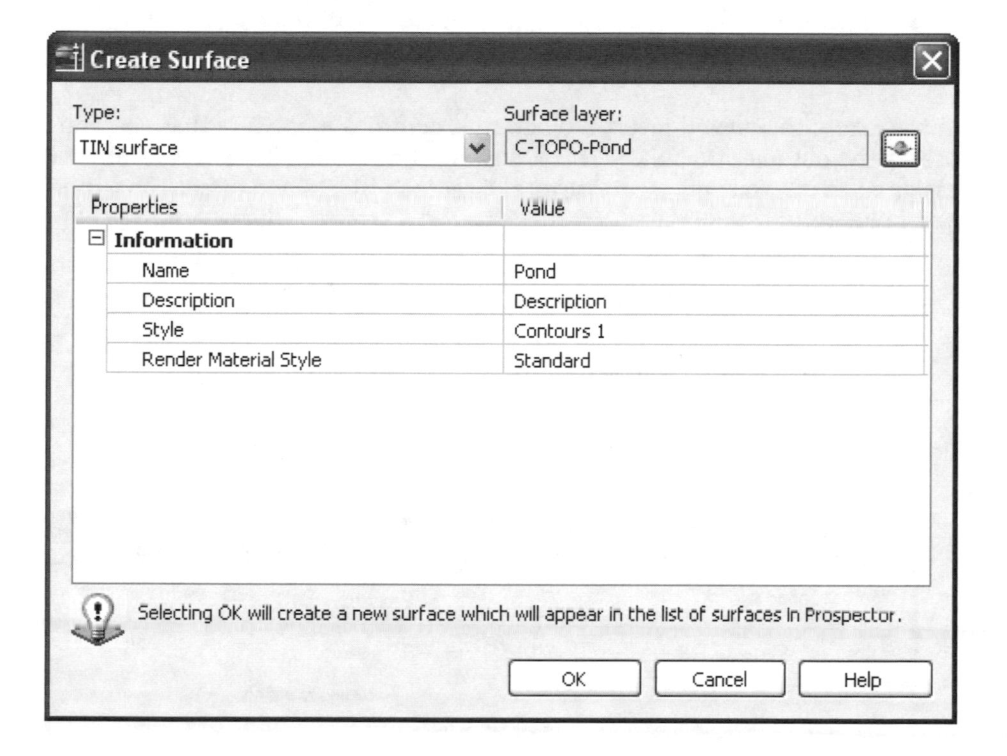

14. Click the **Surface Layer** button to open the *Object Layer* dialog box.

15. Set the **Modifier** to **Suffix**.

16. Enter **-*** as the *Modifier value*.

17. Click **<<OK>>** to close the **Object Layer** dialog box.

18. Confirm that the surface **Name** is set to **Pond**.

19. Confirm that the surface **Style** is set to **Contours 1**.

20. Click **<<OK>>** to create the new surface based on the grading group.

9.2.2 Creating a Grading Object

In this exercise you will insert a block that contains a polyline that you will use as the outline of a pond. This is just a 2D polyline with an elevation assigned to it. You will create several Grading objects and offsets based on this polyline.

1. Select **Insert ⇒ Block.**

2. Browse to the file:

C:\Cadapult Training Data\Civil 3D 2007\Level 1\Drawings\pond.dwg

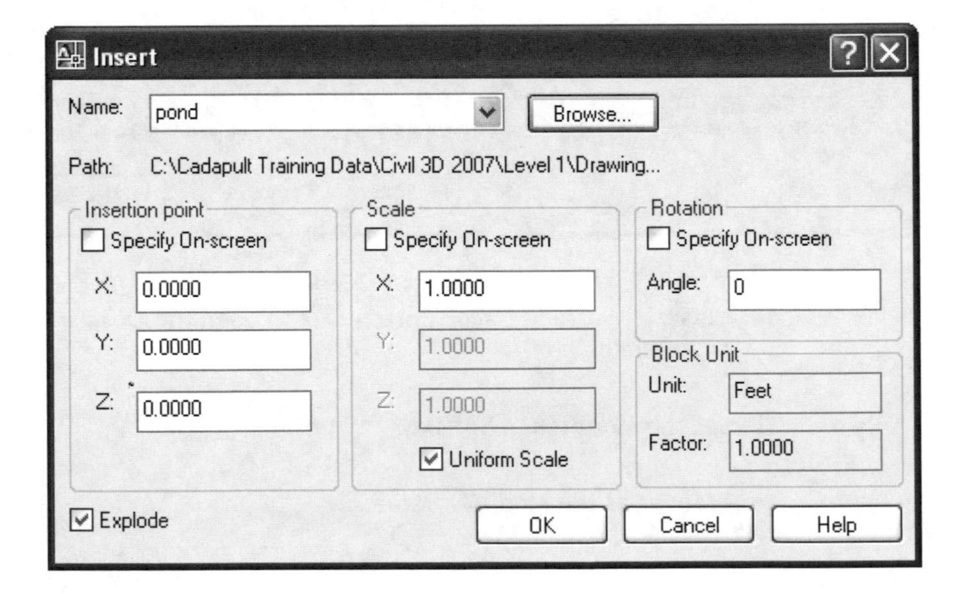

3. **Disable** the option to **Specify the Insertion Point** On-screen.

4. **Enable** the option to **Explode** the block.

5. Click **<<OK>>** to insert the pond outline.

6. Zoom in to the pond outline in the northeast corner of the site.

7. Use the AutoCAD `Offset` command to offset the red pond outline that you just inserted **5** feet to the inside.

You now have an outer line to daylight to the surface and an inner line to use to grade the inside of the pond. This will create a 5 foot wide berm around the pond.

8. Select **Grading ⇒ Grading Creation Tools** to display the *Grading Creation Tools* toolbar.

At the bottom of the *Grading Creation Tools* toolbar you will see that the *Grading Group* is set along with the target *Surface*.

9. Click the **Select a Criteria Set** button on the *Grading Creation Tools* toolbar.

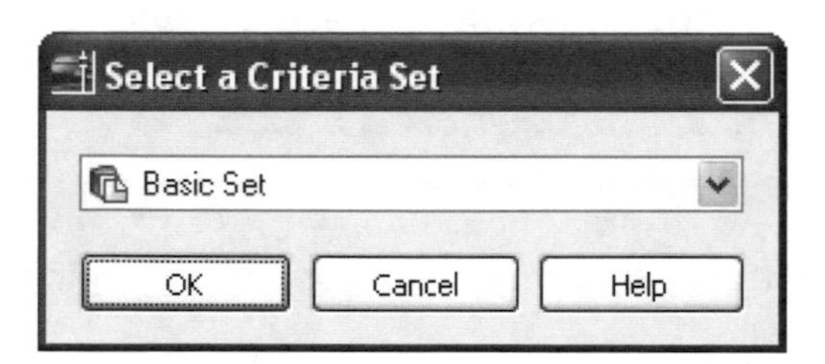

10. Select the **Basic Set** grading criteria.

11. Click **<<OK>>**.

12. Set the **Grading Criteria** to **Surface @ 3-1 Slope**.

13. Click the **Create Grading** button .

14. Pick the outer Pond Outline.

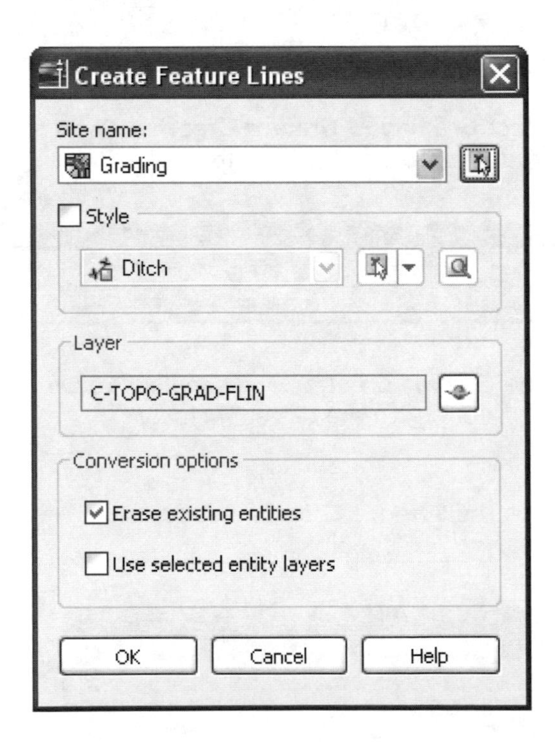

15. Confirm that the **Site name** is set to **Grading**.

16. Confirm that the *Conversion option* to **Erase existing entities** is enabled.

If the *Erase existing entities* option is not enabled the new feature line will be created on top of the existing polyline.

You also have options to apply a feature line style to the object and control the layer of the new object.

17. Click **<<OK>>** to create the feature line.

18. Pick a point outside of the pond outline to grade to the outside.

19. At the command line **<<Enter>>** to apply the grading to the entire length of the feature line.

The *Grading Object* is now created and daylights to the surface *Final* at a 3:1 slope according to the grading criteria.

20. **<<Enter>>** to end the command when asked to select another feature.

The surface *Pond* is now rebuilt with the information from the grading group and displayed as 1 foot contours according to the surface style.

21. Set the **Grading Criteria** to Relative **Elevation @ Slope**.

22. Click the **Create Grading** button .

23. Pick the interior pond outline.

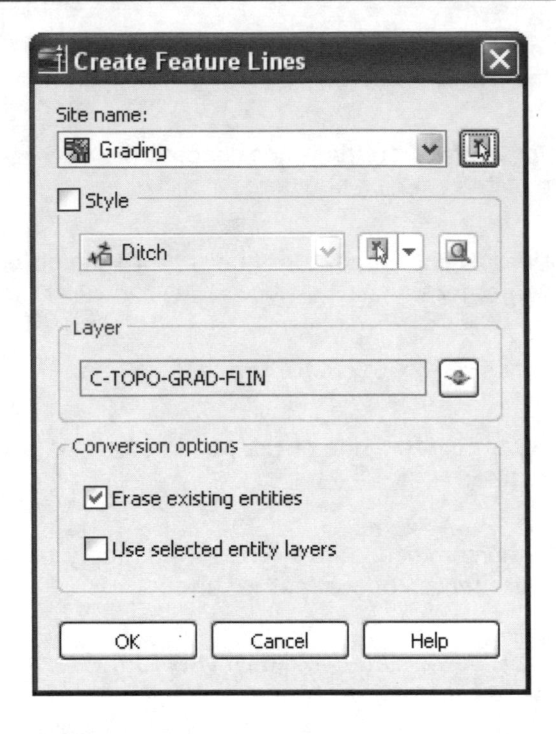

24. Confirm that the **Site name** is set to **Grading**.

25. Confirm that the *Conversion option* to **Erase existing entities** is enabled.

26. Click <<**OK**>> to create the feature line.

27. Pick a point inside of the pond outline you just selected to grade to the inside.

28. At the command line <<**Enter**>> to apply the grading to the entire length of the feature line.

29. Enter **-2** for the **Relative Elevation**.

30. Enter **5** for the **Slope**.

The *Grading Object* is now created projecting down 2 feet into the center of the pond at a 5:1 slope, according to the grading criteria.

31. **<<Enter>>** to end the command when asked to select another feature.

The surface *Pond* is now rebuilt with the information from the grading group and displayed as 1 foot contours according to the surface style.

32. Use the AutoCAD `Offset` command to offset the inner most feature line created during the last grading **5** feet to the **inside**.

This will create a polyline on the current layer that you will use as the next feature line to continue grading the interior of the pond. It will create a 5 foot wide, flat bench in the pond.

33. Confirm that the **Grading Criteria** is set to **Relative Elevation @ Slope**.

34. Click the **Create Grading** button.

35. Pick the feature line you created with the offset command.

36. Pick a point inside of the pond outline you just selected to grade to the inside.

37. At the command line **<<Enter>>** to apply the grading to the entire length of the feature line.

38. Enter **-12** for the Relative Elevation.

39. Enter **2** for the Slope.

The *Grading Object* is now created projecting down 12 feet into the center of the pond at a 2:1 slope, according to the grading criteria.

40. **<<Enter>>** to end the command when asked to select another feature.

The surface *Pond* is now rebuilt with the information from the grading group and displayed as 1 foot contours according to the surface style.

41. Save the drawing.

9.2.3 Creating a Grading Infill

In a grading group, areas between individual grading objects have no data. This creates blank or void areas in the surface that is created from the grading group. These blank areas can be filled in by creating an Infill. The Infill fills the gap between the individual grading objects in the group with a straight slope.

In this exercise you will create three Infills in the Pond grading group to fill the gaps on the benches and at the bottom of the pond.

1. Review the Surface Pond with the Object Viewer to see the blank areas at the benches and the bottom of the surface.

2. If the *Grading Creation Tools* toolbar is not open select **Grading ⇒ Grading Creation Tools**.

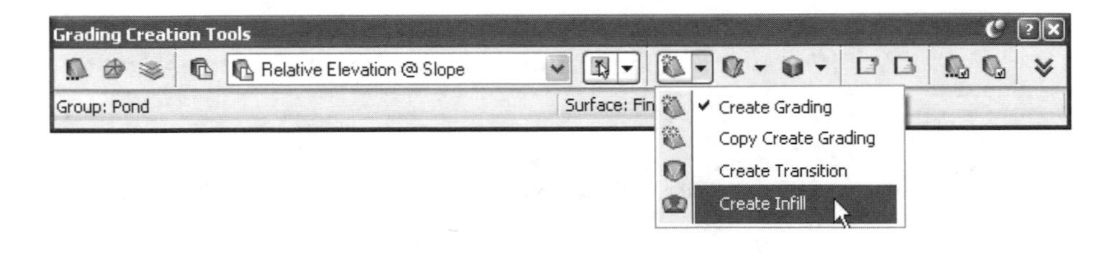

3. Click the **Create Infill** button.

The command line will prompt you to `Select an area to infill.`

4. Move your cursor over the flat part of the berm around the top of the pond. You will see the extents of the flat area highlight.

5. Pick any point in the flat, highlighted, area to create the Infill.

6. Move your cursor over the flat part of the bench on the inside of the pond. You will see the extents of the flat area highlight.

7. Pick any point in the flat, highlighted, area to create the Infill.

8. Move your cursor over the flat bottom of the pond. You will see the extents of the flat area highlight.

9. Pick any point in the flat, highlighted, area to create the Infill.

10. **<<Enter>>** to end the command.

The surface *Pond* is now rebuilt with the information from the infills and displayed as 1 foot contours according to the surface style.

11. Close the *Grading Creation Tools* toolbar.

12. Review the surface with the Object Viewer to confirm that the blank areas are now filled in.

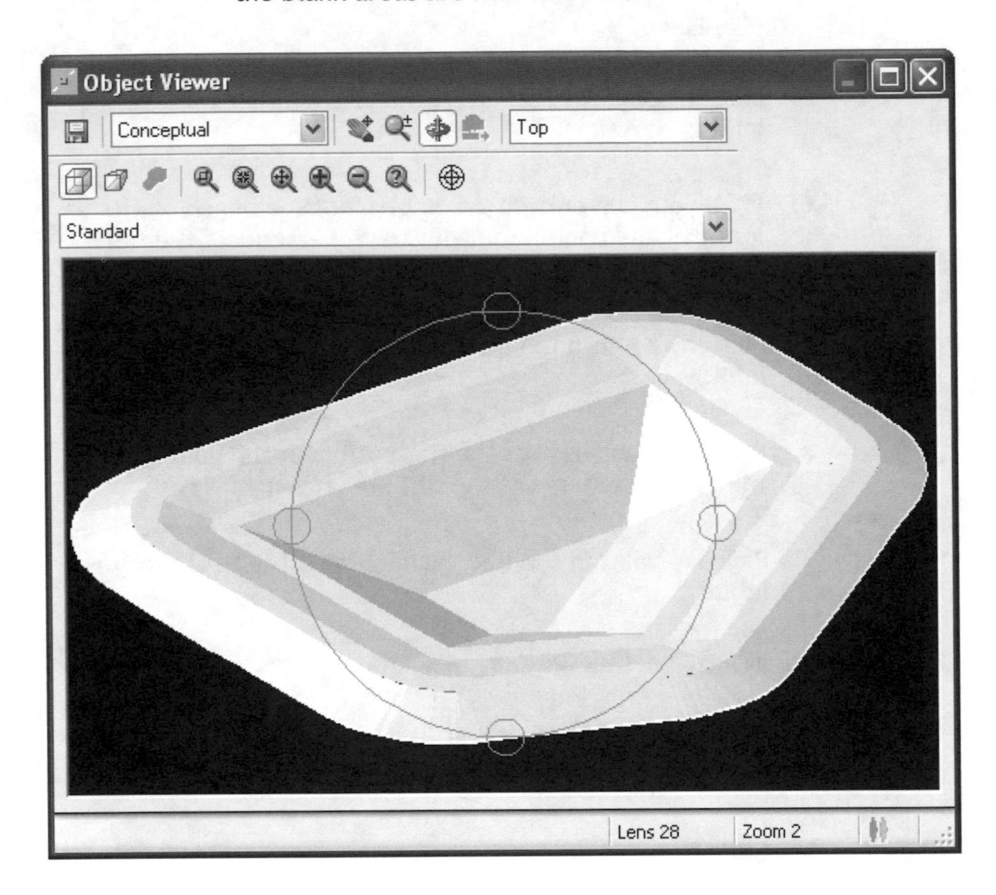

13. Save the drawing.

9.2.4 Reviewing Grading Group Properties

1. On the *Prospector* tab of the *Toolspace*, expand **Sites, Grading,** and **Grading Groups.**

2. Right-click on *Grading Group* **Pond** under the Site *Grading* and select ⇒ **Properties.**

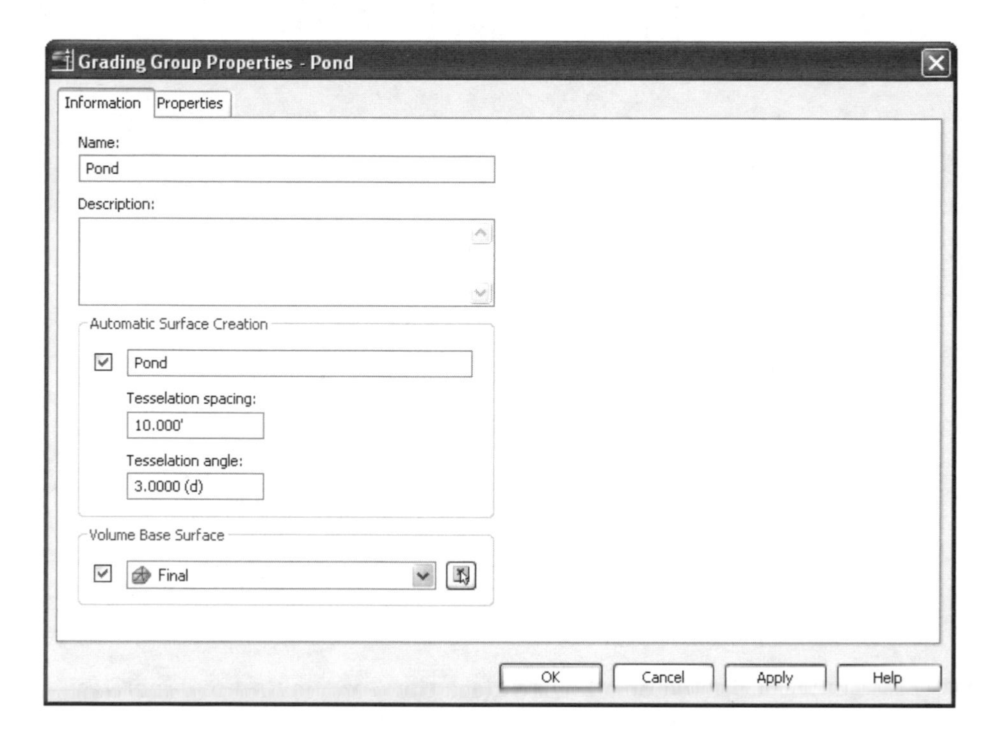

On the *Information* tab of the *Grading Group Properties* dialog box you can control the name, the surface created from the grading group, and the volume base surface. For this exercise you will not make any changes to the *Information* tab.

3. Select the **Properties** tab.

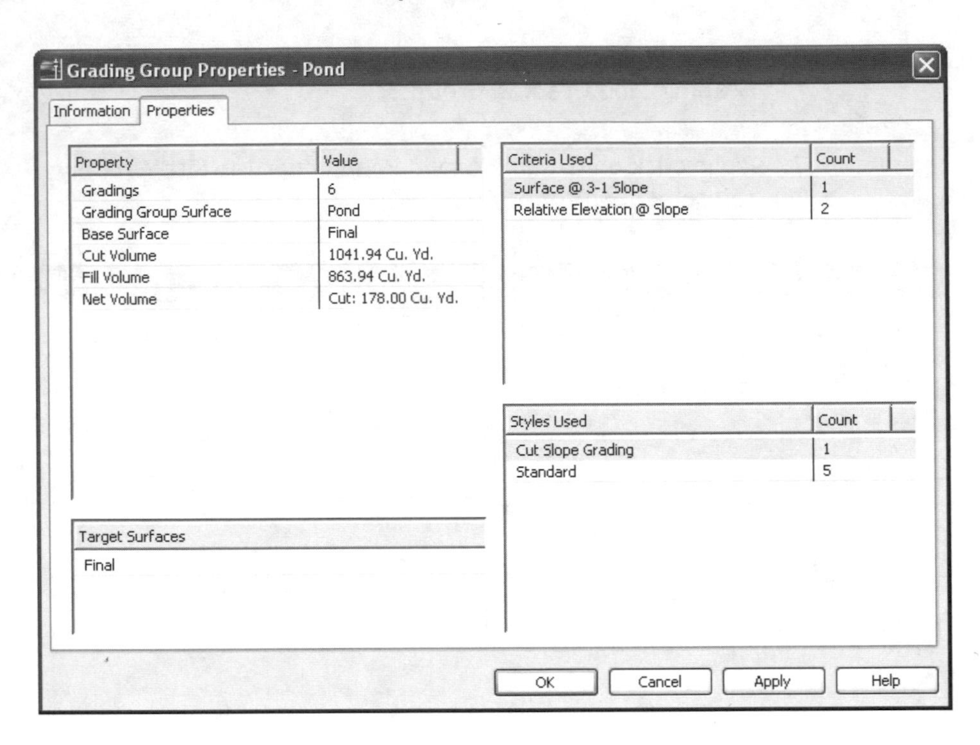

On the *Properties* tab of the *Grading Group Properties* dialog box you can review the current volume of the grading group along with the grading criteria and styles used.

4. Click <<**Cancel** >> to close the *Grading Group Properties* dialog box without making any changes.

9.3 Volume Calculations

In this section you will calculate the volumes between two surfaces by creating a volume surface. Using multiple methods to calculate volumes is a good way to verify and confirm your results. The volume surface can then be used to display the depth of cut and fill.

9.3.1 Creating a Grid Volume Surface

Grid Volume Surfaces compare two surfaces and calculate a volume based on a user defined grid. In this type of a volume calculation the elevation difference is measured between the two surfaces at the center of each grid cell. That difference is then applied to the entire cell. This means that the accuracy is dependant on the grid size.

1. On the *Prospector* tab of the **Toolspace,** right-click on **Surfaces** and select ⇒ **New.**

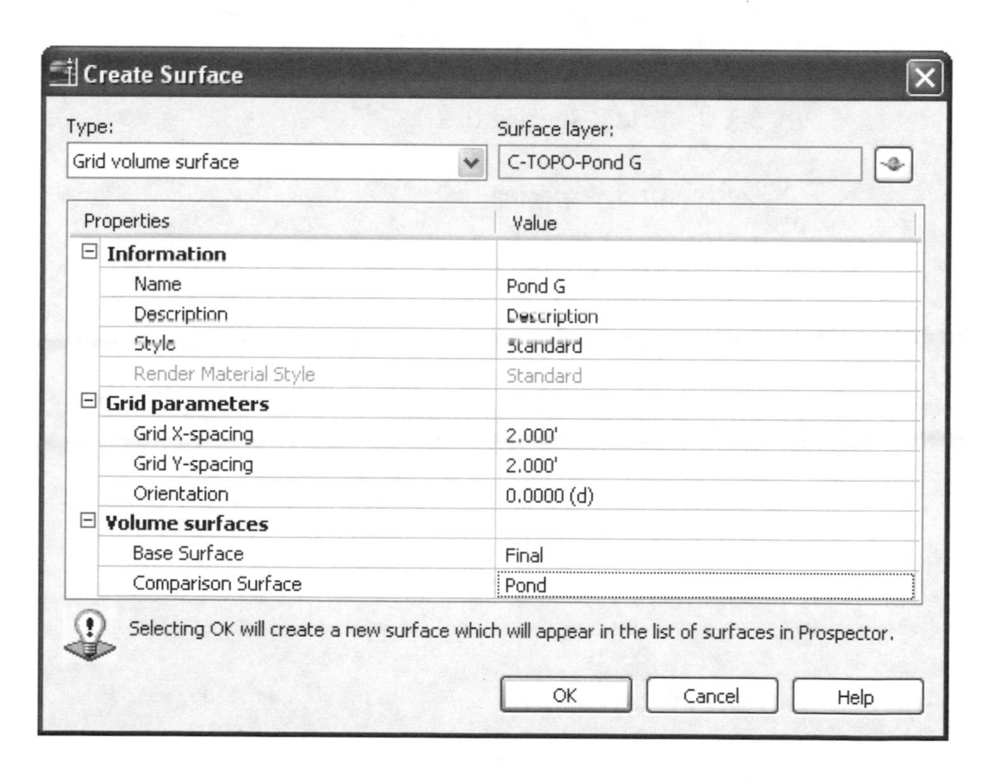

2. Select **Grid Volume surface** as the **Type**.

3. Click the **Surface Layer** button to open the *Object Layer* dialog box.

4. Set the **Modifier** to **Suffix**.

5. Enter **-*** as the *Modifier value*.

6. Click **<<OK>>** to close the **Object Layer** dialog box.

7. Enter **"Pond G"** for the **Name**.

8. Set the surface **Style** to **Standard**.

The surface style *Standard* will display only the border of the *Volume Surface*.

9. Set the **Grid X-spacing** to **2**.

10. Set the **Grid Y-spacing** to **2**.

11. Set the **Base Surface** to **Final**.

12. Set the **Comparison Surface** to **Pond**.

13. Click **<<OK>>** to create the *Volume Surface* and calculate the volume.

14. On the *Prospector* tab of the *Toolspace*, right-click on the *Surface* **Pond G** and select ⇒ **Properties.**

15. Select the **Statistics** tab of the *Surface Properties* dialog box.

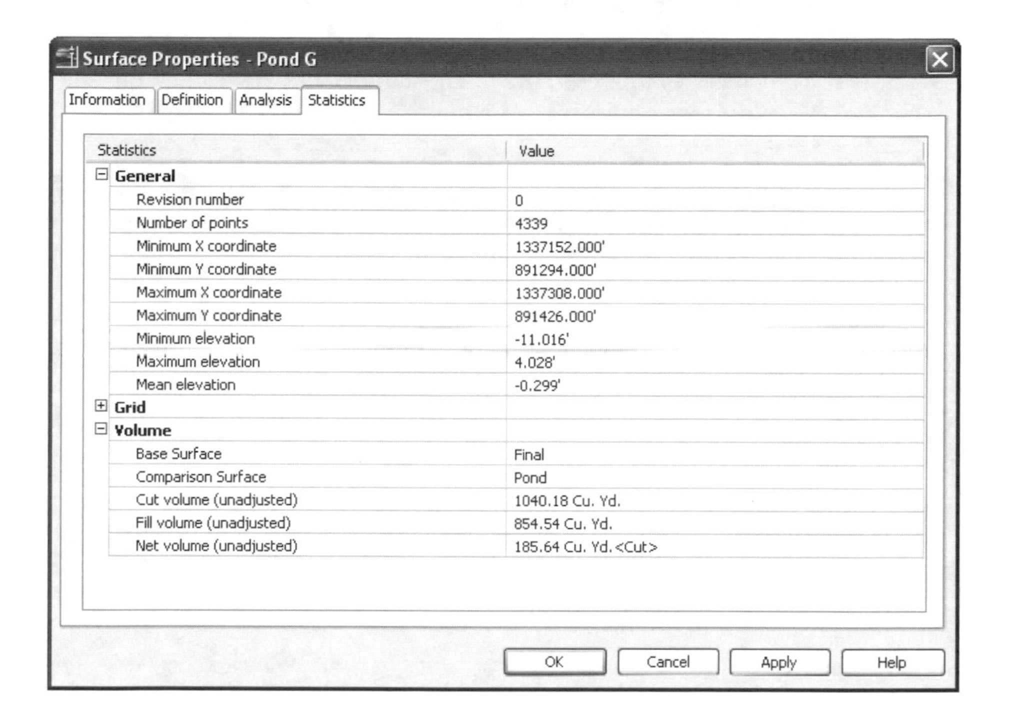

16. Expand the **General** node

Here you can review the minimum and maximum elevations of the surface. For a volume surface that is the maximum depth of cut and fill.

17. Expand the **Volume** node.

Here you can review the surfaces used to calculate the volume and the volume results.

18. Click **<<OK>>** to close the **Surface Properties** dialog box.

9.3.2 Creating a TIN Volume Surface

TIN Volume Surfaces compare two surfaces and calculate a volume. This is a true surface comparison volume where all the points from each surface are used to calculate elevation differences that are used to create the *Volume Surface* and calculate volumes.

1. On the *Prospector* tab of the **Toolspace,** right-click on **Surfaces** and select ⇒ **New**.

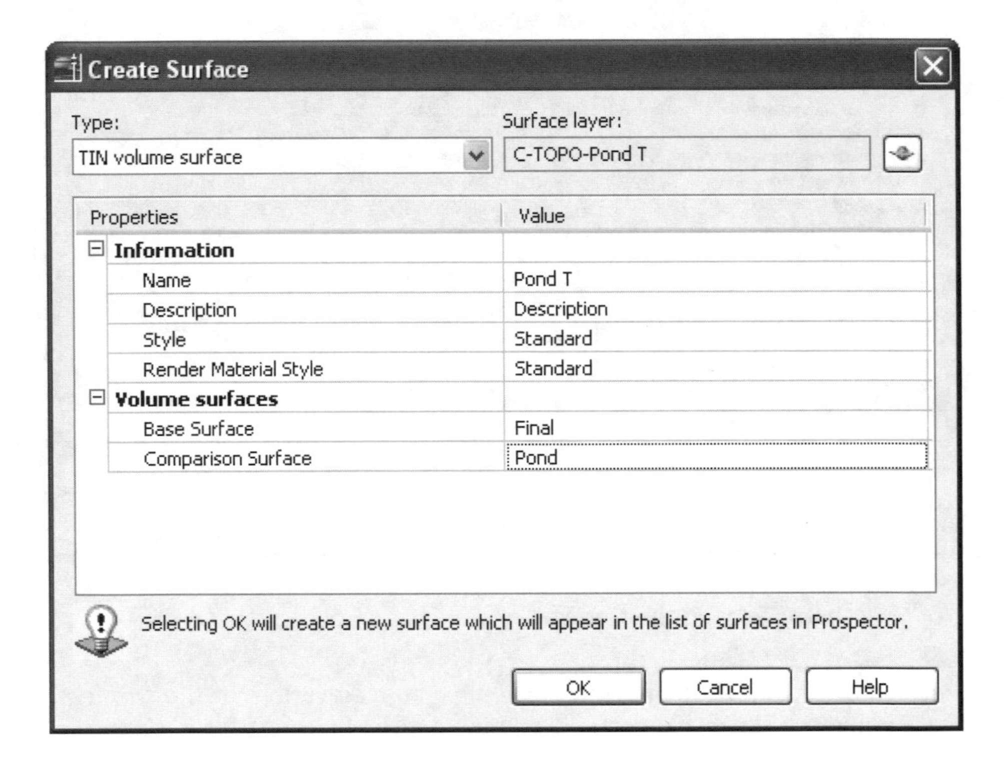

2. Select **TIN Volume surface** as the **Type.**

3. Click the **Surface Layer** button to open the *Object Layer* dialog box.

4. Set the **Modifier** to **Suffix.**

5. Enter **-*** as the *Modifier value*.

6. Click **<<OK>>** to close the **Object Layer** dialog box.

7. Enter **"Pond T"** for the **Name**.

8. Set the surface **Style** to **Standard**.

The surface style *Standard* will display only the border of the *Volume Surface*. Notice the border of the *TIN Volume* surface is much smoother than the *Grid Volume* surface because it is not based on a grid.

9. Set the **Base Surface** to **Final**.

10. Set the **Comparison Surface** to **Pond**.

11. Click **<<OK>>** to create the *Volume Surface* and calculate the volume.

Notice the border of the *TIN Volume* surface is much smoother than the *Grid Volume* surface because it is not based on a grid.

12. On the *Prospector* tab of the *Toolspace*, right-click on the *Surface* **Pond T** and select ⇒ **Properties**.

13. Select the **Statistics** tab of the *Surface Properties* dialog box.

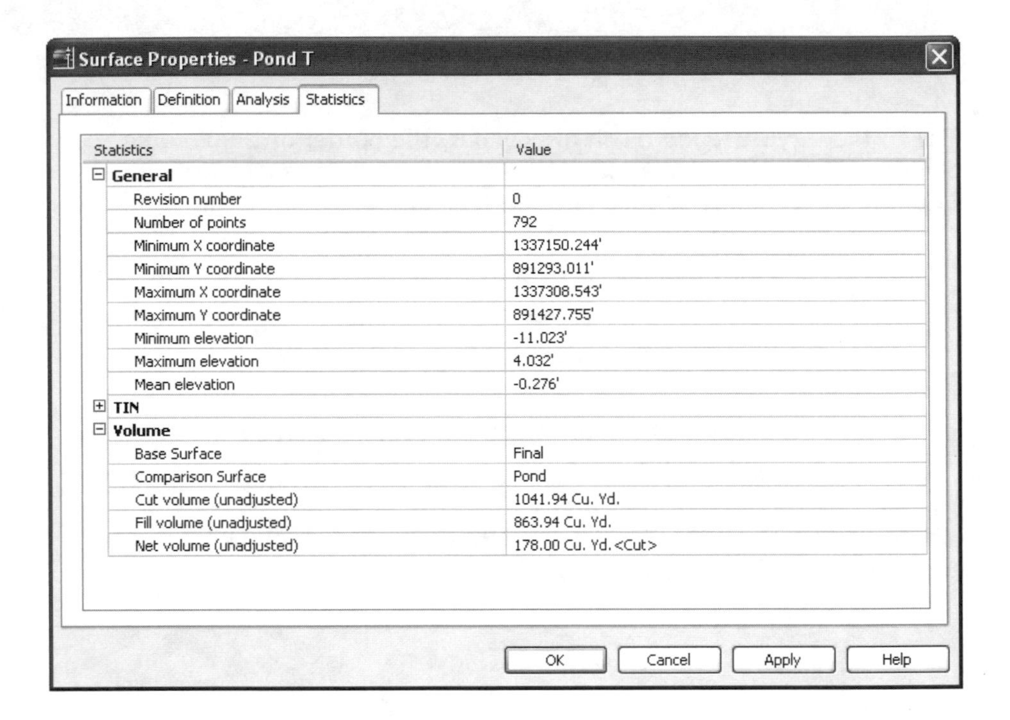

14. Expand the **General** node.

Here you can review the minimum and maximum elevations of the surface. For a volume surface that is the maximum depth of cut and fill.

15. Expand the **Volume** node.

Here you can review the surfaces used to calculate the volume and the volume results.

16. Click <<**OK**>> to close the *Surface Properties* dialog box.

9.3.3 Displaying Cut and Fill with Surface Styles

Volume Surfaces are displayed with styles just like *TIN Surfaces*. In this exercise you will use the style *Border & Elevations* that you created in *Chapter 5* for elevation banding to display depths of cut and fill.

1. On the *Prospector* tab of the *Toolspace*, right-click on the *Surface* **Pond T** and select ⇒ **Properties.**

2. Change the surface style to **Border & Elevations.**

This is the surface style you created in *Chapter 5* for elevation banding.

3. Select the **Analysis** tab of the *Surface Properties* dialog box.

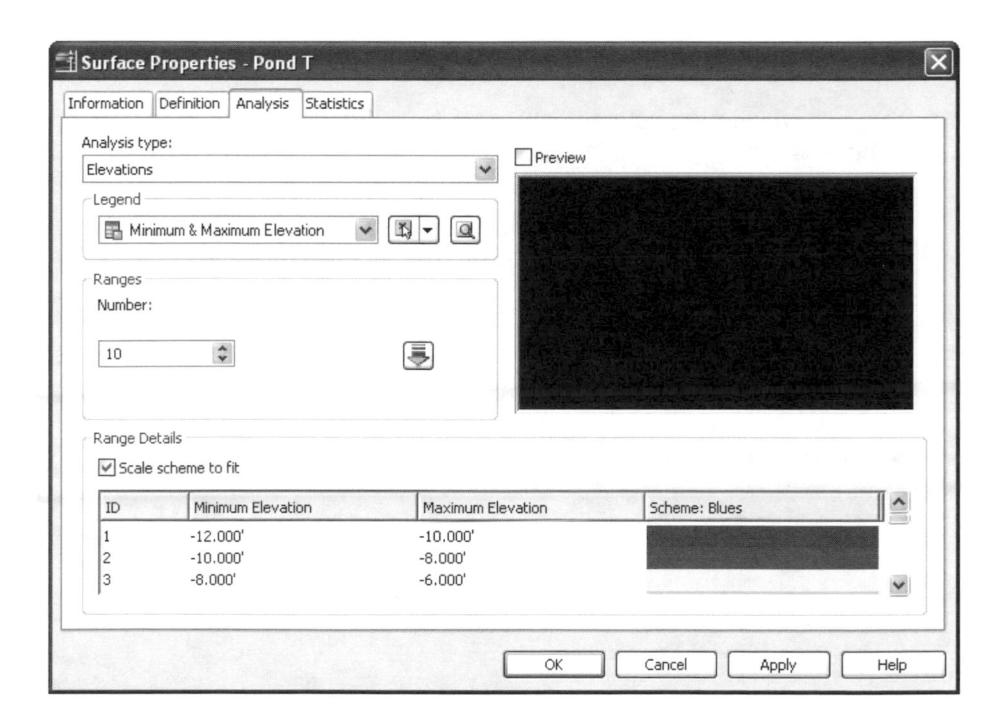

4. Confirm that the **Analysis type** is set to **Elevations.**

5. Set the **Number** of ranges to **10**.

6. Click the **Run Analysis** button to analyze the surface.

7. Double click to edit the **Minimum** and **Maximum Elevation** fields to edit the elevation ranges. Set up elevation ranges in 2 foot increment starting at -12 and ending at 6.

Minimum Elevation	Maximum Elevation
-12	-10
-10	-8
-8	-6
-6	-4
-4	-2
-2	0
0	2
2	4
4	6

You can also change the colors if you want to highlight certain depths of cut or fill.

8. Click **<<OK>>** to display the elevation bands representing the depth of cut and fill in the volume surface.

9.3.4 Creating a Legend for Cut and Fill Depths

1. Select **Surfaces ⇒ Add Legend Table.**

2. When asked to `Select a surface` at the command line, Enter to select it from a list.

You also have the option to pick it graphically from the screen. However, this can be difficult if there are several surfaces displayed that all overlap each other.

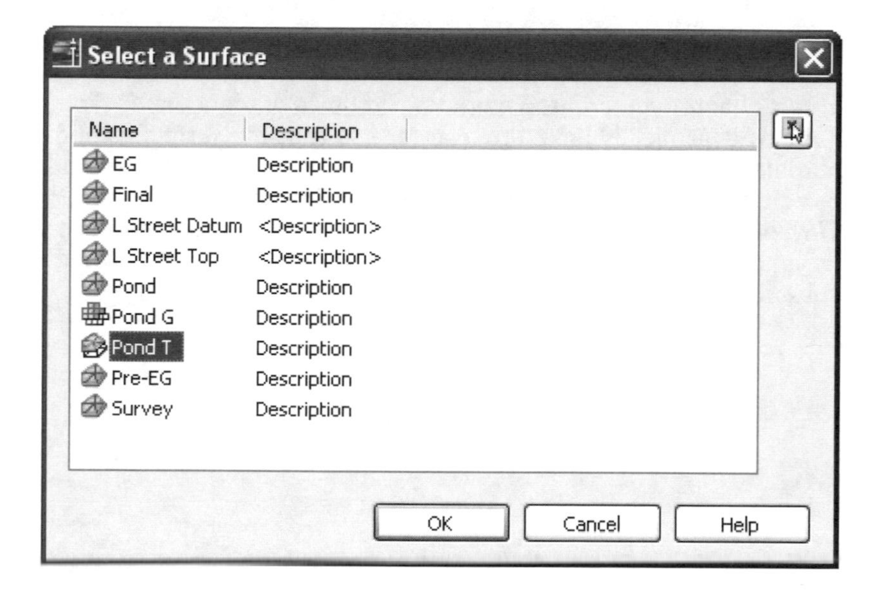

3. Select the surface **Pond T.**

4. Click **<<OK>>**.

5. When asked to `Enter a table type` at the command line, **<<Enter>>** to select the default option of **Elevations.**

6. When asked to select a `Behavior` at the command line, **<<Enter>>** to select the default option of **Dynamic**.

7. Pick a point in a blank area to the right of the pond for the location of the upper left corner of the legend.

A legend table will be created for the elevation bands displaying the depth of cut and fill. This table is a single object so you can easily move it to another location if it is overlapping anything. The display of the table is controlled by the table style as selected on the *Analysis* tab of the *Surface Properties* dialog box.

8. Save the drawing.

9.4 Chapter Summary

In this chapter you created a grading group to grade a pond. The grading group is dynamically linked to a surface so that the surface is updated automatically each time that you add a grading object to the grading group or edit a grading object within the group. Finally, you created *Grid* and *TIN Volume* surfaces to calculate the volume of the pond and explored ways to use surface styles to display the volume surfaces showing depths of cut and fill.

Other Titles Available by this Author Include:

Available at www.cadapult-software.com/store

- **Digging Into Autodesk Civil 3D 2006**
  Level 1 Training
- **Digging Into Autodesk Land Desktop 2006**
  Level 1 Training
- **Digging Into Autodesk Land Desktop 2004**
  Level 1 Training
- **Digging Into Autodesk Map 3D 2005**
  Level 1 Training

Coming Soon:

- **Digging Deeper Into Autodesk Civil 3D 2007**
  Level 2 Training
- **Digging Deeper Into Autodesk Map 3D 2007**
  Level 2 Training